EXAM PREPARATION & PRACTICE

Haese and Harris Publications
specialists in mathematics publishing

Endorsed by
University of Cambridge International Examinations

IGCSE
Cambridge International
Mathematics

(0607)
Extended

Keith Black
Alison Ryan
Michael Haese
Robert Haese
Sandra Haese
Mark Humphries

Keith Black
Alison Ryan
James Foley
Katie Richer
Michael Haese

Haese & Harris Publications

CAMBRIDGE IGCSE INTERNATIONAL MATHEMATICS (0607) EXTENDED
EXAM PREPARATION & PRACTICE GUIDE

Keith Black   B.Sc.(Hons.), Dip.Ed.
Alison Ryan   B.Sc., M.Ed.
James Foley   B.Ma.Comp.Sc.(Hons.)
Katie Richer   B.Sc., B.Ed.(Hons.)
Michael Haese   B.Sc.(Hons.), Ph.D.

Haese Mathematics
152 Richmond Road, Marleston, SA 5033, AUSTRALIA
Telephone: +61 8 8210 4666, Fax: + 61 8 8354 1238
Email: info@haesemathematics.com.au
Web: www.haesemathematics.com.au

National Library of Australia Card Number & ISBN 978-1-921500-16-9

© Haese & Harris Publications 2011

Published by Raksar Nominees Pty Ltd
152 Richmond Road, Marleston, SA 5033, AUSTRALIA

    First Edition   2011
    Reprinted   2014

Artwork by Piotr Poturaj.
Cover design by Piotr Poturaj.

Typeset in Australia by Susan Haese and Charlotte Frost (Raksar Nominees).

Typeset in Times Roman 9/10.

Printed in Singapore by Opus Group.

This Guide has been endorsed by University of Cambridge International Examinations.

This book is copyright. Except as permitted by the Copyright Act (any fair dealing for the purposes of private study, research, criticism or review), no part of this publication may be reproduced, stored in a retrieval system, or transmitted in any form or by any means, electronic, mechanical, photocopying, recording or otherwise, without the prior permission of the publisher. Enquiries to be made to Haese Mathematics.

Copying for educational purposes: Where copies of part or the whole of the book are made under Part VB of the Copyright Act, the law requires that the educational institution or the body that administers it, has given a remuneration notice to Copyright Agency Limited (CAL). For information, contact the Copyright Agency Limited.

Acknowledgements: While every attempt has been made to trace and acknowledge copyright, the authors and publishers apologise for any accidental infringement where copyright has proved untraceable. They would be pleased to come to a suitable agreement with the rightful owner.

FOREWORD

The aim of this Guide is to help you prepare for the Cambridge IGCSE International Mathematics (0607) Extended examinations. It should be used in conjunction with your textbook.

One of the examination objectives is to test your ability to apply mathematical concepts and techniques to solve problems. Using practice or past examination papers is an ideal way to test your knowledge.

This Guide covers all eleven topics in the Cambridge IGCSE International Mathematics (0607) syllabus. Each topic has a summary of key facts and concepts followed by a set of 'Skill Practice' questions. There are over 500 'Skill Practice' questions in total.

Following topics 1 – 11 are three Practice Exams, divided into Paper 2, Paper 4, and Paper 6 as per the Cambridge IGCSE International Mathematics (0607) Extended examination. Each Paper 2 has 10 – 11 short response questions to be completed within 45 minutes without the use of a calculator. Each Paper 4 has 11 – 12 medium to extended response questions to be completed within 2 hours 15 minutes. Each Paper 6 has more open-ended questions divided into two parts: A. Investigation and B. Modelling, to be completed within 1 hour 30 minutes. A graphics calculator may be used for Papers 4 and 6.

Fully worked solutions are provided for all questions in this Guide.

There are a number of graphics calculator requirements for the Cambridge IGCSE International Mathematics (0607) Extended examination. Students should know how to do the following using a graphics calculator, and should seek help before the examination if they are not confident:

- Sketch the graph of a function.
- Produce a table of values for a function.
- Find the zeros and local maxima or minima of a function.
- Find the intersection point of two graphs.
- Find the mean, median, and quartiles of a data set.
- Find the linear regression equation for a data set.

Try to complete the practice exams under examination conditions. Getting into good habits will reduce pressure during the examination.

Become familiar with any formulae that are given at the front of the examination paper.

- It is important that you persevere with a question, but sometimes it is a good strategy to move on to other questions and return later to ones you have found challenging. Time management is very important during the examination, and too much time spent on a difficult question may mean that you do not leave yourself sufficient time to complete other questions.
- Use a pen rather than a pencil, except for graphs and diagrams.

- If you make a mistake draw a single line through the work you want to replace. Do not cross out work until you have replaced it with something you consider better.
- Set out your work clearly with full explanations. Do not take shortcuts.
- Diagrams and graphs should be sufficiently large, well labelled, and clearly drawn.
- Remember to leave answers correct to three significant figures unless an exact answer is more appropriate or a different level of accuracy is requested in the question.
- Get used to reading the questions carefully.
- Rushing into a question may mean that you miss subtle points. Underlining key words may help.
- Questions in the examination are often set so that even if you cannot get through one part, the question can still be picked up in a later part.

After completing a practice exam, identify areas of weakness.

- Return to your notes or textbook and review the topic.
- Ask your teacher or a friend for help if further explanation is needed.
- Summarise each topic. Summaries that you make yourself are the most valuable.
- Test yourself, or work with someone else, to help improve your knowledge of a topic.
- If you have had difficulty with a question, try it again later. Do not just assume that you know how to do it once you have read the solution.
- It is important that you work on areas of weakness, but do not neglect the other areas.

Your graphics display calculator is an essential aid.

- Make sure you are familiar with the model you will be using.
- In trigonometry questions make sure your calculator is in degree mode.
- Become familiar with common error messages and how to respond to them.
- Important features of graphs may be revealed by zooming in or out.
- Asymptotic behaviour is not always clear on a graphics calculator screen; don't just rely on appearances. As with all aspects of the graphics calculator, reflect on the reasonableness of the results.
- Are your batteries fresh?

We hope this guide will help you structure your revision program effectively. Remember that good examination techniques will come from good examination preparation.

We welcome your feedback:

web: *http://haesemathematics.com.au*
email: *info@haesemathematics.com.au*

TABLE OF CONTENTS

1	**NUMBER**	**6**
	Number sets	6
	Primes and composites	6
	Other terms	6
	Order of operations	6
	Rounding numbers	6
	Absolute value	6
	Fractions	6
	Time	6
	Speed, distance and time	6
	Surds and radicals	6
	Exponents and indices	7
	Exponents and radicals	7
	Standard form	7
	Ratio and proportion	7
	Percentage	7
	Simple interest	7
	Compound interest	7
	Skill practice	7

2	**ALGEBRA**	**9**
	Expansion laws	9
	Factorisation	9
	Algebraic fractions	9
	Linear equations	9
	Simultaneous linear equations	10
	Problem solving using equations	10
	Formulae	10
	Quadratic equations	10
	Power equations	10
	Exponential equations	10
	Direct variation	10
	Inverse variation	10
	Variation modelling	10
	Linear inequalities	11
	Other one-variable inequalities	11
	Number sequences	11
	Skill practice	11

3	**FUNCTIONS**	**15**
	Mappings	15
	Functions	16
	Reciprocal functions	16
	Absolute value function	16
	Quadratic functions	16
	Cubic functions	16
	Exponential functions	17
	Logarithms	17
	Trigonometric functions	17
	Using technology	17
	Transforming functions	17
	Skill practice	17

4	**GEOMETRY**	**21**
	Lines and line segments	21
	Symmetry	21
	Angles	22
	Triangles	22
	Polygons	22
	Pythagoras	22
	Congruence	23
	Similarity	23
	Circles	23
	Circle theorems	23
	Theorems involving arcs	24
	Cyclic quadrilaterals	24
	Cyclic quadrilateral theorems	24
	Skill practice	24

5	**TRANSFORMATIONS IN TWO DIMENSIONS**	**29**
	Vectors	29
	Transformations	29
	Skill practice	30

6	**MENSURATION**	**33**
	Units	33
	Length	33
	Area	33
	Surface area	33
	Volume	34
	Capacity	34
	Mass	34
	Skill practice	34

7	**COORDINATE GEOMETRY**	**37**
	The number plane	37
	Distance between two points	37
	Midpoint of a line segment	37
	Gradient	37
	Straight lines	37
	Linear inequality regions	37
	Skill practice	38

8	**TRIGONOMETRY**	**40**
	Right angled triangles	40
	The unit circle	40
	Area of a triangle using sine	41
	The sine rule	41
	The cosine rule	41
	Trigonometric graphs	41
	Skill practice	42

9	**SETS**	**44**
	Venn diagrams	45
	Skill practice	45

10 PROBABILITY — 46
- Experimental probability — 46
- Expectation — 46
- Sample space — 46
- Theoretical probability — 46
- Complementary events — 46
- Independent events — 46
- Dependent events — 46
- Mutually exclusive events — 46
- Skill practice — 46

11 STATISTICS — 48
- Variables — 48
- Statistical graphs — 48
- Census or sample — 49
- The distribution of a data set — 49
- The centre of a discrete data set — 49
- Measuring the spread of discrete data — 49
- Frequency tables — 49
- Grouped discrete data — 49
- Using technology — 49
- Continuous data — 49
- Correlation — 50
- Line of best fit — 50
- Skill practice — 50

PRACTICE EXAM 1 — 54
- Paper 2: 45 minutes / 40 marks — 54
- Paper 4: 2 hours 15 minutes / 120 marks — 55
- Paper 6: 1 hour 30 minutes / 40 marks — 56

PRACTICE EXAM 2 — 57
- Paper 2: 45 minutes / 40 marks — 57
- Paper 4: 2 hours 15 minutes / 120 marks — 58
- Paper 6: 1 hour 30 minutes / 40 marks — 59

PRACTICE EXAM 3 — 61
- Paper 2: 45 minutes / 40 marks — 61
- Paper 4: 2 hours 15 minutes / 120 marks — 61
- Paper 6: 1 hour 30 minutes / 40 marks — 63

SOLUTIONS — 65
- Topic 1: Number — 65
- Topic 2: Algebra — 68
- Topic 3: Functions — 81
- Topic 4: Geometry — 94
- Topic 5: Transformations in two dimensions — 102
- Topic 6: Mensuration — 108
- Topic 7: Coordinate geometry — 112
- Topic 8: Trigonometry — 118
- Topic 9: Sets — 124
- Topic 10: Probability — 126
- Topic 11: Statistics — 129
- Practice exam 1 — 136
- Paper 2 — 136
- Paper 4 — 137
- Paper 6 — 139
- Practice exam 2 — 140
- Paper 2 — 140
- Paper 4 — 141
- Paper 6 — 144
- Practice exam 3 — 146
- Paper 2 — 146
- Paper 4 — 146
- Paper 6 — 149

TOPIC 1: NUMBER

NUMBER SETS

Natural or **counting** numbers $\mathbb{N} = \{0, 1, 2, 3, 4, ...\}$

Integers or **whole** numbers
$\mathbb{Z} = \{...., -3, -2, -1, 0, 1, 2, 3, ...\}$

Rational numbers $\mathbb{Q} = \left\{ \dfrac{p}{q} \text{ where } p, q \in \mathbb{Z}, \ q \neq 0 \right\}$

Real numbers $\mathbb{R} = \{\text{all numbers on the number line}\}$

Irrational numbers are all the real numbers which are not rational.

PRIMES AND COMPOSITES

The **factors** of a positive integer are the positive integers which divide exactly into it, leaving no remainder.

A positive integer is a **prime** number if it has exactly two factors, 1 and itself.

A positive integer is a **composite** number if it has more than two factors.

Any positive integer greater than 1 can be written uniquely as the product of prime factors.

OTHER TERMS

A **common factor** is a number that is a factor of two or more other numbers.

The **highest common factor** is the largest factor that is common to two or more numbers.

A **multiple** of any positive integer is obtained by multiplying it by another positive integer.

The **lowest common multiple** of two or more positive integers is the smallest multiple which is common to all of them.

ORDER OF OPERATIONS

Brackets
Exponents
Division
Multiplication } in the order you come to them
Addition
Subtraction } in the order you come to them

ROUNDING NUMBERS

You should be able to round numbers:

- to a certain number of **decimal places**
- to a certain number of **significant figures**
- to the **nearest integer** or **whole number**

You should be able to **estimate** values using **one figure approximations**.

ABSOLUTE VALUE

The **absolute value** of a number is its size, ignoring its sign.

$$|x| = \begin{cases} x & \text{if } x \geqslant 0 \\ -x & \text{if } x < 0. \end{cases}$$

FRACTIONS

A **fraction** consists of a **numerator** and a **denominator**, separated by a bar.

numerator → 2
bar → —
denominator → 3

In a **proper fraction** the numerator is less than the denominator.

In an **improper fraction** the numerator is more than the denominator. We can write improper fractions as **mixed numbers**.

Equivalent fractions represent the same proportion.
For example, $\frac{1}{2} = \frac{2}{4}$.

The **lowest common denominator** of two or more fractions is the lowest common multiple of their denominators.

You should know how to perform addition, subtraction, multiplication, and division with fractions.

TIME

$$\begin{aligned} 1 \text{ minute} &= 60 \text{ seconds} \\ 1 \text{ hour} &= 60 \text{ minutes} = 3600 \text{ seconds} \\ 1 \text{ day} &= 24 \text{ hours} \\ 1 \text{ week} &= 7 \text{ days} \\ 1 \text{ year} &= 12 \text{ months} = 365 \text{ days} \end{aligned}$$

You should understand how to use 24-hour time.

SPEED, DISTANCE AND TIME

The **average speed** s is the distance travelled divided by the time taken.

$$s = \frac{d}{t}$$

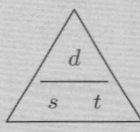

You should be able to interpret **travel graphs**, including:

- times when the thing is stationary
- times when the thing is travelling fastest
- average speeds.

SURDS AND RADICALS

A **radical** is a number written using the radical sign $\sqrt{\ }$.

A **surd** is an **irrational radical**.

$\sqrt{a} \times \sqrt{b} = \sqrt{a \times b}, \quad a \geqslant 0, \ b \geqslant 0$

$\dfrac{\sqrt{a}}{\sqrt{b}} = \sqrt{\dfrac{a}{b}}, \quad a \geqslant 0, \ b \geqslant 0$

A surd is in simplest form when the number under the radical sign is the smallest integer possible.

If an expression involves division by a surd, we can write the expression with an integer denominator which does not contain surds.

If the denominator has the form $a + \sqrt{b}$, we multiply both numerator and denominator by the **radical conjugate** $a - \sqrt{b}$.

EXPONENTS AND INDICES

We use an **exponent** or **power** to display the product of factors in **index form**.

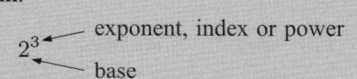

2^3 ← exponent, index or power
← base

A **negative base** raised to an **odd power** is **negative**.
A **negative base** raised to an **even power** is **positive**.

$a^m \times a^n = a^{m+n}$	To **multiply** numbers with the **same base**, keep the base and **add** the indices.
$\dfrac{a^m}{a^n} = a^{m-n}$, $a \neq 0$	To **divide** numbers with the **same base**, keep the base and **subtract** the indices.
$(a^m)^n = a^{m \times n}$	When **raising** a **power** to a **power**, keep the base and **multiply** the indices.
$(ab)^n = a^n b^n$	The power of a product is the product of the powers.
$\left(\dfrac{a}{b}\right)^n = \dfrac{a^n}{b^n}$, $b \neq 0$	The power of a quotient is the quotient of the powers.
$a^0 = 1$ for all $a \neq 0$	Any non-zero number raised to the power zero, is one.
$a^{-n} = \dfrac{1}{a^n}$	a^n and a^{-n} are **reciprocals** of each other.
$\left(\dfrac{a}{b}\right)^{-n} = \left(\dfrac{b}{a}\right)^n$, $a \neq 0$, $b \neq 0$	

EXPONENTS AND RADICALS

If $a^n = b$ then $\sqrt[n]{b} = a$.

$a^{\frac{1}{n}} = \sqrt[n]{a}$

$a^{\frac{m}{n}} = \sqrt[n]{a^m} = (\sqrt[n]{a})^m$

STANDARD FORM

A number in **standard form** or **scientific notation** is written as a number between 1 and 10, multiplied by an integer power of 10.

$a \times 10^n$ where $1 \leqslant a < 10$ and $n \in \mathbb{Z}$.

RATIO AND PROPORTION

In a **ratio** we compare quantities with the **same units**.

To **simplify** a ratio we can multiply or divide each part by the same non-zero number. We do this so that each number is an integer, and they are as small as possible.

Ratios are **equal** if they can be expressed in the same simplest form.

A **proportion** is a statement that two ratios are equal.

PERCENTAGE

$x\%$ means $\dfrac{x}{100}$.

To find $x\%$ of a quantity, multiply by $\dfrac{x}{100}$.

You should understand how percentages are applied to **profit** and **loss**, **mark up** and **mark down**.

You should be able to use a **multiplier** to increase or decrease a quantity by a given percentage.

SIMPLE INTEREST

Simple interest is calculated on the initial amount borrowed for the entire period of the loan.

$I = Prn$ where I is the simple interest
P is the principal or amount borrowed
r is the rate of interest as a decimal
n is the duration of the loan in years.

COMPOUND INTEREST

Compound interest is always calculated on the current balance.

The compounding effect comes from paying interest on interest.

You should be able to use multipliers to help with compound interest problems.

SKILL PRACTICE

1. Copy and complete:

Number	$\mathbb{N}$	$\mathbb{Z}$	$\mathbb{Q}$	$\mathbb{R}$
$\frac{1}{2}$				
-5				
$\sqrt{-3}$				
π				
4.8712				

2. Find the lowest common multiple of:
   a 6 and 8
   b 3, 4 and 5

3. Fifi earns $14.20 per hour as a secretary. She loses 25% of her pay in tax.
   a How much money does Fifi receive for each hour of work?
   b For how many hours must Fifi work to receive $4000?

4. Without using a calculator, find $3^1 + 3^2 + 3^3 + 3^4$.

5. Jessica scored $\frac{3}{8}$ of the goals for her team in the season. If Jessica scored 42 goals in total, how many did the whole team score?

6. Write as ratios in simplest form:
   a €15 is to €4.50
   b 375 ml is to 2 litres
   c 2 kg is to 800 g
   d 3 h is to 100 min.

7. Poj sells photo frames in two sizes. Their dimensions are in the same ratio. The smaller frame is 12 cm × 18 cm.
   If the longer side of the larger frame is 24 cm, find the length of its shorter side.

8. What is the time difference between 1310 on one day and 0706 the next morning?

9. Enrique invests €4500 in an account which pays 5.4% p.a. compounded annually. Find:
   a the value of his account after 4 years
   b the total interest earned after 4 years.

10 Simplify: $4\sqrt{2} - 5\sqrt{3} - 2 + 3\sqrt{3} - 5\sqrt{2}$

11 Thomas built a chest of drawers and sold it for £280. If his profit was 15%, what was the cost of building the chest of drawers?

12 Find the selling price for the following items:
  a a car purchased for £12 000 and sold at a 10% discount
  b a swimsuit purchased for $56 and sold at a 28% markup
  c a computer bought for €950 and sold at a 35% loss
  d a fruit tree bought for $32 and sold at a 20% profit.

13 Simplify the following, giving your answer in standard form:
  a $(3.2 \times 10^4) \times (6.8 \times 10^3)$
  b $(4.5 \times 10^3) \div (6 \times 10^5)$

14 Write down the values of:
  a $\left|-\frac{1}{3}\right|$
  b $|3.6801 - 2.1508|$
  c $\left|\frac{-3.15}{5}\right|$

15 **a** The government spent 236.8 million pounds on highway projects last month. Express this amount in pounds in standard form.
  b A laser emits light of wavelength 6.8×10^{-7} m. Write this as a decimal number.
  c A planet orbits its star at 260 km per second. How far does it travel in 365 days? Write your answer in kilometres, in standard form.

16 Lance has completed 60% of his daily training ride. If he has already covered 75 km, how many kilometres does he still need to ride?

17 Convert 12 years into seconds, rounding your answer to 3 significant figures.

18 Calculate the simple interest on a $4500 loan at 12% p.a. for 5 years.

19 Simplify:
  a $(3\sqrt{7})^3$
  b $-4\sqrt{3} \times 2\sqrt{3}$
  c $3\sqrt{7} \times 2\sqrt{2}$
  d $-4\sqrt{3} + 2\sqrt{3}$

20 Calculate the total amount to be repaid on a $2900 simple interest loan at $6\frac{1}{2}$% p.a. for 3 years 4 months.

21 Write as the product of prime factors in index form:
  a 168
  b 975

22 Selena scored 87% for an examination. If she scored $130\frac{1}{2}$ marks, how many total marks were in the exam?

23 Write in simplest surd form:
  a $\sqrt{72}$
  b $\sqrt{\frac{30}{27}}$
  c $\sqrt{\frac{6}{5}} \times \sqrt{\frac{15}{8}}$

24 How much do I need to invest now, at 5% p.a. compound interest, for the total in 6 years to be $10 000?

25 Find the average speed of a bee which flies 320 m in 7 minutes. Give your answer in:
  a m/s
  b km/h.

26 Find the rate of simple interest per annum if:
  a £3375 is charged on a £45 000 loan over 18 months
  b €1800 is charged on a €2400 loan over 10 years.

27 Simplify using the index laws:
  a $(-3)^4 \times (-3)^3$
  b $(2^4)^3$
  c $\left(\frac{2}{7}\right)^2$
  d $\left(\frac{4}{3}\right)^{-1}$
  e $2^{-2} + 2^{-1} + 2^0$
  f $\left(3\frac{1}{4}\right)^{-2}$

28 Find the percentage change that occurs when:
  a 15 m reduces to 7 m
  b ¥67 000 increases to ¥70 000
  c 80 kg increases to 95 kg
  d 27 tonnes decreases to 20 tonnes.

29 Simplify:
  a $3\sqrt{2}(5 - \sqrt{2})$
  b $-\sqrt{3}(2\sqrt{2} + \sqrt{3})$
  c $(\sqrt{3} - \sqrt{2})^2$
  d $\left(\frac{1}{\sqrt{2}} + \sqrt{2}\right)^2$
  e $(2\sqrt{3} - 1)(1 + 2\sqrt{3})$

30 Evaluate without using a calculator:
  a $81^{\frac{1}{4}}$
  b $125^{-\frac{2}{3}}$
  c $64^{\frac{5}{6}}$

31 Kerrie rides her bicycle for 20 minutes at an average speed of 27 km/h. How far does she travel in this time?

32 An antique chair *appreciates* or increases in value by 7% each year. Ten years ago the chair was worth £2100.
  a What is the value of the chair now?
  b By what percentage has the value of the chair increased?

33 A motor scooter bought for £650 was later sold for £900. Find the profit as a percentage of the cost price.

34 Express with rational denominator:
  a $\dfrac{5 - \sqrt{3}}{2 - \sqrt{3}}$
  b $\dfrac{\sqrt{2} - \sqrt{3}}{1 + 3\sqrt{2}}$

35 How long will it take to earn ¥13 500 on a loan of ¥675 000 at a rate of 8% p.a. simple interest?

36 What is the overall effect of:
  a increases of 4%, $4\frac{1}{2}$%, and 5% over three consecutive years
  b an increase of $3\frac{1}{2}$% over 5 consecutive years
  c an increase of 10%, a decrease of 40%, and an increase of 15%, over three consecutive years?

37 Evaluate if possible:
  a $9^{\frac{3}{2}}$
  b $(-1)^{\frac{3}{2}}$
  c $(-1)^{\frac{2}{3}}$

38 Frank and Betty are hiking up a mountain. Their progress is shown on the distance-time graph.

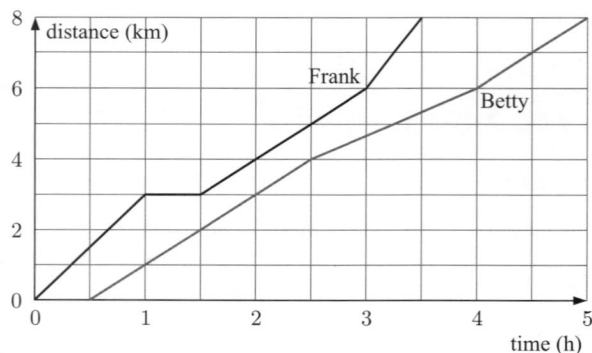

  a How far was the trail?
  b For how long did Frank stop for lunch?
  c Suggest where the mountain was steepest, giving reasons for your answer.
  d Find the average speed of:
    i Frank for the first hour
    ii Betty for the whole hike.

39 By considering $13^1, 13^2, 13^3, ...$, find the last digit of 13^{1313}.

40 Bing has 15 000 yuan to invest. He has been offered a simple interest rate of 4.9% p.a. for a six year period. What rate of compound interest would Bing need to find in order to earn the same amount of interest after 6 years?

41 A racing pigeon is released 420 km from its home. If it flies at an average speed of 80 km/h, how long will it take to get there?

42 Show that $0.\overline{37}$ is a rational number.

43 Write the following in index form:

   a $\sqrt[3]{13}$ **b** $\frac{1}{\sqrt[5]{5}}$ **c** $\sqrt{24}$ **d** $\frac{1}{\sqrt{29}}$

44 At the beginning of the year, Sally started jogging to improve her fitness. She was initially able to run 1.5 km without stopping. However, she has noticed a steady improvement of 9% per month in the distance she can run. If this pattern continues, how far will she be able to run by the end of the year?

45 Write in the form $a + b\sqrt{2}$ where $a, b \in \mathbb{Q}$:

   a $\dfrac{3 - 2\sqrt{2}}{\sqrt{2} - 1}$ **b** $\dfrac{\frac{1}{\sqrt{2}} - 3\sqrt{2}}{\sqrt{2} - \frac{1}{\sqrt{2}}}$

46 Michael takes his dog for a walk to the park. The graph below describes the journey:

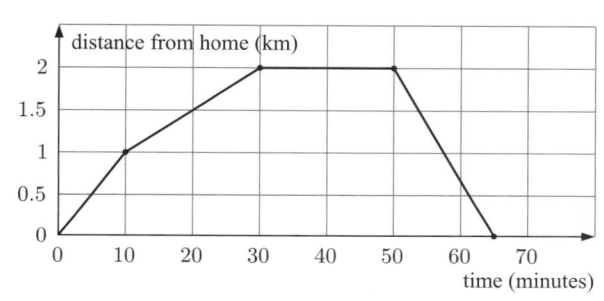

   a How far away is the park from Michael's house?

   b How long did the entire trip take?

   c How long did Michael spend at the park?

   d When were Michael and his dog walking fastest? Explain your answer.

TOPIC 2: ALGEBRA

EXPANSION LAWS

Distributive law $a(b + c) = ab + ac$

Product (FOIL) law $(a + b)(c + d) = ac + ad + bc + bd$

Difference of two squares $(a + b)(a - b) = a^2 - b^2$

Perfect squares $(a + b)^2 = a^2 + 2ab + b^2$

                      $(a - b)^2 = a^2 - 2ab + b^2$

FACTORISATION

Factorisation is the reverse process of expansion.

$$x^2 + (\alpha + \beta)x + \alpha\beta = (x + \alpha)(x + \beta)$$

coefficient of x is the **sum** of α and β

constant term is the **product** of α and β

The factorisation process:

Expression
↓
Remove any common factors.
↓
Look for special types:
- **difference of two squares**
- **perfect squares**

↓
Look for the **sum and product** factorisation.
↓
Look for **splitting the middle terms**.

ALGEBRAIC FRACTIONS

Simplifying fractions

If the numerator and denominator of an algebraic fraction are both written in factored form and common factors are found, we can simplify by **cancelling common factors**.
It is useful to remember that $b - a = -(a - b)$.
Be careful when cancelling!

$\dfrac{a + \cancel{3}^1}{\cancel{3}_1} = \dfrac{a + 1}{1} = a + 1$ is **WRONG** since the factor in the numerator is $(a + 3)$.

Multiplication

$$\frac{a}{b} \times \frac{c}{d} = \frac{ac}{bd}$$

- multiply numerators and multiply denominators
- cancel any common factors
- write the result in simplest form

Division

$$\frac{a}{b} \div \frac{c}{d} = \frac{a}{b} \times \frac{d}{c} = \frac{ad}{bc}$$

To divide by a fraction, multiply by its reciprocal.

Addition and subtraction

$$\frac{a}{c} + \frac{b}{c} = \frac{a + b}{c} \qquad \frac{a}{c} - \frac{b}{c} = \frac{a - b}{c}$$

To add or subtract fractions, first write them with the same lowest common denominator.

LINEAR EQUATIONS

A **linear equation** is an equation which can be written in the form $ax + b = 0$ where x is a variable, and a and b are constants, $a \neq 0$.

You should be able to solve linear equations by:
- expanding any brackets
- collecting like terms
- using inverse operations to isolate the unknown while maintaining the balance of the equation.

To solve equations involving fractions, write the fractions with the lowest common denominator, then equate the numerators.

SIMULTANEOUS LINEAR EQUATIONS

The **simultaneous solution** to two equations is the pair of values of the variables which satisfy both equations.

You should be able to solve linear simultaneous equations by:
- equating values of y
- substitution
- elimination

PROBLEM SOLVING USING EQUATIONS

- Identify the unknown quantity and label it using a variable.
- Decide which operations are involved.
- Translate the problem into an equation and check that your translation is correct.
- Solve the equation by isolating the variable.
- Check that your solution does satisfy the original problem.
- Write your answer in sentence form.

FORMULAE

A **formula** is an equation which connects two or more variables. We usually write one variable by itself on the left hand side as the **subject** of the formula, with the other variables and constants in an expression on the right hand side.

We can **substitute** the values of any known variables into the formula. If there is only one unknown we can solve the equation to find its value.

We can **rearrange** formulae using the same processes which we use to solve equations. We do this to make a different variable the subject of the formula.

QUADRATIC EQUATIONS

A **quadratic equation** can be written in the form $ax^2 + bx + c = 0$ where $a \neq 0$.

Quadratic equations may have zero, one, or two solutions.

If $x^2 = k$ then $\begin{cases} x = \pm\sqrt{k} & \text{if } k > 0 \\ x = 0 & \text{if } k = 0 \\ \text{there are no real solutions} & \text{if } k < 0. \end{cases}$

The **Null Factor law** states:
> When the product of two or more numbers is zero, then at least one of them must be zero.
> If $ab = 0$ then $a = 0$ or $b = 0$.

Solving quadratic equations:
- rearrange the equation so one side is zero
- fully factorise the other side
- use the Null factor law
- solve the resulting linear equations
- check at least one of the solutions

Do not cancel a common factor that involves a variable unless you know that factor cannot be zero.

The quadratic formula

If $ax^2 + bx + c = 0$ where $a \neq 0$,

then $x = \dfrac{-b \pm \sqrt{b^2 - 4ac}}{2a}$.

POWER EQUATIONS

A **power equation** is an equation of the form $x^n = k$, where $n = 2, 3, 4, \ldots$.

If $x^n = k$ where n is odd, then $x = \sqrt[n]{k}$.
If $x^n = k$ where n is even and $k \geqslant 0$, then $x = \pm\sqrt[n]{k}$.

EXPONENTIAL EQUATIONS

If $a^x = a^k$ then $x = k$.

So, if the base numbers are the same, then we can equate indices.

You may first need to rearrange the equation using index laws.

You should be able to solve exponential equations using a graphics calculator.

DIRECT VARIATION

Two variables x and y are **directly proportional** if multiplying one of them by a number results in the other being multiplied by the same number.

We write $y \propto x$ or $y = kx$ where k is called the **proportionality constant**.

When y is graphed against x, k is the gradient of the graph, and the line passes through the origin.

INVERSE VARIATION

Two variables are **inversely proportional** if, when one is multiplied by a constant, the other is *divided* by the same constant.

If y is **inversely proportional** to x, then y is **directly proportional** to $\dfrac{1}{x}$. So, $y = \dfrac{k}{x}$ or $xy = k$.

VARIATION MODELLING

Direct variation

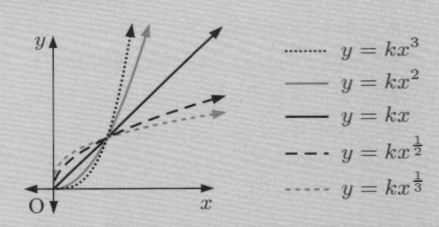

.... $y = kx^3$
——— $y = kx^2$
——— $y = kx$
- - - $y = kx^{\frac{1}{2}}$
----- $y = kx^{\frac{1}{3}}$

The graph always passes through the origin $(0, 0)$.

Inverse variation

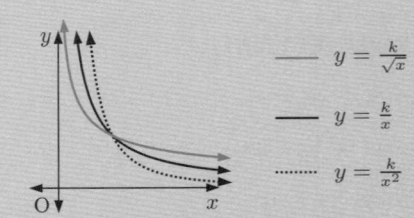

——— $y = \dfrac{k}{\sqrt{x}}$
——— $y = \dfrac{k}{x}$
.... $y = \dfrac{k}{x^2}$

The graph is always asymptotic to both the x and y-axes.

If $y \propto x^n$ then:
- $y = kx^n$ where k is the proportionality constant
- the graph of y against x^n is a straight line with gradient k.

Power modelling

Power models have the form $y = ax^b$.
- If $b > 0$, we have **direct variation**.
- If $b < 0$, we have **inverse variation**.

You should know how to use the power regression function on your calculator to fit a power model to a data set.

LINEAR INEQUALITIES

We can represent inequalities on a **number line**.

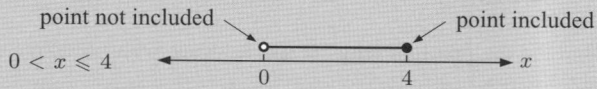

$0 < x \leqslant 4$ point not included ⟶ 0, point included ⟶ 4

When solving linear inequalities we use the same method of inverse operations as for linear equations except:

- when interchanging sides, we reverse the inequality sign.
- when multiplying or dividing both sides by a negative number, we reverse the inequality sign.

OTHER ONE-VARIABLE INEQUALITIES

You should be able to use technology to graph functions and hence solve one-variable inequalities.

For example:

- if $f(x) > 0$ we seek values of x for which the function is above the x-axis
- if $f(x) > g(x)$ we seek values of x for which $y = f(x)$ is *above* $y = g(x)$.

NUMBER SEQUENCES

A **number sequence** is a set of numbers listed in a specific order, where the numbers can be determined from a rule.

The first term is denoted u_1, the second u_2, and so on.

The **nth term** is u_n.

You should be able to describe a number sequence in words, giving the rule that connects successive terms.

Algebraic rules for sequences

A sequence can often be described by an algebraic formula for u_n. In this case, u_n is often called the **general term**.

An algebraic rule for u_n is only valid for $n \in \mathbb{Z}^+$.

You should be able to:

- find terms of a sequence, given the formula for the general term
- deduce the formula for the general term of a sequence, given its first few terms.

Geometric sequences

In a **geometric sequence**, each term is found by multiplying the previous one by the same constant.

Its general term will have the form $u_n = u_1 \times r^{n-1}$ for some constant r.

The difference method for sequences

You should be able to generate the difference tables for:

- Linear sequences $u_n = an + b$
- Quadratic sequences $u_n = an^2 + bn + c$
- Cubic sequences $u_n = an^3 + bn^2 + cn + d$.

We place the values of n (1, 2, 3, 4, ...) in the first row, and the corresponding values of u_n in the second row.

We then generate rows of **differences** until the differences in a given row are all the same.

- *Linear sequences*:
  The third row gives the **first difference** $\Delta 1$, the difference between successive terms.
- *Quadratic sequences*:
  We include a fourth row for the **second difference** $\Delta 2$, the difference between successive $\Delta 1$s.
- *Cubic sequences*:
  We include a fifth row for the **third difference** $\Delta 3$, the difference between successive $\Delta 2$s.

By comparing the 'general' table with the table for the sequence being considered, the values of a, b, c and d can be calculated.

SKILL PRACTICE

1 Expand and simplify:

   a $-x(1 + 2x)$ **b** $4x(1 - x^2)$

   c $-x^2 \left(3 - \dfrac{1}{x}\right)$

2 Expand and simplify:

   a $5x(1 - x) + 3x$ **b** $6(x^2 - 1) - x(1 + x)$

   c $3(2 - x) - x(2x + 1)$ **d** $5x(1-x) + x\left(\dfrac{1}{x} - 2\right)$

3 Consider the figure below:

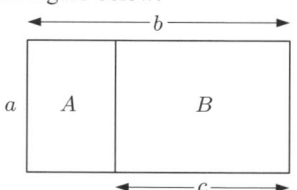

[figure: rectangle divided into rectangle A (with side a) and rectangle B; top width b, overall width c]

   a Write down an expression for the area of:

      i rectangle A **ii** rectangle B

      iii the overall rectangle

   b Use your expressions in **a** to explain why $a(b - c) = ab - ac$.

4 Expand and simplify:

   a $(2 + 3x)(2 - 3x)$ **b** $(2x - 5)^2$

   c $(3 - 2x)(1 + 3x)$ **d** $(x + 2y)^2$

5 Simplify, if possible:

   a $\dfrac{3a^2 b}{6ab^2}$ **b** $\dfrac{4x^2 y^2}{10xy}$ **c** $\dfrac{4 - x}{4}$

6 Factorise:

   a $x^2 y - 2xy$ **b** $3xy + 6x^3$

   c $8x^2(5 - 2x) + 12x(2x - 5)$

7 Represent the following on a number line:

   a $-4 < x \leqslant 2$ **b** $x \leqslant -1$ or $x > 3$

8 Simplify, if possible:

   a $\dfrac{6(x + 1)}{2(x + 1)}$ **b** $\dfrac{(x - 1)(x + 2)}{1 - x}$ **c** $\dfrac{8(x + 1)^2}{2(x - 3)(x + 1)}$

9 Expand and simplify:

   a $(2 - x)^2 - (x + 1)^2$

   b $(2x + 3)^2 - (x - 1)(x + 5)$

10 Expand and simplify: $(x - 1)(2x + 1)(3 - x)$

11 Write down a rule to describe the sequence and hence find its next two terms:
   a 2, −1, −4, −7, ...
   b 6, 3, $1\frac{1}{2}$, $\frac{3}{4}$, ...
   c 4, 5, 7, 10, 14, ...
   d 5, 15, 45, 135, ...

12 Factorise:
   a $3(1+x) - x(1+x)$
   b $(x-2)^2 + (x-1)(x-2)$
   c $4(1-x)^2 - x + 1$
   d $2(a-3)(a+4) + a(a+4)$

13 Fully factorise using the difference between two squares:
   a $3x^2 - 27$ **b** $-8x^2 + 32$ **c** $(a-5)^2 - 4a^2$

14 Use the difference between two squares to show that $27 \times 15 = 21^2 - 6^2$.

15 Solve for x:
   a $\frac{4x-1}{5} = 2$
   b $\frac{1}{3}(3-x) = -4$

16 Fully factorise:
   a $x^2 + 10x + 25$
   b $-18x^2 - 12x - 2$

17 Solve for x:
   a $3x^2 = 48$
   b $3 - 4x^2 = -13$
   c $5x^2 - 7 = 58$

18 Explain why $x^2 + 16 \geqslant 8x$ for all real x.

19 Fully factorise:
   a $x^2 + x - 20$
   b $2x^2 + 2x - 4$
   c $6x^2 - 3x - 3$
   d $-9t^2 - 30t + 24$

20 Simplify:
   a $\frac{3x - x^2}{x^2 + x}$
   b $\frac{2x^2 - 8}{2x^2 + 4}$
   c $\frac{3x^2 - 3y^2}{6y^2 - 6xy}$

21 Solve for x:
   a $3(2x-1) = 5 - x$
   b $2(1-3x) + x = 4(2x+1)$

22 Simplify:
   a $(2x^2)^3$
   b $\left(\frac{3}{4a}\right)^2$
   c $\left(\frac{x}{y^2}\right)^{-2}$

23 Solve for x:
   a $\frac{2x-5}{3} = \frac{x}{4}$
   b $\frac{2+x}{5} = \frac{x-1}{2}$

24 Write down the inequality used to describe the set of numbers:
   a [number line from −1 (closed) to 4 (open)]
   b [number line with −6 (open) and 0 (closed), arrows outward]

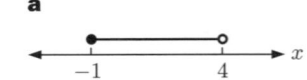

25 Simplify:
   a $\frac{x}{3} \times \frac{y}{2}$
   b $2 \div \frac{3}{x}$
   c $\frac{3n}{2} - \frac{4n}{5}$

26 Solve for x:
   a $(x+5)^2 = 9$
   b $(2x-1)^2 = 18$
   c $3(2-x)^2 = 192$

27 Solve for x using the Null Factor law:
   a $(1-2x)(3+x) = 0$
   b $(4x-1)(2-x) = 0$
   c $2(2x-3)^2 = 0$

28 Simplify:
   a $\frac{x}{2} \div \frac{x}{3}$
   b $\frac{2}{x} \times \frac{x}{y} \times \frac{y}{4}$
   c $3 - \frac{a}{4} + \frac{2a}{3}$

29 Solve for x:
   a $\frac{1}{3x} = \frac{7}{5}$
   b $\frac{2}{x+2} - \frac{3x}{x-1} = -\frac{10}{3}$

30 Find the first four terms of the sequence with nth term:
   a $u_n = 4n - 1$
   b $u_n = 3n^2 - 2$
   c $u_n = n(n^2 - 3)$

31 Simplify:
   a $\frac{x^2 - 2x - 3}{x^2 - 5x + 6}$
   b $\frac{x^2 - x - 2}{2x^2 - 4x}$
   c $\frac{2x^2 + 5x - 3}{-2x^2 + 7x - 3}$

32 Simplify by writing as a single fraction:
   a $\frac{x}{3} + x - 2$
   b $\frac{2}{x} - 3 + \frac{x}{3}$

33 The sum of three consecutive odd integers is 51. Find the smallest of the integers.

34 Solve for x:
   a $3x - 2x^2 = 0$
   b $x^2 - 25 = 0$
   c $x^2 + 8x + 16 = 0$

35 Make y the subject of:
   a $4x - 2y = 3$
   b $ax + by + c = 0$

36 Solve for x:
   a $x^2 + 7x + 12 = 0$
   b $x^2 = 10x - 25$
   c $x^2 - 5x = 24$

37 Write down a rule to describe the sequence and hence find its next two terms:
   a 5, 11, 17, 23, ...
   b 2, 5, 10, 17, 26, ...
   c 4, 12, 36, 108, ...
   d 3, 12, 21, 30, ...

38 Solve for x and graph the solutions:
   a $6 - 5x \leqslant 1$
   b $4(3-x) > 2$

39 Solve for x: $\frac{3x-1}{4} + \frac{x+3}{2} = \frac{4x+3}{3}$

40 Match each graph to its rule:
   A [curve increasing steeply] **B** [straight line through origin] **C** [decreasing hyperbola]
   a $y = \frac{k}{x}$
   b $y = kx^2$
   c $y = kx$

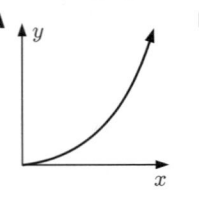

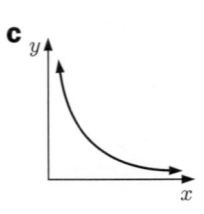

41 Solve for x:
   a $3x^2 - 5x - 2 = 0$
   b $11x = 2x^2 + 15$
   c $16x^2 = 24x - 9$

42 Simplify:
   a $\frac{a}{x} \times \frac{x}{b}$
   b $\frac{b}{m^2} \div \frac{b^2}{m}$
   c $\frac{x}{3} + \frac{x}{7} + \frac{x}{5}$

43 Make l the subject of $T = \frac{k\sqrt{l}}{\pi}$. Hence determine l when $k = 2\pi$ and $T = 6$.

44 Draw the next two matchstick figures in these sequences and write the number of matchsticks used as a number sequence:
  a
  b

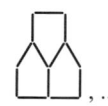

45 When 11 is added to a number and the result is divided by 3, the answer is 1 less than the original number. Find the number.

46 Solve for x:
  a $x(x-1) - 3(x+7) = 0$
  b $2x(x+1) - 3(x+2) = 0$

47 Solve for x, giving answers correct to 3 significant figures:
  a $x^3 - 26 = 0$
  b $\dfrac{x}{3} = \dfrac{4}{x}$
  c $x^4 = 40$
  d $\dfrac{x}{5} = -\dfrac{3}{x}$

48 **a** Write $\dfrac{3x}{(x+1)(x-2)} - \dfrac{2}{x+1}$ as a single fraction.
  b Find the values of x when this expression is:
    i undefined   **ii** zero.

49 Use technology to solve, correct to 3 decimal places:
  a $x^2 - \sqrt{2}x - 5 = 0$
  b $\pi x^2 - x = 6$

50 Elliott bought Fiona flowers for her birthday. Lilies cost €5 each and orchid spikes cost €2 each. Elliott bought four more orchid spikes than lilies, and in total he spent €43. How many of each flower did Elliott buy?

51 Simplify:
  a $(3p^2q)^{-1}$
  b $(3x^2y^3)^3$
  c $\left(\dfrac{4x^{-1}}{3y^2}\right)^3$

52 **a** Find a formula for the general term u_n of: 3, 6, 9, 12, 15, ...
  b *Hence* find a formula for the general term u_n of:
    i 7, 10, 13, 16, 19, ...   **ii** −1, 2, 5, 8, 11, ...

53 The kinetic energy E of an object with mass m and speed v is given by $E = \frac{1}{2}mv^2$ joules.
  a Find the kinetic energy of a 2.4 kg basketball travelling at 9.3 m/s.
  b Find the speed of a 0.15 kg bullet with kinetic energy 270 J.

54 Solve for x:
  a $5^x = \dfrac{1}{125}$
  b $8 \times 3^x = 72$
  c $32^x = \left(\dfrac{1}{4}\right)^{x-1}$

55 A half of a number plus twice its reciprocal is two. Find the number.

56 The table below shows the amount €C charged for using p kWh of power.

p (kWh)	0	1	2	3	4	5
C (€)	0	0.23	0.46	0.69	0.92	1.15

  a Draw a graph of C against p.
  b Explain why C and p are directly proportional, using the features of your graph.
  c Find:
    i the proportionality constant
    ii the law connecting C and p.

57 Write a formula for the total wages €W of
  a 12 employees for 38 hours at €21 per hour
  b 12 employees for h hours at €21 per hour
  c 12 employees for h hours at €p per hour
  d n employees for h hours at €p per hour.

58 Solve by equating values of y:
  a $y = 2x - 3$
     $y = 3x + 1$
  b $y = x - 5$
     $y = -2x + 1$

59 Solve for x:
  a $2x - 1 \geqslant 0$
  b $\dfrac{x-2}{5} \leqslant 3$
  c $2x - 3 < 3x + 5$

60 If $y \propto x$ and $y = 211.5$ when $x = 9$, find:
  a y when $x = 17.2$
  b x when $y = 329$.

61 Use the quadratic formula to solve for x, giving exact answers:
  a $2x^2 - 11x + 13 = 0$
  b $3x + \dfrac{1}{x} = -6$

62 **a** Find a formula for the general term of 1, 8, 27, 64, ...
  b *Hence* find a formula for the general term u_n of:
    i −1, 6, 25, 62, ...
    ii 1, $\frac{1}{8}$, $\frac{1}{27}$, $\frac{1}{64}$, ...
    iii $\frac{1}{8}$, $\frac{2}{27}$, $\frac{3}{64}$, $\frac{4}{125}$, ...

63 The volume of stormwater flowing through a pipe is directly proportional to the square of the pipe's radius. If the radius is increased by 25%, by what percentage would the volume of stormwater flow increase?

64 **a** Simplify: $\dfrac{\frac{3}{1-x} - \frac{x}{2}}{x - 5}$
  b Find the values of x when this expression is:
    i undefined   **ii** zero.

65 According to Newton's Second Law of Motion, the force F newtons applied to an object equals its mass m kilograms multiplied by its acceleration a m/s^2.
  a Write this formula with a as the subject.
  b Find the acceleration on a 2.4 kg object under a force of 14.3 N.

66 Solve for x and graph the solutions:
  a $2x + 5 \geqslant 3 - 4x$
  b $2 - (3 - x) < 4(x - 1) - 3$

67 For each of the following tables:
    i find values of $\dfrac{y}{\sqrt{x}}$
    ii establish that $y \propto \sqrt{x}$
    iii find the rule connecting y and x
    iv find the value of a.

  a
x	1	16	81	196
y	3	12	27	a

  b
x	25	36	49	a
y	2.5	3	3.5	6.5

68 Solve by substitution:
  a $y = 2x - 1$
   $2x - 4y = -4$
  b $x = -2y - 3$
   $3y - 4 = 2x$

69 Use the quadratic formula to solve for x, if possible:
  a $6x^2 - 17x - 57 = 0$
  b $2x^2 + 5x + 4 = 0$
  c $\frac{13}{x} - 7 = 2x$

70 Consider the formula $H = \frac{d}{2x^2 y}$.
  a Make y the subject of the formula.
  b Find the value of y when:
   i $H = 25$, $d = 100$, $x = 2$
   ii $H = 18$, $d = 144$, $x = 2$

71 Solve for x without using a calculator:
  a $3^{2x+1} = 81$
  b $7^x = 1$
  c $3^{x^2 - 3} = 9^x$

72 Calculate the value of xy for each point in the following tables. Hence determine for each table whether x and y are inversely proportional. If an inverse proportionality exists, determine the law connecting the variables and draw the graph of y against x.

  a
x	15	3	12	30
y	4	20	5	2

  b
x	2	4	6	12
y	7.5	7	6.5	6

  c
x	4	20	80	10
y	25	5	1.25	10

73 Solve for x:
  a $\frac{x}{x-1} = \frac{8}{x+2}$
  b $\frac{5}{x+3} - \frac{x}{x+1} = 3$

74 Solve by elimination:
  a $5x - 3y = 12$
   $-2x + 3y = 3$
  b $3x - 4y = 6$
   $4x + y = 27$

75 Use the method of differences to find the general term u_n of:
  a $1, 7, 13, 19, 25, 31, ...$
  b $-1, 0, 3, 8, 15, 24, ...$
  c $-2, -1, -14, -53, -130, -257, ...$

76 Solve for x using your calculator:
  a $3^{-x} \geqslant 2^x - 4$
  b $x^3 \leqslant \sqrt{x}$
  c $\frac{1}{x^2 - 1} > 2$

77 The length of a rectangle is 3 cm more than its width. Find the length of the rectangle given its area is 270 cm².

78 List the first five terms of the geometric sequence defined by:
  a $u_n = 4 \times 3^n$
  b $u_n = 5 \times (-\frac{1}{2})^{n-1}$

79 If $a \times 3^n = 108$ and $a \times 6^n = 864$, find a and n.

80 Find the variation model for these data sets:

  a
x	2	5	10	25
y	62.5	4	0.5	0.032

  b
t	2	4	6	8	10
p	12	48	108	192	300

81 The surface area of a cylinder with radius r and height h is $A = 2\pi r^2 + 2\pi r h$.
  a Make h the subject of this formula.
  b Find the value of h when:
   i $A = 240\pi$ m² and $r = 6$ m
   ii $A \approx 1385.2$ cm² and $r \approx 8.9$ cm.

82 Write down a rule to describe the sequence and hence find its next two terms:
  a $1, \frac{1}{4}, \frac{1}{9}, \frac{1}{16}, \frac{1}{25}, ...$
  b $100, 50, 25, 12\frac{1}{2}, ...$
  c $10, 13\frac{1}{2}, 17, 20\frac{1}{2}, ...$
  d $1, 4, 10, 19, 31, ...$

83 Q varies inversely to the cube of t. When $t = 5$, $Q = 50$.
  a Find Q when $t = 17$.
  b Find t when $Q = 10\,800$.

84 The area of a kite is given by the formula
$$\text{Area} = \tfrac{1}{2}(\text{width} \times \text{length}).$$
A kite is 25 cm longer than it is wide. Its area is 3838 cm². Find the dimensions of the kite.

85 Draw the next two figures in these sequences and write the number of dots used as a number sequence:
  a
  b

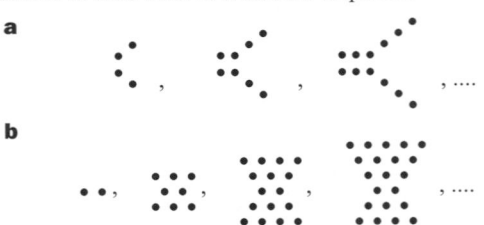

86 The intensity of light below sea level varies inversely to the depth at which the measurement is taken. When the depth is 3 m, the light intensity is 45 units. Find:
  a the light intensity when the depth is 12 m
  b the depth at which the light intensity is 5 units.

87 Use your graphics calculator to solve for x, giving answers correct to 3 decimal places:
  a $3^x = 100$
  b $3 \times 2^{-3x} = \frac{1}{2}$
  c $2^{3x-2} = 10$

88 Find x and y:

(diagram: quadrilateral with sides $3y$ cm, $(x+4y)$ cm, $(5x-y)$ cm, $(2x+2)$ cm)

89 Use your graphics calculator to find the power model which best fits the following data:

  a
t	1	2	3	4	5
a	0.45	7.2	36.45	115.2	281.25

  b
x	10	12	14	16	18
Q	50.2	54.0	57.5	60.6	63.6

90 Find the next two terms and a formula for the nth term of:
  a $9, 18, 36, 72, ...$
  b $96, -24, 6, -\frac{3}{2}, ...$
  c $1, \frac{1}{2}, \frac{1}{4}, \frac{1}{8}, ...$

91 The volume of a cylinder is given by $V = \pi r^2 h$ where r is the radius and h is the height.

  a If we consider cylinders of fixed height but variable radius, what proportionality exists between V and r?

  b What would happen to the volume of a cylinder if the radius was increased by 40%?

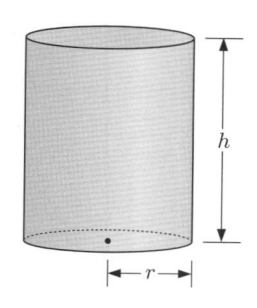

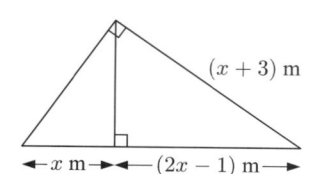

92 Find x correct to 3 significant figures.

(triangle with sides $(x+3)$ m, x m, and $(2x-1)$ m)

93 Solve for x: $\quad |x^2 - x - 6| > |x - 3|$

94 The force between two particles is inversely proportional to the square of the distance between them. If the distance between the two particles is decreased by 55%, how does the force between the particles change?

95 Use the method of differences to find the general term u_n of:

  a 22, 15, 8, 1, −6, −13, ...

  b −3, 3, 19, 45, 81, 127, ...

  c 0, 5, 20, 51, 104, 185, ...

96 It is suspected that for two variables x and y, y is directly proportional to the square root of x.

Use the data from the graph to find a model connecting x and y.

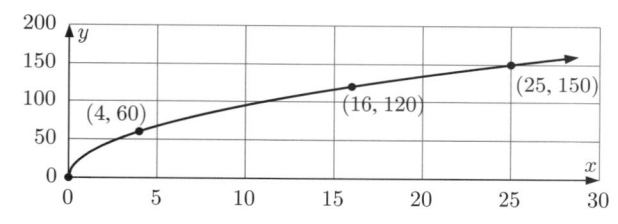

(graph with points (4, 60), (16, 120), (25, 150))

97 a Make x the subject of $y = \dfrac{2x - 3}{1 - x}$.

  b Hence determine the value of x when y equals:

  i $-\frac{4}{3}$ **ii** $-\frac{5}{3}$

98 Jo needs to wrap 150 drinking glasses in paper to protect them during transport. Jo can wrap x glasses each hour. If Jo was able to wrap 10 more glasses each hour, the job would take 6 minutes less. How many glasses can Jo wrap each hour?

99 Consider the dot pattern:

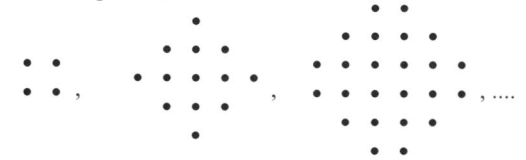

Suppose u_n is the number of dots in the nth figure.

  a Find u_n for $n = 1, 2, 3, 4, 5$.

  b Find a formula for the general term u_n.

  c How many dots are needed to make up the 45th pattern?

100 A CPU manufacturer wishes to find out the relationship between the number of CPUs sold per year, C, and the number of years since 1992, t.

A search of production records yields the following data:

t	2	4	6	8	10
C (×1000)	33.5	74.9	119.9	167.4	216.8

  a Obtain a power model for this data.

  b Is the power model appropriate? Explain your answer.

  c Estimate the number of CPUs sold in 2007.

101 Consider the sequence u_n where:

$u_1 = 2 \times 3$
$u_2 = 2 \times 3 + 4 \times 6$
$u_3 = 2 \times 3 + 4 \times 6 + 6 \times 9$
$u_4 = 2 \times 3 + 4 \times 6 + 6 \times 9 + 8 \times 12$
.... and so on.

  a Find the values of u_1, u_2, u_3, u_4, u_5 and u_6.

  b Use the difference method to find a formula for u_n.

  c Hence, find the value of u_{75}.

102 A supermarket sells dried apricots in 500 g and 1 kg packs. The 500 g packs cost €3.25 each and the 1 kg packs cost €5.85 each. Over a week, 66 kg of dried apricots were sold, with total revenue €404.95. How many of each size pack did the supermarket sell?

103 Anya makes spherical glass ornaments of diameter d cm. She believes that the volume of glass used, V cm^3, is directly proportional to some power of d.

Anya takes the following measurements:

d cm	5	8	10
V cm^3	65.4	268.1	523.6

  a Use technology to find k and n for which $V = kd^n$.

  b Check your model by substituting in the first data point.

  c Verify your value of k by substituting $r = \dfrac{d}{2}$ into the volume formula $V = \frac{4}{3}\pi r^3$.

  d Find:

  i V when $d = 4$ **ii** d when $V = 500$.

104 Consider the matchstick pattern

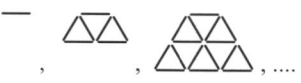

Suppose u_n is the number of matches in the nth figure.

  a Find u_n for $n = 1, 2, 3, 4, 5$, and 6.

  b Use the method of differences to find a formula for u_n.

  c How many matchsticks are in the 20th figure?

TOPIC 3: FUNCTIONS

MAPPINGS

A **mapping** is used to map the members or elements of one set called the **domain**, onto the members of another set called the **range**.

You should be able to draw **mapping diagrams** to illustrate mappings. You should understand what is meant by **one-one**, **many-one**, **one-many**, and **many-many** mappings.

FUNCTIONS

A **function** is a mapping in which each element of the domain maps onto *exactly one* element of the range.

One-one and many-one mappings can be functions.
One-many and many-many mappings cannot be functions.

If a function maps set A onto set B then:
- A is the **domain** of the function
- B is the **range** of the function.

Geometric or vertical line test for functions

If we draw all possible vertical lines on the graph of a relation, the relation is a function if each line cuts the graph no more than once.

Function notation

$$f : x \mapsto 2x - 1 \quad \text{or} \quad f(x) = 2x - 1$$

function f such that x maps onto $2x - 1$

Composite functions

The **composite function** of f and g is the function which maps x onto $f(g(x))$.

Inverse functions

The **inverse function** f^{-1} of a function f is the function such that, for every value of x that f maps to $f(x)$, f^{-1} maps $f(x)$ back to x.

The inverse of $y = f(x)$ can be found algebraically by interchanging x and y, and then making y the subject of the resulting formula. The new y is $f^{-1}(x)$.

The **horizontal line test**:

For a function to have an inverse function, no horizontal line can cut it more than once.

RECIPROCAL FUNCTIONS

A **reciprocal function** has the form $f(x) = \dfrac{k}{x}$ where k is a constant.

The graph of a reciprocal function is called a **rectangular hyperbola**.

It has a **horizontal asymptote** which it approaches as x gets very large.

It also has a **vertical asymptote** which occurs when the denominator of the fraction becomes zero.

ABSOLUTE VALUE FUNCTION

The **absolute value** function $|x| = \begin{cases} x & \text{if } x \geqslant 0 \\ -x & \text{if } x < 0. \end{cases}$

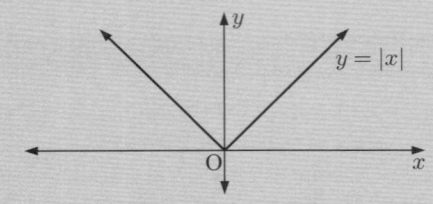

You should be able to graph composite functions involving the absolute value function, such as $y = |2 - x|$, $y = 5 - |x|$, and $y = |3x + 2|$. To do this you need to consider intervals where the expression within the modulus signs is positive, and where it is negative.

QUADRATIC FUNCTIONS

A **quadratic function** has the form $f(x) = ax^2 + bx + c$ where a, b, and c are constants, $a \neq 0$. The graphs of all quadratic functions are **parabolas**.

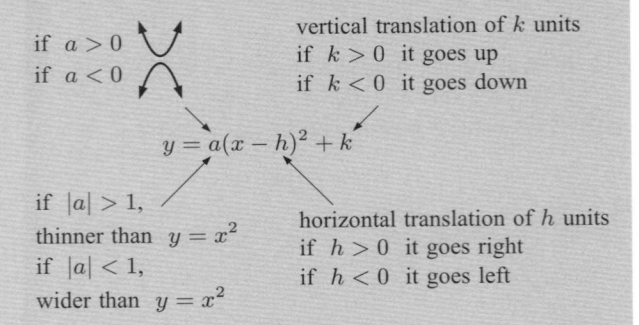

if $a > 0$
if $a < 0$

vertical translation of k units
if $k > 0$ it goes up
if $k < 0$ it goes down

$$y = a(x - h)^2 + k$$

if $|a| > 1$, thinner than $y = x^2$
if $|a| < 1$, wider than $y = x^2$

horizontal translation of h units
if $h > 0$ it goes right
if $h < 0$ it goes left

If $a > 0$, $y = ax^2$ is a vertical stretch of $y = x^2$ with invariant x-axis and scale factor a.

If $a < 0$, $y = ax^2$ is a vertical stretch of $y = -x^2$ with invariant x-axis and scale factor $-a$.

The x-**intercepts** of a quadratic are the values of x where the graph meets the x-axis.

For the function $y = ax^2 + bx + c$, the x-intercepts are found by solving $ax^2 + bx + c = 0$.

The number of x-intercepts corresponds to the number of solutions of the quadratic equation.

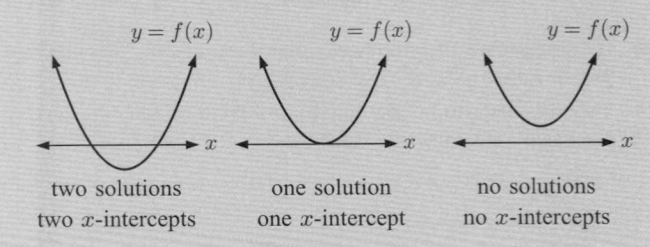

two solutions
two x-intercepts

one solution
one x-intercept

no solutions
no x-intercepts

The y-**intercept** is the value of y where the graph meets the y-axis. It is found by letting $x = 0$.

The graphs of all quadratic functions have a vertical **line of symmetry**.

The **turning point** or **vertex** of the graph lies on the line of symmetry.

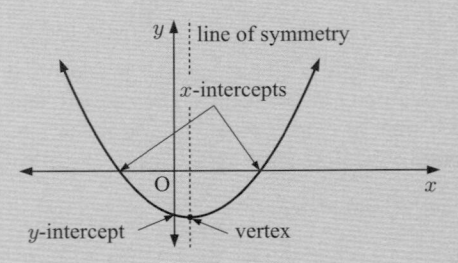

CUBIC FUNCTIONS

A **cubic function** has the form $f(x) = ax^3 + bx^2 + cx + d$ where a, b, c, and d are constants, $a \neq 0$.

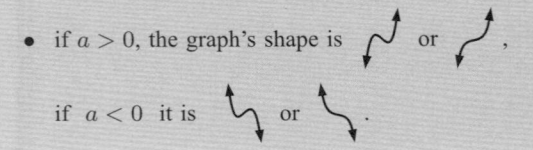

- if $a > 0$, the graph's shape is
- if $a < 0$ it is

- $y = (x-h)^3 + k$ is the translation of $y = x^3$ through $\binom{h}{k}$

EXPONENTIAL FUNCTIONS

An **exponential function** is a function in which the variable occurs as part of the exponent or index.

Graphs of the form $f(x) = a^x$ where $a > 0$, $a \neq 1$:
- have a horizontal asymptote $y = 0$ (the x-axis)
- pass through $(0, 1)$.

The **initial value** of an exponential function is the value when the independent variable (usually n or t) is zero.

Exponential Modelling

You should be able to:
- use a graphics calculator to fit an exponential model of the form $y = a \times b^x$ to a data set
- use your model to predict values
- comment on the reliability of your model.

LOGARITHMS

If $y = a^x$ then x is the logarithm of y in base a.

$$y = a^x \iff x = \log_a y$$

Rules for logarithms

$$\log_a(xy) = \log_a x + \log_a y$$

$$\log_a\left(\frac{x}{y}\right) = \log_a x - \log_a y$$

$$\log_a(x^n) = n \log_a x$$

You should be able to use these rules to simplify logarithmic expressions.

The logarithmic function

The **logarithmic function** is $f(x) = \log_a x$ where $a > 0$, $a \neq 1$.

The inverse function of $f(x) = \log_a x$ is $f^{-1}(x) = a^x$.

Logarithmic equations

You should be able to use the logarithm laws to:
- write an equation as a logarithmic equation in base 10
- write a logarithmic equation without logarithms.

TRIGONOMETRIC FUNCTIONS

You should be able to recognise trigonometric functions of the form $y = a\sin(bx)$, $y = a\cos(bx)$, $y = \tan x$.

Function	Period	Amplitude
$y = a\sin(bx)$	$\dfrac{360°}{b}$	$\|a\|$
$y = a\cos(bx)$	$\dfrac{360°}{b}$	$\|a\|$
$y = \tan x$	$180°$	

USING TECHNOLOGY

You should be able to use your calculator to obtain:
- a table of values for a function
- a sketch of the function
- the zeros or x-intercepts of the function
- the y-intercept of the function
- any asymptotes of the function
- the turning points of the function
- the points of intersection of two functions.

TRANSFORMING FUNCTIONS

- $y = f(x)$ maps onto $y = f(x) + k$ under a **vertical translation** of $\binom{0}{k}$
- $y = f(x)$ maps onto $y = f(x + k)$ under a **horizontal translation** of $\binom{-k}{0}$
- $y = f(x)$ maps onto $y = kf(x)$ under a stretch with invariant x-axis and scale factor k.

SKILL PRACTICE

1 Consider the relation $y = 2x^2$ on the domain $\{-2, -1, 0, 1, 2\}$.
   a Construct a mapping diagram to display the relation.
   b State the range of the mapping.
   c State whether the mapping is one-one, many-one, one-many, or many-many.
   d Is the relation a function?

2 For each of the following graphs, state the domain and range, and decide whether it is the graph of a function:

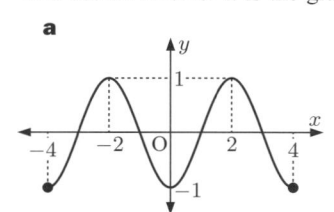

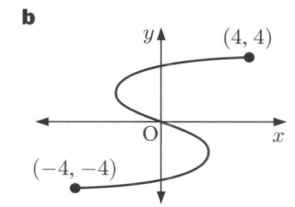

   a

   b

3 Given $f(x) = \dfrac{3x - 2}{x^2}$, find:
   a $f(1)$ **b** $f(2)$ **c** $f(3)$

4 Which of the following sets of ordered pairs are functions? Give reasons for your answers.
   a $\{(0, 3), (1, 1), (2, 2), (3, -1), (4, 1)\}$
   b $\{(3, 0), (1, 1), (2, 2), (-1, 3), (1, 4)\}$

5 Given $m(x) = x^2 - 5$, find x such that:
   a $m(x) = -1$ **b** $m(x) = 11$

6 If $f(x) = 3 - 4x$, find in simplest form:
   a $f(a)$ **b** $f(-a)$ **c** $f(1 - x)$
   d $f(x^2)$ **e** $f(x^2 + \frac{3}{4})$

7 Consider the relation $x^2 + y^2 = 1$ on the domain $\{-1, 0, 1\}$.
   a Construct a mapping diagram to display the relation.
   b State the range of the mapping.
   c State whether the mapping is one-one, one-many, many-one, or many-many.
   d Is the relation a function?

8 If $f(x) = 3^{-x} + 1$, find the value of:
  a $f(0)$ **b** $f(-2)$ **c** $f(3)$

9 If $f(x) = 2x + 1$ and $g(x) = x^2$, find:
  a $f(2)$ **b** $g(2)$ **c** $f(g(2))$
  d $g(f(2))$ **e** $f(f(2))$ **f** $g(g(2))$

10 Consider the function $y = \dfrac{16}{x}$.
  a Find the asymptotes of the function.
  b Find y when: **i** $x = 256$ **ii** $x = -256$
  c Find x when: **i** $y = 256$ **ii** $y = -256$
  d Graph $y = \dfrac{16}{x}$.

11 Write an equivalent logarithmic statement for:
  a $81^{\frac{1}{2}} = 9$ **b** $2^{10} = 1024$
  c $32^{-\frac{1}{5}} = \frac{1}{2}$ **d** $8\sqrt{2} = 2^{3.5}$

12 If $x = -2$, evaluate:
  a $|2x - x^2|$ **b** $\dfrac{|x|}{3 - 2x}$ **c** $\dfrac{-x^2}{|2 + 3x|}$

13 For the function $f(x) = -2x^2 + 5x + 1$, find:
  a $f(3)$ **b** a such that $f(a) = 4$.

14 Without using a calculator, find the value of:
  a $\log_2 \sqrt{2}$ **b** $\log_{10} 10\,000\,000$
  c $\log_5 3125$ **d** $\log_7 1$
  e $\log_3 \left(\dfrac{1}{81\sqrt{3}}\right)$ **f** $\log_{\sqrt{2}} 64$

15 From a table of values for $x = -3, -2, -1, 0, 1, 2, 3$, draw the graph of:
  a $2x^2 - 3x - 1$ **b** $-x^2 + x + 5$

16 Write an equivalent exponential statement for:
  a $\log_3 81 = 4$ **b** $\log_7 7\sqrt{7} = 1.5$
  c $\log_{\sqrt{5}} 125 = 6$

17 Solve for x:
  a $|3 - x| = 5$ **b** $3 - |x| = 2$

18 Suppose $f(x) = 2 - x$ and $g(x) = 3 + 2x$.
  a Find, in simplest form:
    i $f(g(x))$ **ii** $g(f(x))$
    iii $f(f(x))$ **iv** $g(g(x))$
  b Show that $f(g(x)) \neq g(f(x))$ for all x.
  c Find x such that $f(f(x)) = g(g(x))$.

19 For the following functions:
    i sketch the graph
    ii find the y-intercept
    iii find the equation of any asymptote.
  a $f(x) = 2^{x-2}$ **b** $f(x) = 2 \times 3^x - 1$
  c $f(x) = 5 - 5^x$

20 Rewrite as logarithmic equations:
  a $y = 3^{\frac{1}{2}x}$ **b** $y = 5^{-a}$ **c** $y = 4 \times 2^n$

21 Rewrite as exponential equations:
  a $P = \log_3 t$ **b** $y = \frac{1}{4} \log_2 x$ **c** $Q = \log_5 \left(\frac{t}{3}\right)$

22 Sketch each of the following functions on the same set of axes as $y = x^2$. In each case state the coordinates of the vertex.
  a $y = (x - 1)^2 - 2$ **b** $y = 2(x + 2)^2 + 1$
  c $y = \frac{1}{2}(x - 3)^2 - 4$ **d** $y = -3(x + 1)^2 + 6$

23 Rewrite the following, making x the subject:
  a $y = 7^{x+1}$ **b** $d = \log_5(3x)$
  c $z = \log_6\left(\frac{x}{2}\right)$ **d** $T = \frac{2}{3} \times 2^{10x}$

24 Write as powers of 10 using $a = 10^{\log_{10} a}$:
  a 40   **b** 0.004   **c** 500   **d** 0.05

25 State the period and amplitude of the following functions. Draw freehand sketches of the functions for $0° \leqslant x \leqslant 720°$.
  a $y = -2\sin(\frac{1}{2}x)$ **b** $y = 3\cos(3x)$

26 Find the inverse function of:
  a $f(x) = 2x - 5$ **b** $f(x) = -x^3 + 2$
  c $f(x) = \dfrac{3}{1 - x}$

27 Use your graphics calculator to solve the following equations correct to 3 significant figures:
  a $\log_{10} x = x - 1$ **b** $\log_{10}(x - 1) + 1 = 8 - 3x$
  c $\log_{10}\left(\frac{x}{2}\right) + 1 = 3^{-x}$

28 Find the axes intercepts of the following functions:
  a $y = -x^2 + 6x - 5$ **b** $y = 2(x - 1)(x + 5)$

29 Write the following functions without the modulus sign, and hence graph each function:
  a $y = 3 - |x|$ **b** $y = 2|x| - 4$

30 The average speed of a professional cyclist on a mountain ascent is given by
$$s(g) = 45 - g - \dfrac{g^2}{8} \text{ km/h}$$
where g is the angle of incline of the mountain in degrees. Find the angle of incline of a 13.6 km mountain which takes him 48 minutes to climb.

31 Write as a single logarithm:
  a $\log_3 7 + \log_3 4$ **b** $5 + \log_2 3$
  c $\log_2 11 - \log_2 7$ **d** $3\log_5 a + 4\log_5 b$

32 Sketch the graph of the quadratic function with:
  a x-intercepts -4 and 2, and y-intercept -6
  b x-intercept 3 and y-intercept 5.

33 Which of the following functions have inverses? Explain your answers.

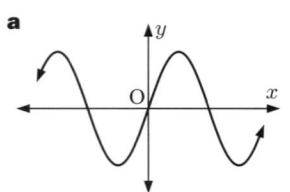

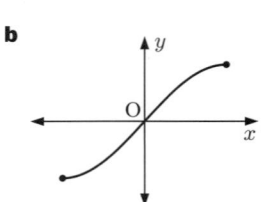

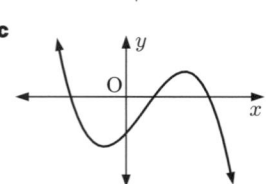

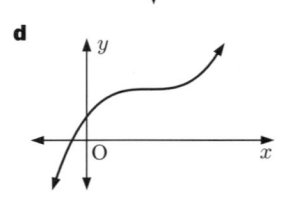

34 If $\log_2 5 = p$ and $\log_2 7 = q$, write in terms of p and q:
   a $\log_2 35$ **b** $\log_2(1.4)$ **c** $\log_2 125$

35 A plate of chicken stew is left on a kitchen bench. Its temperature after x minutes is given by
$T = 65 \times (0.95)^x$ °C.
   a Find the initial temperature of the stew.
   b Find the temperature after:
      i 3 minutes   **ii** 10 minutes.
   c Once the stew's temperature drops to $20\,°C$, it must be thrown away for health reasons. How long will this take?

36 Sketch the graphs of the following by considering the axes intercepts:
   a $y = (x+4)(x-2)$ **b** $y = -2(x-1)^2$
   c $y = 2x^2 + 3x - 5$ **d** $y = -6x^2 + 13x - 5$

37 Write y in terms of a and b:
   a $\log_3 y = \frac{1}{2} \log_3 a$ **b** $\log_{10} y = a - \log_{10} b$
   c $\log_7 y = -\log_7 b$
   d $\log_2 y = \log_2 a + 3\log_2 b$

38 Consider $f(x) = -2x + 3$.
   a On the same grid, graph $y = f(x)$, $y = f(x-1)$, and $y = -f(x)$.
   b What transformation has occurred to obtain:
      i $y = f(x-1)$ **ii** $y = -f(x)$?

39 Determine the equations of the following reciprocal graphs:
   a graph passing through $(3, -4)$
   b graph passing through $(-5, -2)$

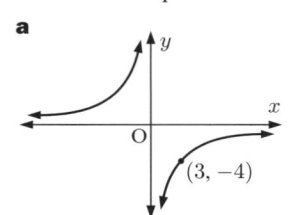

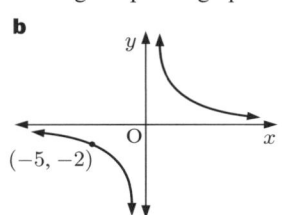

40 Without using a calculator, simplify:
   a $\dfrac{\log_3 81}{\log_3 \sqrt{3}}$ **b** $\dfrac{\log_2(\frac{1}{4})}{\log_2 16}$

41 Using your calculator, find the vertex of each of the following quadratic functions:
   a $f(x) = 2x^2 - 3x + 1$ **b** $f(x) = -\frac{1}{2}x^2 + 4x + 2$

42 Sketch the graphs of:
   a $f(x) = |x+3|$ **b** $f(x) = |3-2x|$

43 Find the inverse function $f^{-1}(x)$ for
   a $f(x) = 2^{-x}$ **b** $f(x) = 4\log_5 x$
   c $f(x) = \log_{\sqrt{3}} x$ **d** $f(x) = 4 \times 10^x$

44 Use axes intercepts to sketch the graphs of:
   a $y = (x-2)(x+1)(x-5)$
   b $y = -(x+2)^2(x-3)$
   c $y = 2x(x-1)(x-2)$
   d $y = -3(2x+1)(2x-1)(2x-5)$

45 Show that an exponential model is appropriate for the following data, and state the equation of the exponential model.
   a
t	1	2	5	10
P	5.5	14.1	315	49 472

   b
x	3	6	9	12
Q	10.5	4.45	1.88	0.792

46 For each of the following quadratic functions, find:
   i the axes intercepts
   ii the equation of the line of symmetry
   iii the coordinates of the vertex
   iv and hence sketch the graph.
   a $y = x^2 - 7x + 10$ **b** $y = -2x^2 + 8x - 8$
   c $y = 3x^2 - 15$ **d** $y = 6x^2 - 15x - 9$

47 Consider $p(x) = x^2$.
   a Graph $y = p(x)$, $y = p(x+3)$, and $y = p(x) - 2$ on the same set of axes.
   b Describe fully the single transformation which maps the graph of $y = p(x+3)$ onto $y = p(x) - 2$.

48 Suppose $f(x) = 1 + x$ and $g(x) = |x|$.
   a Graph $y = f(x)$ and $y = g(x)$ on the same set of axes.
   b Find x such that $f(x) = g(x)$.
   c Find $f(g(x))$ and $g(f(x))$ in simplest form.
   d Graph $y = f(g(x))$ and $y = g(f(x))$ on the same set of axes.
   e Find x such that $f(g(x)) = g(f(x))$.

49 Write as a single logarithm and simplify if possible:
   a $\log 4 + \log 5$ **b** $3 - \log_{10} 5$
   c $2\log 3 + 3\log 4$ **d** $\frac{1}{4}\log 81 + \log(\frac{1}{3})$
   e $5\log 2 - 3\log 5$ **f** $\log_{10} 25 + \log_{10} 4$

50 For each of the following quadratic functions:
   i sketch the graph using axes intercepts
   ii hence find the line of symmetry and the vertex.
   a $y = -3(x-2)(x+3)$ **b** $y = 2(x+1)(7-x)$
   c $y = 4(x-3)^2$ **d** $y = 2x^2 - 7x + 3$

51 Without using a calculator, show that:
   a $\log\left(\sqrt[5]{2}\right) = \frac{1}{5}\log 2$ **b** $\log\left(\frac{1}{81}\right) = -4\log 3$
   c $\log 20 = \log 2 + 1$

52 Find a and b if $y = a\sin(bx)$ has graph:
   a (graph with amplitude 5, period indicated with markings at $-270°$ and $270°$)
   b (graph with amplitude 3, period indicated with markings at $-360°$ and $360°$)

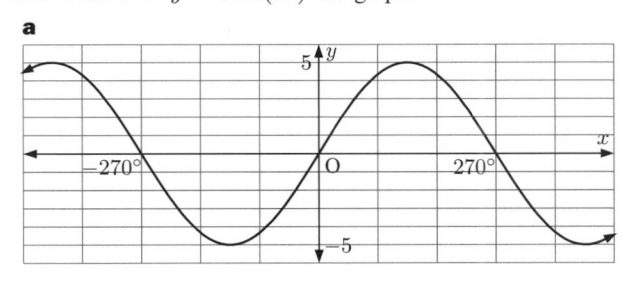

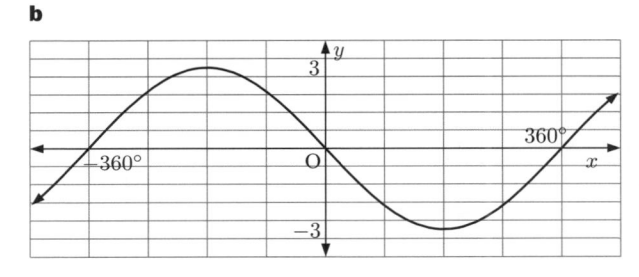

53 Find the quadratic function with:
 a vertex $(-1, 2)$ and y-intercept 3
 b x-intercepts $-\frac{3}{2}$ and $\frac{1}{2}$, and y-intercept -3
 c vertex $(3, 2)$ and which passes through $(5, -2)$
 d x-intercepts 0 and 8, and which passes through $(7, -5)$.

54 Find the function $f(x) = |ax + b|$ which has the graph:
 a
 b

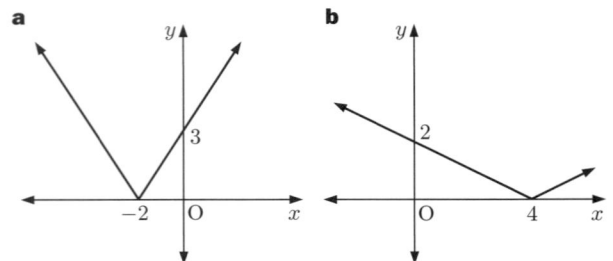

55 Write the following as logarithmic equations in base 10:
 a $y = a^2 b$
 b $Q = \dfrac{m\sqrt{n}}{p}$
 c $P = \sqrt[3]{s^2 t}$

56 Find the function corresponding to each of these graphs, giving your answer in the form $y = a(x - h)^2 + k$.
 a
 b
 c
 d

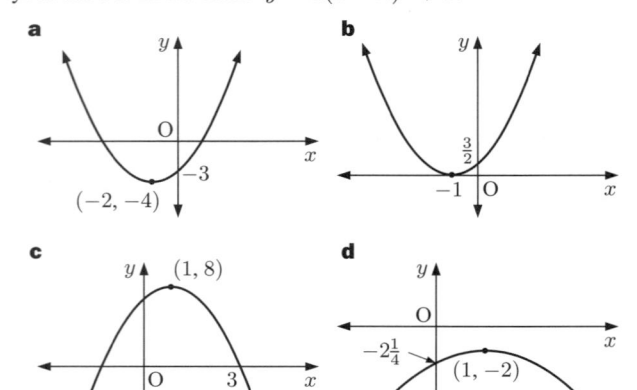

57 Write these equations without logarithms:
 a $\log_{10} T = x - 2$
 b $\log_{10} K = \frac{1}{3} x - 1$
 c $\log_{10} Q \approx 0.903 + x$
 d $\log p = -\frac{1}{2} \log q$
 e $2 + \log_{10} y = \log_{10} x$
 f $\log A = \log c + \frac{1}{3} \log d$

58 For each of the following:
 i use technology to graph the function
 ii find the equations of the asymptotes.
 a $y = -\dfrac{3}{x - 1}$
 b $y = -1 + \dfrac{2}{1 - x}$

59 The value of an investment after n years at 7.6% p.a. compound interest is given by $F = 7500 \times (1.076)^n$ euros.
 a What was the original investment?
 b Find the value of the investment after:
  i 4 years
  ii 10 years.
 c Find the time taken for the value of the investment to reach €25 000.

60 Solve for x using logarithms, giving your answers to 3 significant figures:
 a $10^x = 30$
 b $10^x = 0.000\,031\,416$
 c $9^x = 45$
 d $7^x = 0.004\,17$
 e $(0.375)^x = 8$
 f $(0.9)^x = 7.29$

61 In each of the following, $g(x)$ is mapped onto $h(x)$ using a single transformation.
 i Describe the transformation fully.
 ii Write $h(x)$ in terms of $g(x)$.
 a 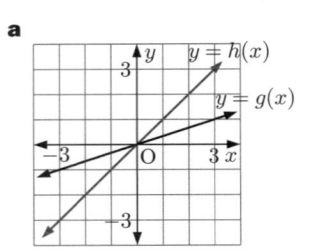
 b

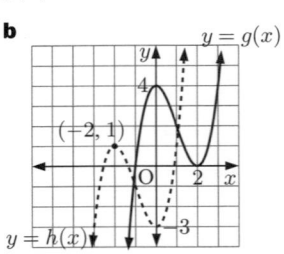

62 Find the quadratic function with:
 a vertex $(4, 3)$ and which passes through the origin
 b vertex $(6, 16)$ and which has 8 as one of its x-intercepts
 c x-intercepts 3 and 5, and which passes through $\left(\frac{9}{2}, 2\right)$.

63 **a** This cubic function has the form $y = \frac{1}{10} x^3 + cx + d$. Find c and d. 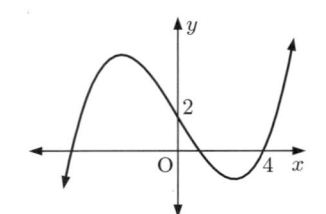
 b This cubic function has the form $y = ax^3 + bx^2 + 2x + 3$. Find a and b.

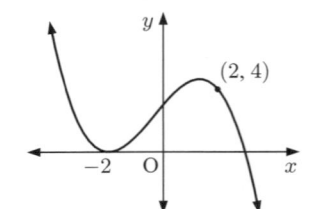

64 Use your calculator to solve correct to 1 decimal place, for $0° \leqslant x \leqslant 360°$:
 a $2 \cos x = -1.3$
 b $\tan x = -2$
 c $\frac{1}{2} \cos(2x) = \frac{1}{3}$
 d $-\sin(3x) = 0.2168$

65 Consider the function $f(x) = \dfrac{x - 2}{x + 1}$.
 a Find $f^{-1}(x)$.
 b Graph $y = f(x)$ and $y = f^{-1}(x)$ on the same set of axes.
 c What do you notice about the graphs of $y = f(x)$ and $y = f^{-1}(x)$?

66 Solve for x:
 a $\log_7 x = 50$
 b $\log_3 x = 2$
 c $\log_2 \left(\frac{x}{2}\right) = -2$

67 The height of a projectile t seconds after its launch is given by $h(t) = 20.2t - 4.9t^2$ m/s.
 a Calculate the height of the projectile after:
  i 2 seconds
  ii 4 seconds.
 b How long will it take for the projectile to return to the ground?

68 Copy and complete the following graphs and transformations:
 a Draw $y = \frac{3}{2} f(x)$
 b Draw $y = f(x + 1)$

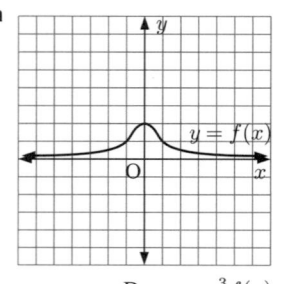

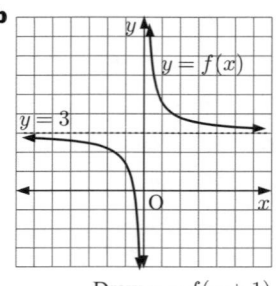

69 Consider $f(x) = \dfrac{x^2 - 4}{x^2 + 2x - 3}$.

   a Use your calculator to help sketch the function.
   b State the equations of any asymptotes.
   c Find the axes intercepts.
   d Find and classify any turning points.

70 Consider the cubic function $y = x^3 + x^2 + x - 3$.

   a Show that the function has an x-intercept of 1.
   b Hence write the function in the form $y = (x - 1)(x^2 + bx + c)$.
   c Show that the function has no other x-intercepts.
   d Use technology to help sketch the function.

71 The weight of a sample of sodium-24 after t hours is given by $W = 5 \times 2^{-0.067t}$ grams.

   After what time will the weight reach:
   a 4 grams   **b** 0.5 grams?

72 Consider $f(x) = x^4 - 2x^3 + 7x - 1$.

   a Use technology to help sketch the graph of $f(x)$.
   b Find the zeros of $f(x)$.
   c Find and classify the turning points of the function.

73 John has his car valued for insurance purposes every 2 years. The valuations were:

Year	2000	2002	2004	2006	2008
Value (£V)	15 000	9600	6150	3900	2500

   a Let the year 2000 be time $t = 0$.
      Show that an exponential model fits the data well, and find the model.
   b Estimate the value of the car in the years:
      i 2005   **ii** 2015.
   c Comment on the reliability of your estimates in **b**.

74 Find the image of:

   a $y = 2^x$ under the translation $\begin{pmatrix} -3 \\ 2 \end{pmatrix}$
   b $y = 3^x$ under a reflection in the line $y = x$.

75 The value of a wooden cabinet t years after its purchase is given by $V(t) = t^2 - 40t + 500$ euros.

   a Find $V(10)$ and explain what this represents.
   b Sketch the function, identifying its key features.
   c Hence identify when the cabinet's value is a minimum. State this minimum value.
   d Do you think this formula will be valid for all $t > 0$? Explain your answer.

76 Consider $f(x) = 3^{-x} - 2$ and $g(x) = -\dfrac{x}{x + 2}$.

   a Sketch the graphs of $y = f(x)$ and $y = g(x)$ on the same set of axes for $-5 \leqslant x \leqslant 5$.
   b Find the coordinates of the points of intersection of the two graphs.
   c Find the values of x for which $3^{-x} \geqslant 2 - \dfrac{x}{x + 2}$.

77 Find a and b if $y = a\cos(bx)$ has graph:

   a

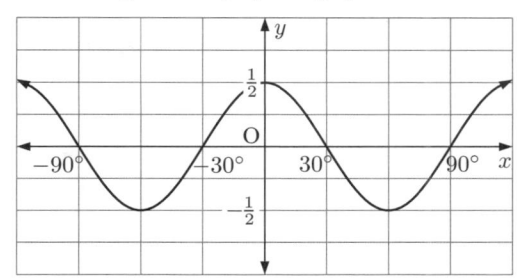

   b

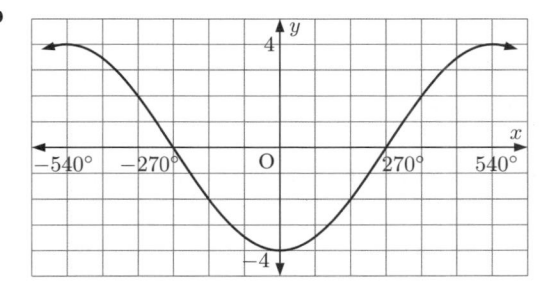

78 The graph below has the form $y = 2x^3 + bx^2 + cx + 18$. Find the values of b and c.

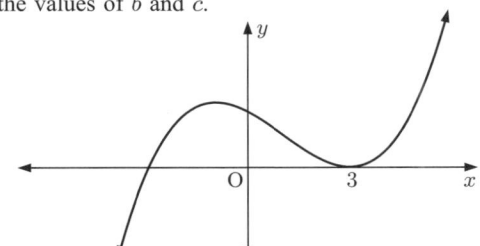

TOPIC 4: GEOMETRY

LINES AND LINE SEGMENTS

Line AB is the endless straight line passing through the points A and B.

Line segment AB is the part of the straight line AB that connects A with B.

The **distance AB** is the length of the line segment AB.

Collinear points lie on a straight line.

Perpendicular lines intersect at right angles.

Parallel lines are lines which never intersect.

SYMMETRY

A figure has **line symmetry** if it can be reflected in a line so that each half of the figure is reflected onto the other half of the figure.

A figure has **rotational symmetry** if it can be mapped onto itself more than once as it rotates through 360° about a point called the **centre of symmetry**.

The number of times an object will fit onto itself when rotated through 360° is called its **order of rotational symmetry**.

ANGLES

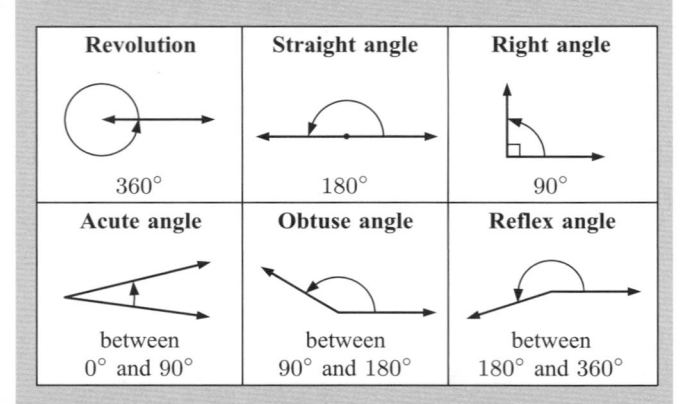

Revolution	Straight angle	Right angle
360°	180°	90°
Acute angle	Obtuse angle	Reflex angle
between 0° and 90°	between 90° and 180°	between 180° and 360°

Complementary angles have sizes which add to 90°.

Supplementary angles have sizes which add to 180°.

Two angles which have the same vertex and share a common arm are called **adjacent angles**.

Property	Figure
The sum of the sizes of the **angles at a point** is 360°.	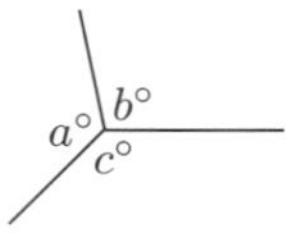 $a + b + c = 360$
The sum of the sizes of the **angles on a line** is 180°.	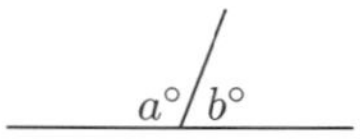 $a + b = 180$
The sum of the sizes of the **angles in a right angle** is 90°.	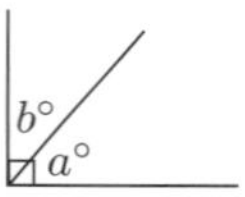 $a + b = 90$
Vertically opposite angles are equal in size.	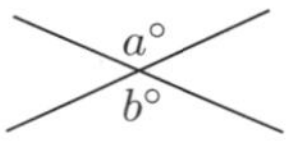 $a = b$
When two *parallel* lines are cut by a third line, the angles in **corresponding** positions are equal in size.	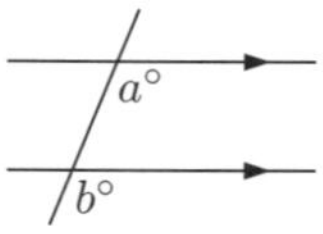 $a = b$
When two *parallel* lines are cut by a third line, the angles in **alternate** positions are equal in size.	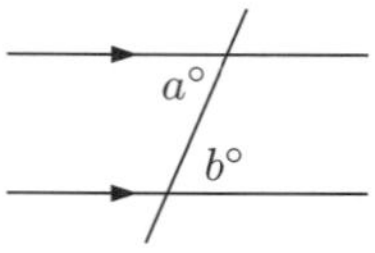 $a = b$
When two *parallel* lines are cut by a third line, the sum of the sizes of the **co-interior angles** is 180°.	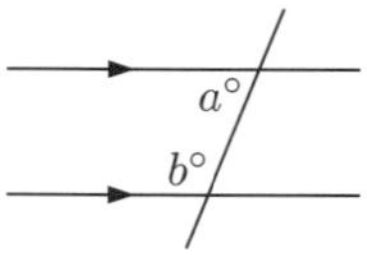 $a + b = 180$

TRIANGLES

Property	Figure
The sum of the interior angles of a triangle is 180°.	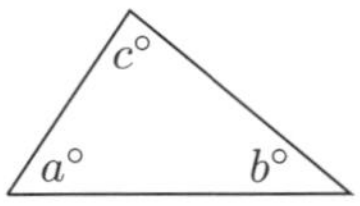 $a + b + c = 180$
The size of the exterior angle of a triangle is equal to the sum of the interior opposite angles.	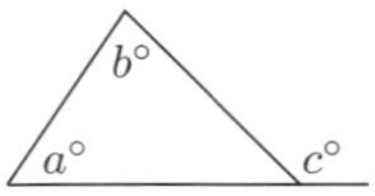 $c = a + b$

The longest side of a triangle is opposite the largest angle.

An **isosceles triangle** is a triangle in which:
- two sides are equal in length
- base angles are equal
- the line joining the apex to the midpoint of the base bisects the vertical angle and meets the base at right angles.

You should understand which properties of a triangle are necessary to indicate it is isosceles.

An **equilateral triangle** is a triangle in which:
- all three sides are equal in length
- all angles are 60°.

POLYGONS

A **polygon** is any closed figure with straight line sides which can be drawn on a flat surface.

A **regular polygon** is a polygon in which all sides have the same length and all interior angles are equal.

You should understand the special properties of the polygons: parallelogram, rectangle, rhombus, square, trapezium, kite.

The sum of the **interior angles** of any n-sided polygon is $(n-2) \times 180°$.

The sum of the **exterior angles** of any polygon is 360°.

PYTHAGORAS

Pythagoras' theorem

In a right angled triangle with hypotenuse c and legs a and b, $c^2 = a^2 + b^2$.

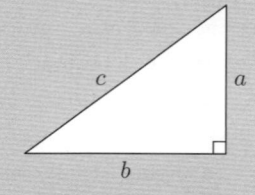

The set of positive integers $\{a, b, c\}$ is a **Pythagorean triple** if it obeys the rule $a^2 + b^2 = c^2$.

If a triangle has sides of length a, b, and c units, and $a^2 + b^2 = c^2$, then the triangle is right angled.

You should recognise the presence of right angled triangles in geometric figures such as rectangles, rhombuses, and isosceles and equilateral triangles.

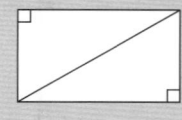

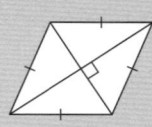

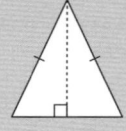

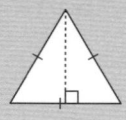

You should recognise the presence of right angled triangles in circle problems (see Circle Theorems).

CONGRUENCE

Two objects are **congruent** if they have the same size and shape.

Two triangles are congruent if any one of the following is true:
- All corresponding sides are equal in length. (**SSS**)

- Two sides and the **included** angle are equal. (**SAS**)

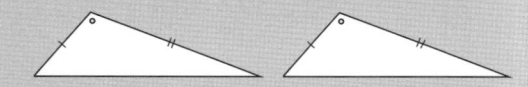

- Two angles and a pair of **corresponding sides** are equal. (**AAcorS**)

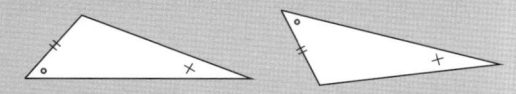

- For right angled triangles, the hypotenuses and one pair of sides are equal. (**RHS**)

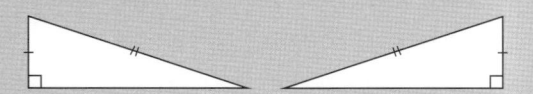

SIMILARITY

Two figures are **similar** if one is an enlargement of the other, regardless of their orientation.

If two polygons are similar then:
- the figures are equiangular
- the corresponding sides are in proportion.

Either of these properties is sufficient to prove that two triangles are similar.

You should be able to show whether two triangles are similar, and hence find unknown angles and side lengths. Remember it is often useful to:
- draw a labelled sketch showing all information
- write the information in a table, showing the equal angles and the sides opposite these angles
- answer problems with a sentence.

If a figure is **enlarged** with **scale factor** k to produce a similar figure, then the new area = $k^2 \times$ the old area.

If a solid is **enlarged** with **scale factor** k to produce a similar solid, then the new volume = $k^3 \times$ the old volume.

CIRCLES

- A **circle** is the set of all points which are equidistant from a fixed point called the **centre**.
- The **circumference** is the distance around the entire circle boundary.

- An **arc** of a circle is any continuous part of the circle.
- A **chord** of a circle is a line segment joining any two points on the circle.
- A **semi-circle** is a half of a circle.
- A **diameter** of a circle is any chord passing through its centre.
- A **radius** of a circle is any line segment joining its centre to any point in the circle.
- A **tangent** to a circle is any line which touches the circle in exactly one point.

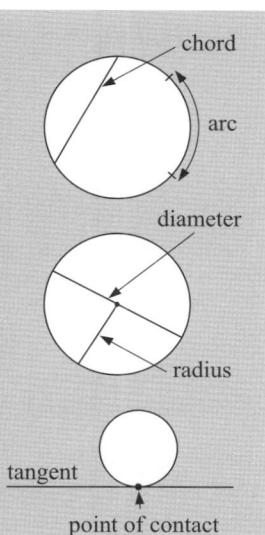

CIRCLE THEOREMS

Angle in a semi-circle

The angle in a semi-circle is a right angle.

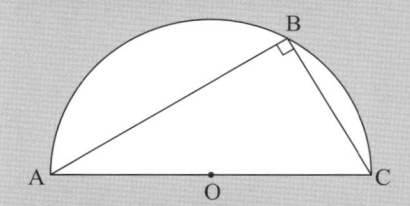

$$A\widehat{B}C = 90°$$

Chords of a circle

The perpendicular from the centre of a circle to a chord bisects the chord.

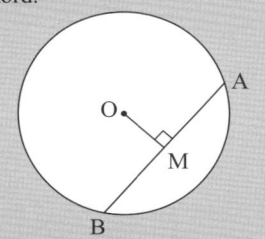

$$AM = BM$$

Radius-tangent

The tangent to a circle is perpendicular to the radius at the point of contact.

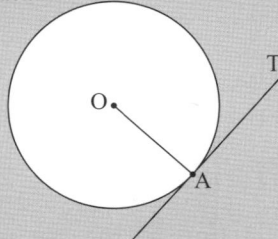

$$O\widehat{A}T = 90°$$

Tangents from an external point

Tangents from an external point are equal in length.

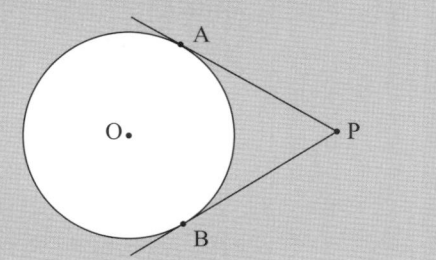

$$AP = BP$$

Two useful **converses** are:

- If line segment AB subtends a right angle at C then the circle through A, B and C has diameter AB.

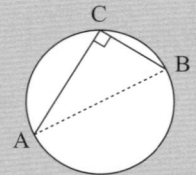

- The perpendicular bisector of a chord of a circle passes through its centre.

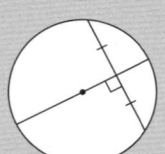

THEOREMS INVOLVING ARCS

Angle at the centre

The angle at the centre of a circle is twice the angle on the circle subtended by the same arc.

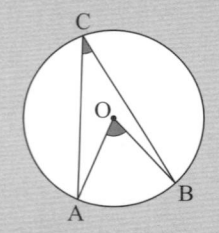

$$A\widehat{O}B = 2 \times A\widehat{C}B$$

The diagrams below show special cases of **the angle at the centre** theorem:

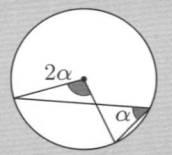

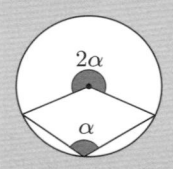

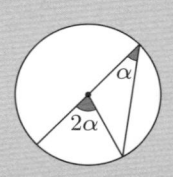

Angles subtended by the same arc

Angles subtended by an arc on the circle are equal in size.

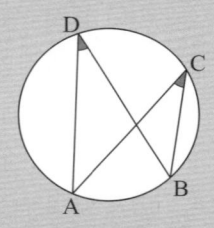

$$A\widehat{D}B = A\widehat{C}B$$

CYCLIC QUADRILATERALS

If a circle can be drawn through four points, we say that the points are **concyclic**.

If any four points on a circle are joined to form a convex quadrilateral then the quadrilateral is said to be a **cyclic quadrilateral**.

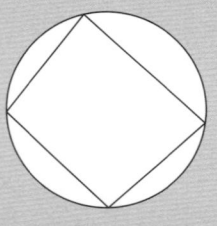

CYCLIC QUADRILATERAL THEOREMS

Opposite angles of a cyclic quadrilateral

The opposite angles of a cyclic quadrilateral are supplementary.

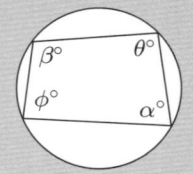

$$\alpha + \beta = 180$$
$$\theta + \phi = 180$$

Exterior angle of a cyclic quadrilateral

The exterior angle of a cyclic quadrilateral is equal to the opposite interior angle.

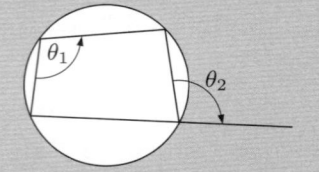

$$\theta_1 = \theta_2$$

Tests for cyclic quadrilaterals

A quadrilateral is a **cyclic quadrilateral** if one of the following is true:

- one pair of opposite angles is supplementary

If $\alpha + \beta = 180°$ then ABCD is a cyclic quadrilateral.

- one side subtends equal angles at the other two vertices

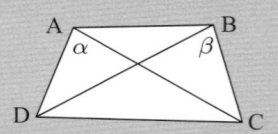

If $\alpha = \beta$ then ABCD is a cyclic quadrilateral.

- an exterior angle is equal to the opposite interior angle

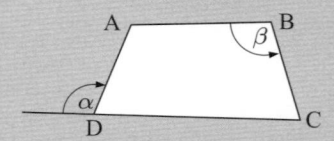

If $\alpha = \beta$ then ABCD is a cyclic quadrilateral.

SKILL PRACTICE

1 Classify these angles:

 a $142°$ **b** $212°$
 c   **d**

2 Classify these polygons:

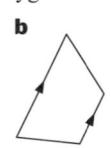

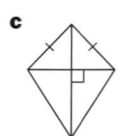

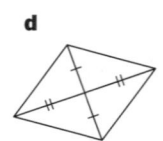

 a   **b**   **c**   **d**

3 Draw a diagram which shows collinear points A, B and C. Also draw a perpendicular line segment BD.

4 State the order of rotational symmetry of:

 a   **b**

5 Draw a rhombus PQRS, clearly marking its key features.

6 Find x, giving your answers correct to 3 significant figures:

a [triangle with legs x cm and 4 cm, right angle]

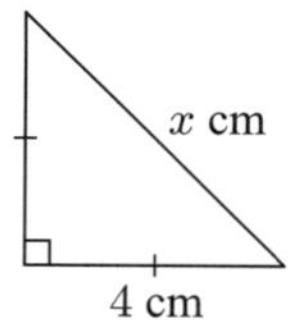

b [right triangle with hypotenuse x m, legs 4.1 m and 2.6 m]

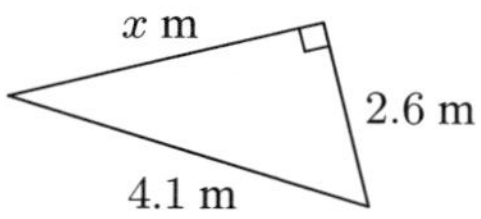

c [right triangle with sides 5.8 km, 3.2 km and x km]

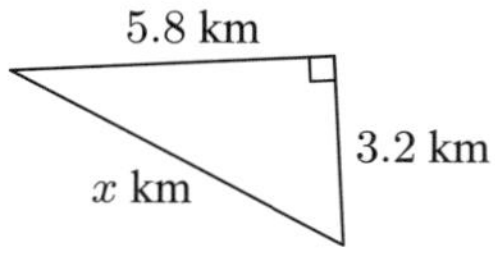

7 Determine if the following are Pythagorean triples:
  a $\{4, 7, 8\}$ **b** $\{5, 12, 13\}$

8 A triangle has sides of length 2 cm, 2 cm, and 3 cm. Describe the symmetry of this triangle.

9 Discuss the congruence of these pairs of triangles:
  a   **b**

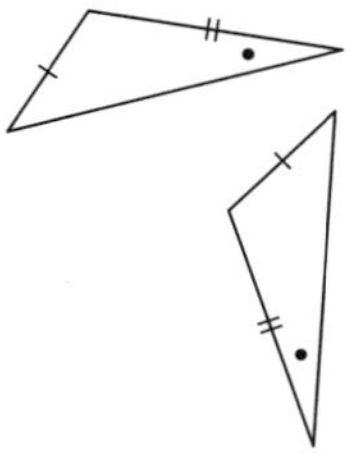

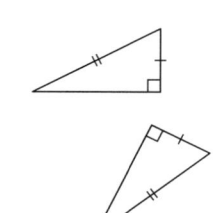

10 Find the values of the unknowns, giving brief reasons:
  a [parallel lines with angles $x°$, $y°$, $115°$]
  b [crossing lines with $100°$, $x°$, $y°$]
  c [angles $20°$, $x°$, $y°$, $40°$]
  d [angles $2x°$, $4x°$, $3x°$]

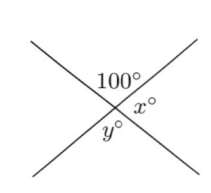

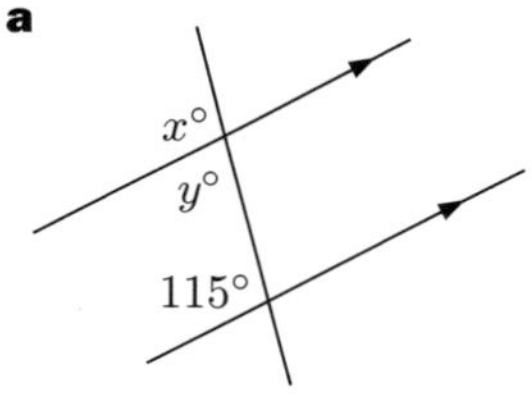

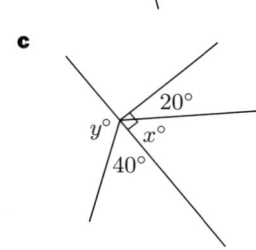

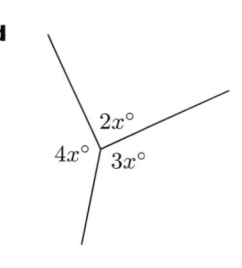

11 [Two similar quadrilaterals: small with sides 3 cm, 6 cm; large with sides x cm, 4.5 cm]

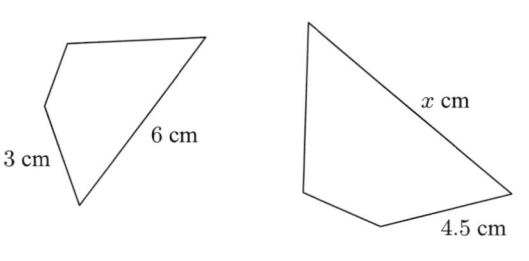

The given figures are similar.
  a Find the scale factor k needed to enlarge the small figure into the large one.
  b Find x.

12 State whether AB is parallel to CD, giving a brief reason for your answer. Note that the diagrams are not drawn to scale.
  a [angles $110°$ and $110°$]
  b [angles $110°$ and $110°$]

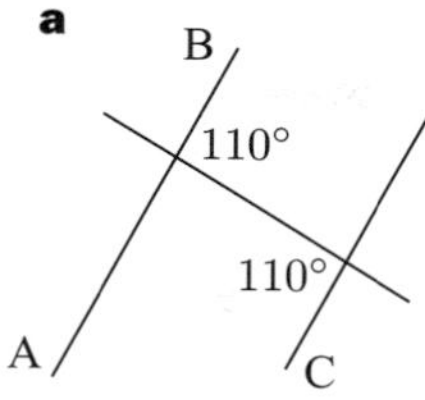

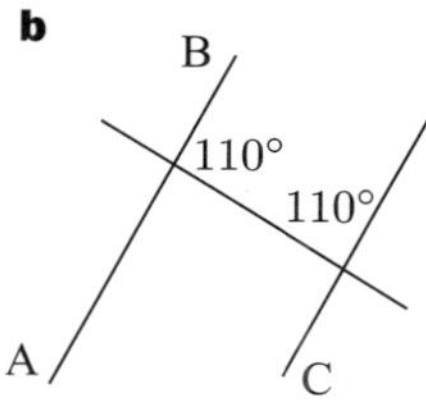

13 Find x, giving your answers in simplest surd form:
  a [right triangle with hypotenuse 85 m, legs $2x$ m and x m]
  b [right triangle with hypotenuse 12 cm, legs $(x+3)$ cm and $(x-3)$ cm]
  c [quadrilateral with sides x cm, 2 cm, 5 cm]

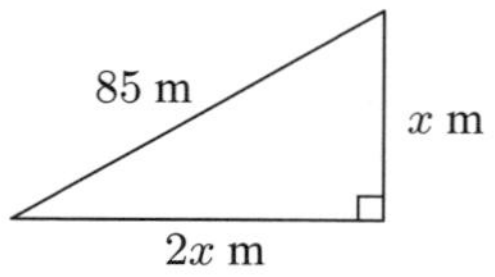

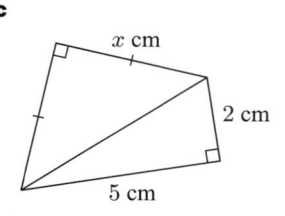

14 Find the unknown in the following, giving brief reasons:
  a [right triangle with angles $36°$ and $x°$]
  b [triangle with angles $35°$, $65°$, $x°$]
  c [isosceles triangle with $110°$ and $x°$]
  d [angles $x°$, $70°$, $30°$, $110°$]

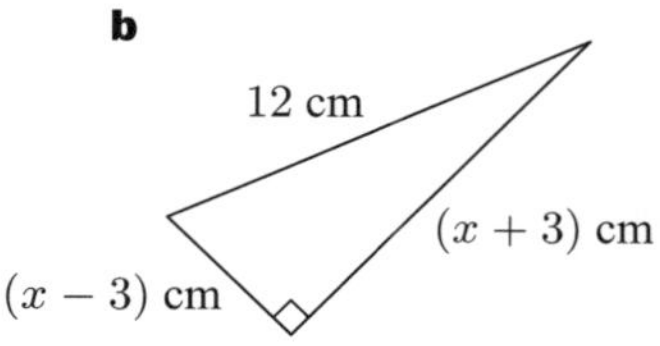

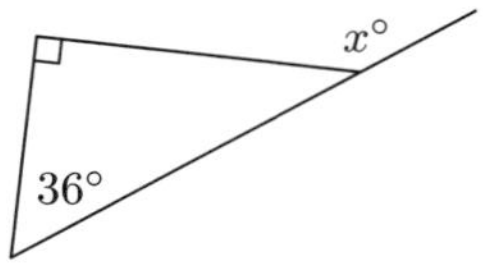

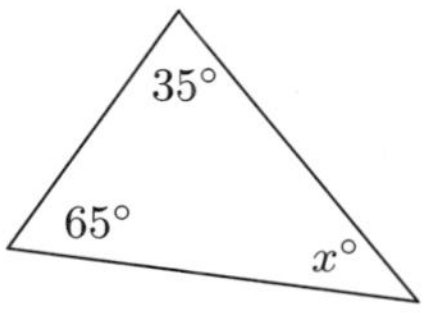

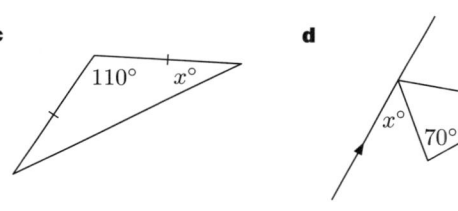

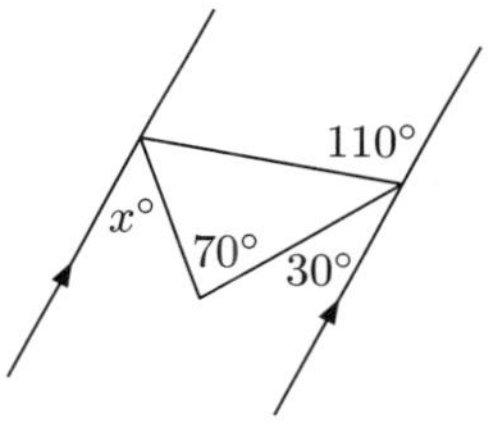

15 Find x given that the figures are similar: [7 cm, 5 cm; 8 cm, x cm]

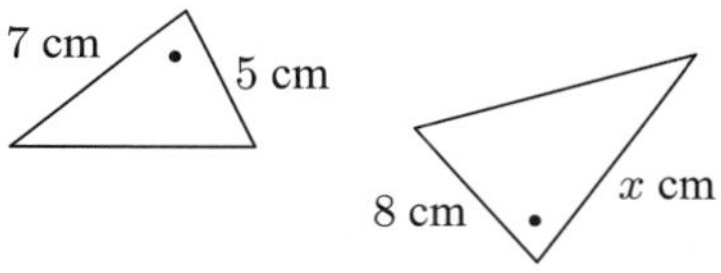

16 Find the length of AD correct to 3 significant figures.
[Triangle ABD with right angle at B, AB = 1 m, BC = CD, AC = CD marked]

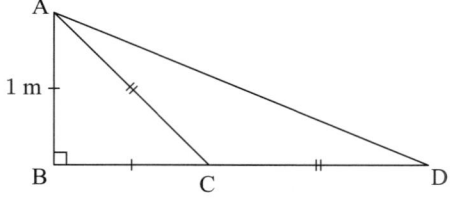

17 [Triangle with angles $2x°$ at A, $(3x-10)°$ at B, $(x+40)°$ at C]
  a Find the value of x.
  b Classify the triangle.
  c State the shortest side of the triangle.

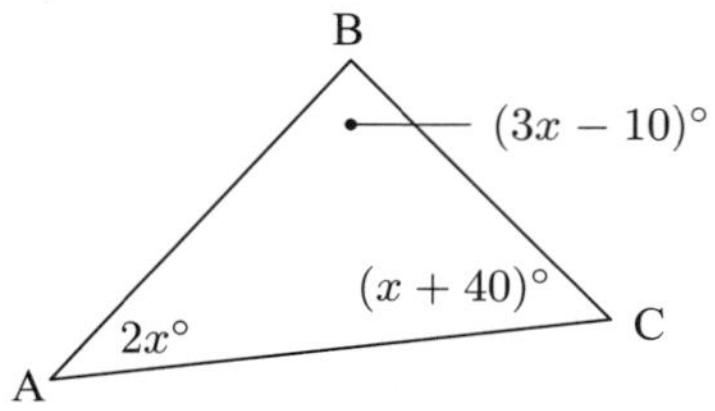

18 Comment on the truth of the following statements, giving reasons for your answers.
  a All equilateral triangles are similar.
  b All rhombuses are similar.

19 Find x, giving brief reasons:
  a
  b

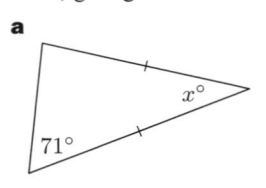

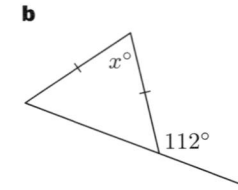

20 Explain why $\{7k, 24k, 25k\}$ is a Pythagorean triple for all $k \in \mathbb{Z}^+$.

21 Classify the following triangle:

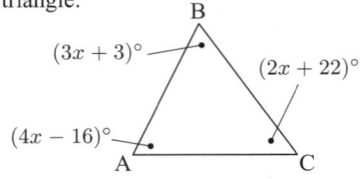

22 Paul has just constructed a window frame, and wants to check it is rectangular. The frame is 54.2 cm wide and 38.7 cm high. Paul measures the diagonal to be 66.6 cm. Do you think the frame is rectangular? Explain your answer.

23 Find the sum of the interior angles of a hexagon.

24 Show that the following figures possess similar triangles:
  a
  b

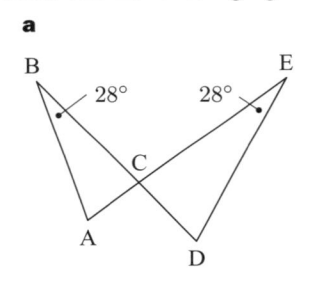

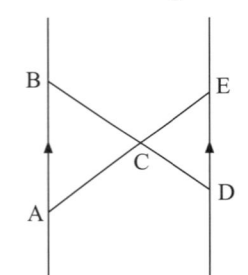

25 Find x, giving brief reasons:
  a
  b

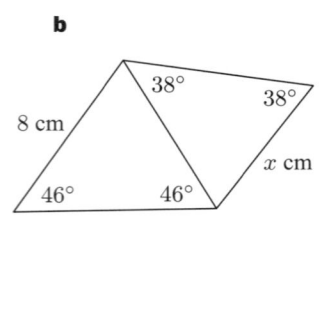

26 A road sign in the shape of a square has the dimension shown.
Find the length of its sides.

27 Find the value of the unknown:
  a
  b

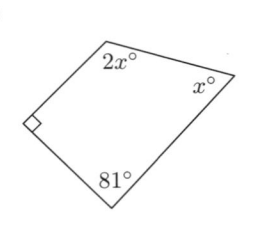

28 Show that the following figures possess similar triangles, and hence find x:
  a
  b

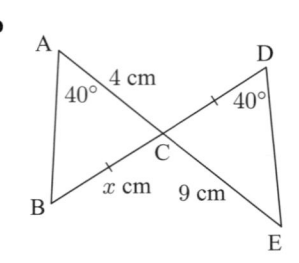

29 Find the values of any unknowns, giving brief reasons for your answers:
  a
  b
  c

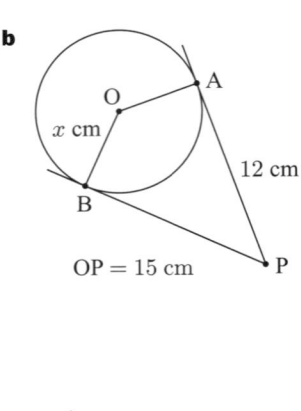

30 Find x:

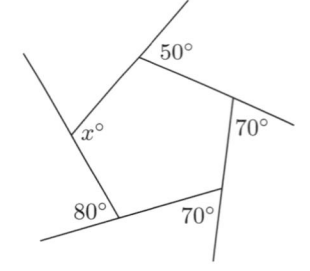

31 Are these triangles similar? Justify your answer.

32 The following triangles are not drawn to scale. Are either of them right angled?
  a
  b

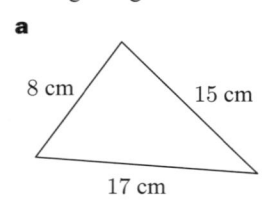

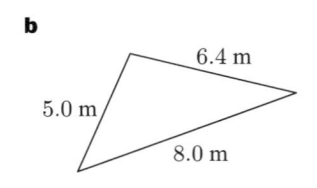

33 The following figures are similar. Find the unknowns.

a

16 cm² , 6 cm

A cm² , 15 cm

b

36 mm, 144 mm²

196 mm², x mm

34 The equal sides of an isosceles triangle are each 5 cm, and the third side is 6 cm long. Find the area of the triangle.

35 Find, giving reasons, the value of x in each of the following:

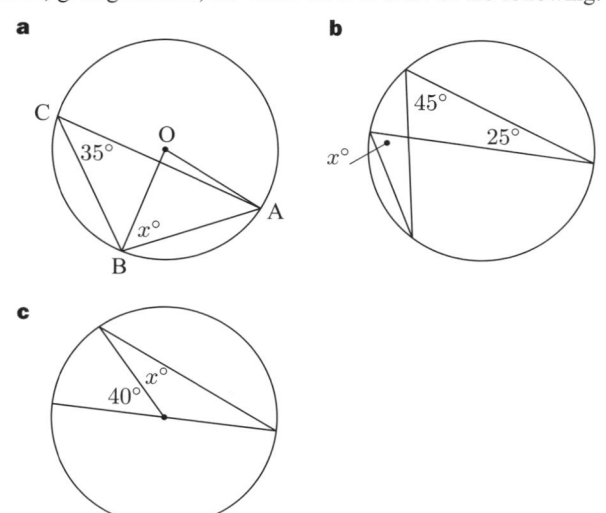

a

C, 35°, O, x°, A, B

b

45°, 25°, x°

c

40°, x°

36 Two triangles are similar if either they are equiangular or their sides are in proportion. Explain with examples why *both* conditions need to be satisfied for other polygons to be similar.

37 Abby's kite has the dimensions shown. Find its area.

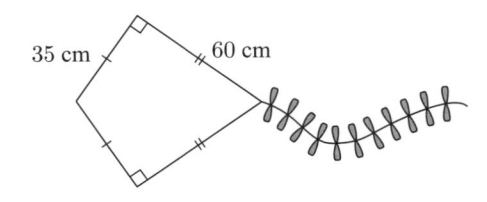

35 cm, 60 cm

38 Find the values of the unknowns in this regular pentagon:

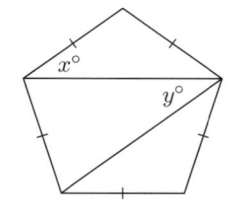

$x°$, $y°$

39 A circle of radius 12 cm contains a chord of length 17 cm. Find the shortest distance from the centre of the circle to the chord.

40 Consider the following *similar* solids. Find the unknowns.

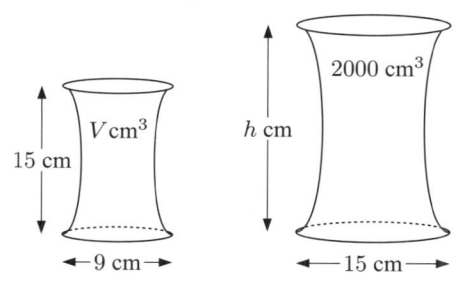

V cm³, 15 cm, 9 cm

2000 cm³, h cm, 15 cm

41 Find x:

6 cm, x cm, 4 cm

42 Is PQRS a cyclic quadrilateral? Give reasons for your answer:

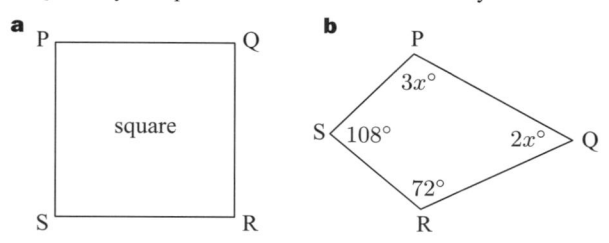

a

P, Q, square, S, R

b

P, $3x°$, S, 108°, $2x°$, Q, 72°, R

43 A courier service charges customers €0.60 per cm of the diagonal of a parcel. Find the cost to courier a parcel which is 26 cm × 18 cm × 16 cm.

44 Find, giving reasons, the values of the unknowns in the following:

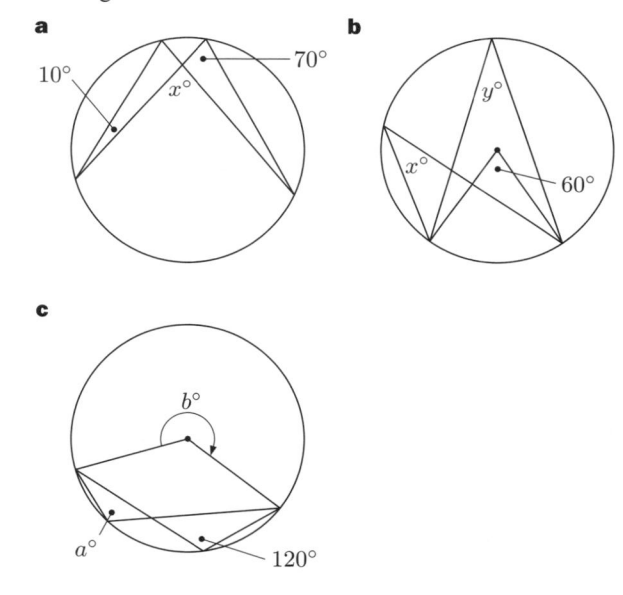

a

10°, $x°$, 70°

b

$y°$, $x°$, 60°

c

$b°$, $a°$, 120°

45 A concrete boat ramp 20 m long slopes into the water at a constant angle. The water reaches 16 m up the concrete, and the water is 1.2 m deep at the bottom of the ramp.

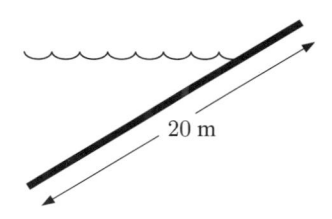

20 m

If Chris' boat needs a depth of 80 cm to be launched, how far down the ramp does he need to reverse?

46 Joseph stands at point A on the edge of a circular pond of radius 6 m. He walks a third of the way around the pond to point B. Without using trigonometry, find the straight line distance AB exactly.

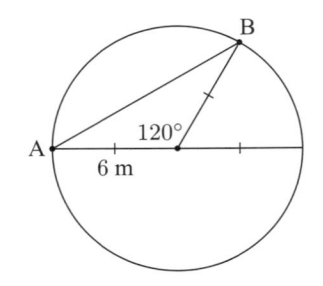

47 Find the values of the unknowns, giving reasons:

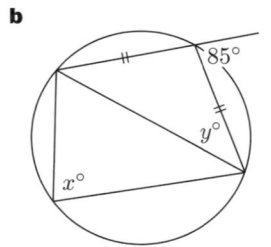

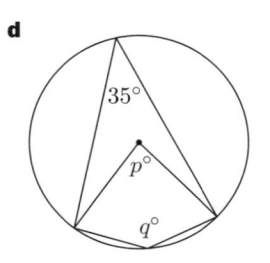

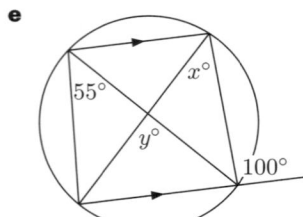

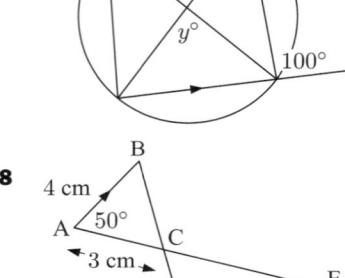

a
b
c
d
e

48 Find the area of triangle:
a ABC
b CDE

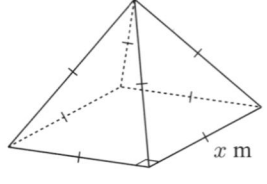

49 All of the edges of a pyramid have the same length x m. Find the height of the pyramid in terms of x.

50 A symmetrical trough 2 m long has the uniform cross-section shown.
a Find the capacity of the trough.
b 0.2 m³ of grain is placed in the trough and spread out to uniform depth. How deep is the grain?

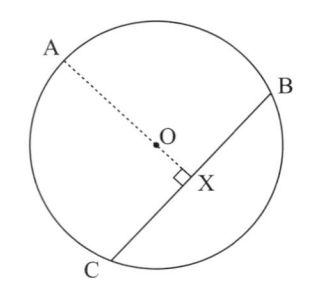

51 Find the radius of the circle, given that $OX = \frac{3}{8} AX$ and $CB = 18$ cm.

52 A circle has diameter 24 cm. Tangent CP of length 16 cm is drawn at point P. Chord AB is then constructed parallel to the tangent, so that A lies on CO.

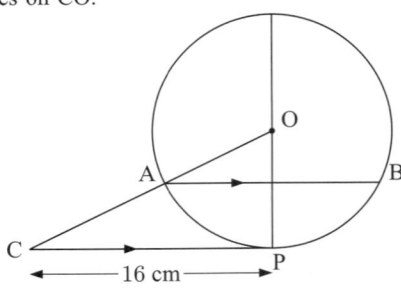

a Find the length CO.
b Find the length of the chord AB.
c Find the distance between the parallel line segments.

53 A solid has surface area 56 cm² and volume 30 cm³. Find the surface area of a similar solid with volume 50 cm³.

54 Consider the greenhouse construction shown. Which is further, AB or AC?

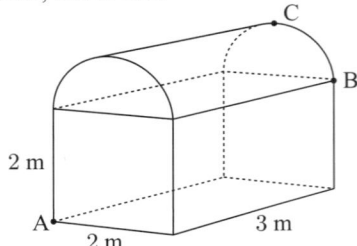

55 AB and AC are chords of a circle with centre O. X and Y are the midpoints of AB and AC respectively. Prove that OXAY is a cyclic quadrilateral.

56 Three circular cans fit exactly in a rectangular box of length 60 cm. Find the width of the box.

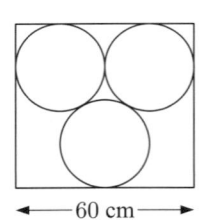

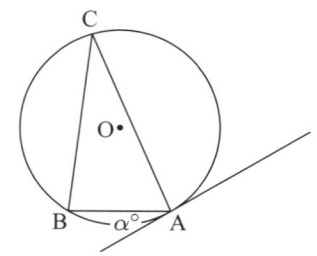

57
a Write $A\hat{O}B$ in terms of α.
b Hence find the size of $A\hat{C}B$ in terms of α.
c Explain the significance of your answer to **b**.

58 A lamp filament 1 cm wide is placed 5 cm from a small aperture as shown. A screen is placed 2.4 m from the far side of the aperture. How wide is the beam of light on the screen?

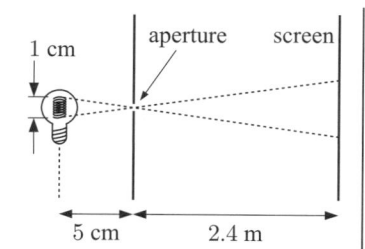

59 OABC is a parallelogram. A circle with centre O and radius OA is drawn. BA produced meets the circle at D. Prove that DOCB is a cyclic quadrilateral.

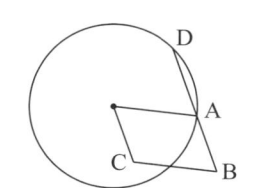

TOPIC 5: TRANSFORMATIONS IN TWO DIMENSIONS

VECTORS

Quantities which have only magnitude are called **scalars**.

Quantities which have both magnitude and direction are called **vectors**.

The vector alongside is

$\overrightarrow{OA}$ or **a** or $\underset{\sim}{a}$

used in textbooks — used by students

Its **magnitude** or **length** is

$|\overrightarrow{OA}|$ or OA or $|\mathbf{a}|$ or $|\underset{\sim}{a}|$.

$\begin{pmatrix} x \\ y \end{pmatrix}$ is a **column vector**. It is a vector in **component form**.

If point A is (a_1, a_2) and point B is (b_1, b_2) then

$$\overrightarrow{AB} = \begin{pmatrix} b_1 - a_1 \\ b_2 - a_2 \end{pmatrix}.$$

Vector equality

Two vectors are **equal** if they have the same magnitude *and* direction.

$\begin{pmatrix} p \\ q \end{pmatrix} = \begin{pmatrix} r \\ s \end{pmatrix}$ if and only if $p = r$ and $q = s$.

Vector addition

To add **a** and **b** geometrically:

Step 1: Draw **a**.

Step 2: At the arrowhead end of **a**, draw **b**.

Step 3: Join the beginning of **a** to the arrowhead end of **b**. This is vector **a** + **b**.

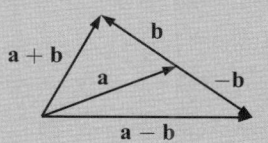

If $\mathbf{a} = \begin{pmatrix} a_1 \\ a_2 \end{pmatrix}$ and $\mathbf{b} = \begin{pmatrix} b_1 \\ b_2 \end{pmatrix}$ then $\mathbf{a} + \mathbf{b} = \begin{pmatrix} a_1 + b_1 \\ a_2 + b_2 \end{pmatrix}$.

The zero vector

The **zero vector**, **0**, is a vector of length 0. It is the only vector with no direction.

For any vector **a**:
- $\mathbf{a} + \mathbf{0} = \mathbf{0} + \mathbf{a} = \mathbf{a}$
- $\mathbf{a} + (-\mathbf{a}) = (-\mathbf{a}) + \mathbf{a} = \mathbf{0}$

Negative vectors

a and −**a** are parallel and equal in length, but opposite in direction.

If $\mathbf{a} = \begin{pmatrix} a_1 \\ a_2 \end{pmatrix}$ then $-\mathbf{a} = \begin{pmatrix} -a_1 \\ -a_2 \end{pmatrix}$.

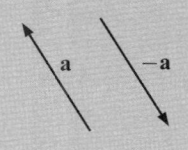

The negative of vector $\overrightarrow{AB}$ is $\overrightarrow{BA}$. $\overrightarrow{BA} = -\overrightarrow{AB}$.

Vector subtraction

To subtract one vector from another, we simply **add its negative**.

So, $\mathbf{a} - \mathbf{b} = \mathbf{a} + (-\mathbf{b})$.

If $\mathbf{a} = \begin{pmatrix} a_1 \\ a_2 \end{pmatrix}$ and $\mathbf{b} = \begin{pmatrix} b_1 \\ b_2 \end{pmatrix}$ then $\mathbf{a} - \mathbf{b} = \begin{pmatrix} a_1 - b_1 \\ a_2 - b_2 \end{pmatrix}$.

The magnitude of a vector

The **magnitude** or **length** of $\mathbf{a} = \begin{pmatrix} a_1 \\ a_2 \end{pmatrix}$

is $|\mathbf{a}| = \sqrt{a_1^2 + a_2^2}$.

Scalar multiplication

If k is a scalar then $k \begin{pmatrix} a \\ b \end{pmatrix} = \begin{pmatrix} ka \\ kb \end{pmatrix}$.

Parallel vectors

Two vectors are parallel $\Leftrightarrow$ one vector is a scalar multiple of the other.

If **a** is parallel to **b** then we write **a** ∥ **b**.

Consider the vector $k\mathbf{a}$ which is parallel to **a**.

- If $k > 0$ then $k\mathbf{a}$ has the same direction as **a**.
- If $k < 0$ then $k\mathbf{a}$ has the opposite direction to **a**.
- $|k\mathbf{a}| = |k||\mathbf{a}|$.

Vectors in geometry

Vectors can be used to establish relationships between the line segments in geometric shapes.

You should be able to use vector relationships to prove geometrical facts.

TRANSFORMATIONS

A change in the size, shape, orientation, or position of an object is called a **transformation**.

The original figure is called the **object** and the new figure is called the **image**.

Translations

If $P(x, y)$ is **translated** h units in the x-direction and k units in the y-direction to become $P'(x', y')$, then $x' = x + h$ and $y' = y + k$.

$$P(x, y) \xrightarrow{\begin{pmatrix} h \\ k \end{pmatrix}} P'(x + h, y + k)$$

$\left. \begin{array}{l} x' = x + h \\ y' = y + k \end{array} \right\}$ are called the **transformation equations**.

$\begin{pmatrix} h \\ k \end{pmatrix}$ is called the **translation vector**.

Rotations

You should be able to rotate points and objects about a fixed point through some multiple of 90° (either clockwise or anticlockwise).

Reflections

The most common reflections are:
- in the x-axis or y-axis
- in lines parallel to the axes
- in the lines $y = x$ and $y = -x$.

Enlargements and reductions

Under an enlargement with centre $O(0, 0)$ and scale factor k, $(x, y) \to (kx, ky)$.

If $k > 1$, the image is an **enlargement** of the object.
If $0 < k < 1$, the image is a **reduction** of the object.

Stretches

In a **stretch** we enlarge or reduce an object in one direction only.

For a stretch with invariant x-axis and scale factor k, $(x, y) \to (x, ky)$.

For a stretch with invariant y-axis and scale factor k, $(x, y) \to (kx, y)$.

The inverse of a transformation

If a transformation maps an object onto its image, then the **inverse** transformation maps the image back onto the object.

Combinations of transformations

We represent 'transformation G followed by transformation H' as HG.

SKILL PRACTICE

1 Using a scale of 1 cm represents 10 units, sketch a vector to represent:

  a 30 m/s in an easterly direction

  b a missile being launched at an angle of 40° to the horizontal at a speed of 40 km/h

  c a force of 29 Newtons downwards

  d a displacement of 36 m in a NW direction.

2 If P is at $(2, -1)$, Q is at $(-4, 5)$, and R is at $(3, 1)$, find:

  a $\overrightarrow{QO}$ **b** $\overrightarrow{RQ}$ **c** $\overrightarrow{PR}$

3 Find the image point when:

  a $(4, 3)$ is translated through $\begin{pmatrix} -2 \\ 6 \end{pmatrix}$

  b $(-4, 1)$ is translated through $\begin{pmatrix} 3 \\ 2 \end{pmatrix}$.

4 **a** Write in terms of vectors **r**, **s** and **t**:

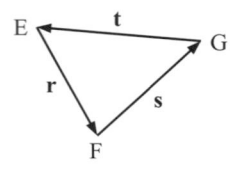

  i $\overrightarrow{EF}$ **ii** $\overrightarrow{GF}$ **iii** $\overrightarrow{EG}$

  b Write an equation connecting **r**, **s** and **t**.

5 Copy the vectors **a** and **b**, and hence show how to find:

  i $\mathbf{a} + \mathbf{b}$ **ii** $\mathbf{a} - \mathbf{b}$

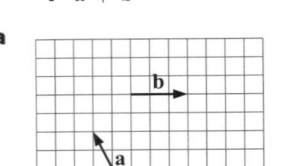

6 Find, by graphical means, the image of the point $(2, 3)$ under a reflection in:

  a the x-axis **b** the line $y = -x$
  c the y-axis **d** the line $y = 5$.

7 What point has image $(1, 5)$ under the translation $\begin{pmatrix} 3 \\ 1 \end{pmatrix}$?

8 Write in component form:

  a $\overrightarrow{WX}$
  b $\overrightarrow{XY}$
  c $\overrightarrow{ZV}$
  d $\overrightarrow{VX}$
  e $\overrightarrow{XV}$
  f $\overrightarrow{WZ}$

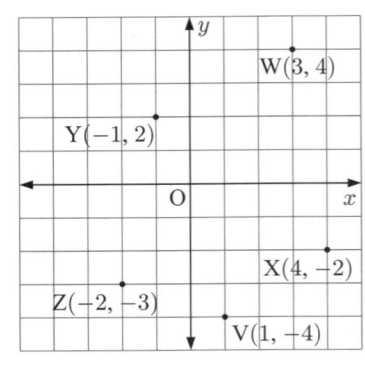

  $W(3, 4)$, $Y(-1, 2)$, $X(4, -2)$, $Z(-2, -3)$, $V(1, -4)$

9 Find the image of $A(-2, 5)$ under these rotations about the origin $O(0, 0)$:

  a 180° **b** anticlockwise through 90°
  c clockwise through 90°.

10 The figure alongside consists of a rhombus with $\overrightarrow{AD} = \mathbf{p}$ and $\overrightarrow{AB} = \mathbf{q}$. Which of the following statements are true?

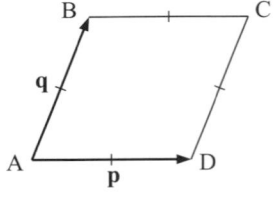

  a $\overrightarrow{AB} = \overrightarrow{CD}$ **b** $\overrightarrow{BC} = \overrightarrow{AD}$ **c** $|\overrightarrow{CB}| = |\mathbf{q}|$
  d $\overrightarrow{BD} = \mathbf{q}$ **e** $\overrightarrow{BA} \parallel \overrightarrow{DC}$

11 Consider $\mathbf{a} = \begin{pmatrix} 4 \\ 7 \end{pmatrix}$, $\mathbf{b} = \begin{pmatrix} -2 \\ 1 \end{pmatrix}$, and $\mathbf{c} = \begin{pmatrix} -5 \\ -8 \end{pmatrix}$.

  Find exactly:

  a $\mathbf{a} + \mathbf{b}$ **b** $\mathbf{a} + \mathbf{a} + \mathbf{c}$ **c** $\mathbf{c} - \mathbf{a}$
  d $|\mathbf{b} + \mathbf{c}|$ **e** $\mathbf{b} - \mathbf{c} + \mathbf{a}$ **f** $|\mathbf{a} + \mathbf{b} - \mathbf{c}|$

12 Find the image of the point:

  a $(1, 4)$ under a reduction with centre $O(0, 0)$ and scale factor $k = \frac{1}{2}$

  b $(5, -6)$ under an enlargement with centre $C(3, 4)$ and scale factor $k = 2\frac{1}{2}$.

13 For $u = \begin{pmatrix} 3 \\ 5 \end{pmatrix}$ and $v = \begin{pmatrix} -1 \\ 3 \end{pmatrix}$, find geometrically:

  a $u - v$ **b** $-2v$ **c** $2(v - u)$

14 Triangle PQR has vertices $P(-1, 4)$, $Q(1, -1)$ and $R(2, 5)$. It is rotated $90°$ clockwise about Q.

  a Draw triangle PQR and draw and label its image P'Q'R'.
  b Write down the coordinates of P', Q' and R'.

15 Mark hikes for 12 km in the direction $120°$ and then 5 km in the direction $030°$. Find Mark's displacement from his starting point.

16 Simplify the following vector expressions:

  a $\overrightarrow{BC} + \overrightarrow{CD}$
  b $\overrightarrow{DA} + \overrightarrow{AB} - \overrightarrow{CB}$
  c $\overrightarrow{BC} + \overrightarrow{CD} - \overrightarrow{AD} - \overrightarrow{BA}$
  d $\overrightarrow{AB} - \overrightarrow{CB} - \overrightarrow{DC}$

17 Square ABCD has vertices $A(-3, 2)$, $B(0, 4)$, $C(2, 1)$, and $D(-1, -1)$.

  a Draw square ABCD on a set of axes.
  b Translate the figure by the translation vector $\begin{pmatrix} 2 \\ -4 \end{pmatrix}$.
  c State the coordinates of the image vertices A', B', C' and D'.
  d Through what distance has each point moved?

18 Find exactly the magnitudes of these vectors:

  a $\begin{pmatrix} 3 \\ 4 \end{pmatrix}$ **b** $\begin{pmatrix} 2 \\ -1 \end{pmatrix}$ **c** $\begin{pmatrix} 5 \\ 0 \end{pmatrix}$
  d $\begin{pmatrix} -2 \\ 7 \end{pmatrix}$ **e** $\begin{pmatrix} 0 \\ -9 \end{pmatrix}$ **f** $\begin{pmatrix} -2 \\ -6 \end{pmatrix}$

19 Reflect P in:

  a the x-axis, and label the image Q
  b the line $y = x$, and label the image R
  c the line $x = 2$, and label the image S.

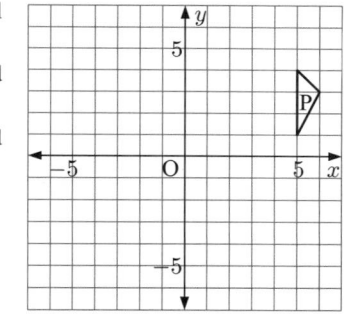

20 If $a = \begin{pmatrix} 2 \\ 1 \end{pmatrix}$ and $b = \begin{pmatrix} -3 \\ -2 \end{pmatrix}$, find algebraically:

  a $-3b$ **b** $2a + b$
  c $\frac{1}{2}(5a + 3b)$ **d** $\frac{1}{4}a - b$

21 Enlarge or reduce the following diagrams with centre C and scale factor k given:

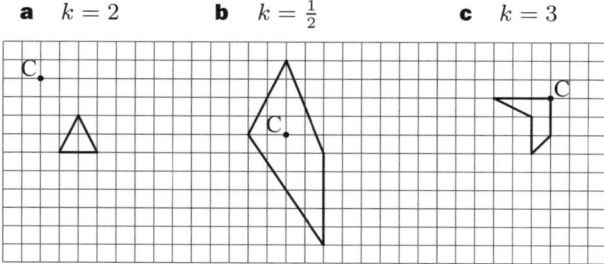

  a $k = 2$ **b** $k = \frac{1}{2}$ **c** $k = 3$

22 For the following pairs of points, find:
  i the vector $\overrightarrow{AB}$ **ii** the exact distance AB.

  a $A(1, 1)$ and $B(-5, -6)$
  b $A(7, 4)$ and $B(7, -9)$
  c $A(-2, 5)$ and $B(3, 6)$

23 Draw the image of the line $y = 2x - 1$ when it is reflected in the line $x = 2$. State the equation of the image.

24 Consider $s = \begin{pmatrix} 4 \\ 6 \end{pmatrix}$ and $t = \begin{pmatrix} -3 \\ 4 \end{pmatrix}$. Find $u = \begin{pmatrix} x \\ y \end{pmatrix}$ such that:

  a $s - 2t + u = 0$ **b** $u + \frac{1}{2}s = 3t$

25 Find the equation of the image of $y = -\frac{1}{2}x$ when it is reflected in:

  a the line $y = 1$ **b** the x-axis
  c the line $y = x$.

26 What two facts can be deduced about **a** and **b** if:

  a $a = -\frac{1}{2}b$ **b** $a = 3b$?

27 Perform the stretch on the following diagrams, using the given invariant line IL and scale factor k:

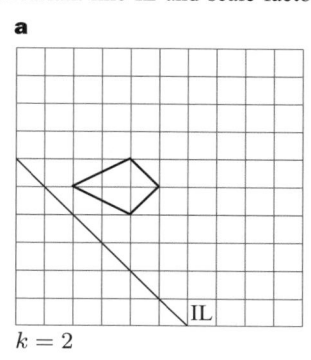

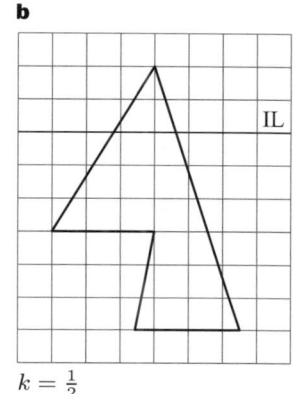

  a $k = 2$
  b $k = \frac{1}{2}$

28 Consider triangle EFG with vertices $E(2, 6)$, $F(4, 5)$, and $G(3, 3)$. If T is a translation of $\begin{pmatrix} 3 \\ -1 \end{pmatrix}$ and R is a $90°$ clockwise rotation about the origin $(0, 0)$, draw the images of:

  a TR   **b** RT.

29 Find the two values of q for which $a = \begin{pmatrix} -5 \\ q \end{pmatrix}$ and $|a| = 13$.

30 Find the equation of the image line when:

  a $y = -2x + 3$ is translated $\begin{pmatrix} 3 \\ -2 \end{pmatrix}$
  b $y = -\frac{5}{2}x$ is translated $\begin{pmatrix} 1 \\ 2 \end{pmatrix}$.

31 Use vector methods to find the remaining vertex of parallelogram STUV: $U(3, 6)$, V, $T(0, 4)$, $S(5, 1)$

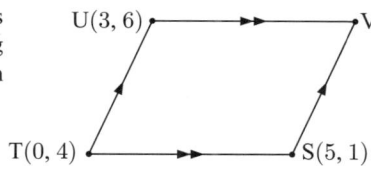

32 Find the image of:

  a $(2, 6)$ under a stretch with invariant y-axis and scale factor $k = \frac{1}{3}$
  b $(2, 3)$ under a stretch with invariant line $y = x$ and scale factor $k = 2$.

33 The diagram below shows a biathlon course skied by Ingmar.

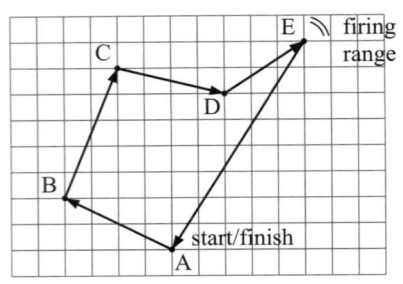

  a Write a column vector to describe each leg of the course.
  b Write a single vector to describe the displacement from the start/finish line to the firing range.
  c Ingmar believes he can find the length of one lap by adding up all the vectors in **a**.
   i Why is Ingmar incorrect?
   ii What does Ingmar need to do instead?
   iii Find the length of one lap of the course.

34 Find the image of $(1, -5)$ under a reflection in the x-axis, followed by a translation of $\begin{pmatrix} 1 \\ -1 \end{pmatrix}$, followed by a reflection in the line $y = x$.

35 Consider the points $A(10, 1)$, $B(4, -1)$ and $C(-2, -3)$.
  a Show that A, B and C are collinear using vector methods.
  b Hence, find k such that $\vec{AC} = k\vec{AB}$.

36 The object triangle ABC is mapped onto the image triangle A'B'C'.
Describe fully the single transformation which has occurred.

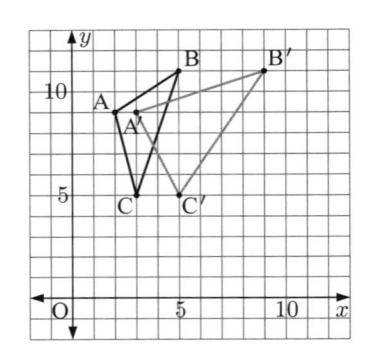

37 A kayaker is paddling south across a river at a speed of 9 km/h. The kayak is subject to a current of 3.75 km/h due east.
  a In what direction is the kayaker actually facing?
  b At what speed would the kayaker be paddling if the water was still?

38 Consider triangle ABC with vertices $A(3, 7)$, $B(6, 5)$, and $C(4, 2)$. If M_1 is a reflection in the line $x = 2$, and M_2 is a reflection in the line $y = -x$, draw the images of:
  a $M_1 M_2$ **b** $M_2 M_1$.

39 In the parallelogram PQRS, $\vec{PQ} = \mathbf{a}$ and $\vec{PS} = \mathbf{b}$.
X and Y are the midpoints of SR and QR respectively.
Find, in terms of **a** and **b**:
  a $\vec{PR}$ **b** $\vec{PX}$ **c** $\vec{QX}$
  d $\vec{PY}$ **e** $\vec{XY}$

40 Find the image equation of the line $x + 2y = -2$ under an anticlockwise rotation of $90°$ about $O(0, 0)$.

41 In the given figure, $CD : DE = 5 : 7$. Deduce a vector expression for $\vec{PD}$ in terms of $\vec{PC}$ and $\vec{PE}$.

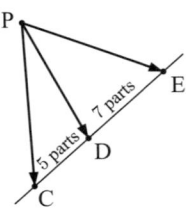

42 Describe fully the inverse transformation for each of the following transformations:
  a a translation of $\begin{pmatrix} 4 \\ -7 \end{pmatrix}$
  b a rotation about point P, anticlockwise through $120°$
  c a stretch with invariant line $y = x$ and scale factor 3
  d a reflection in the line $x = -4$
  e a reduction with centre $(1, 2)$ and scale factor $\frac{1}{5}$.

43 Figure ABCD is a rhombus. P, Q, R, S are the midpoints of its sides.

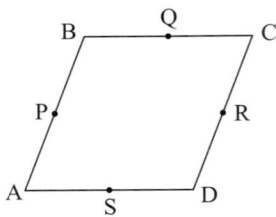

  a If $\vec{AB} = \mathbf{a}$ and $\vec{AD} = \mathbf{b}$, find in terms of **a** and **b**:
   i $\vec{PS}$ **ii** $\vec{QR}$ **iii** $\vec{PQ}$ **iv** $\vec{SR}$
  b Find in terms of **a**: **i** $|\vec{PR}|$ **ii** $|\vec{SQ}|$.
  c What can be deduced about quadrilateral PQRS?

44 Use triangle 0 as the object shape and consider the following transformations T_1 through T_7.

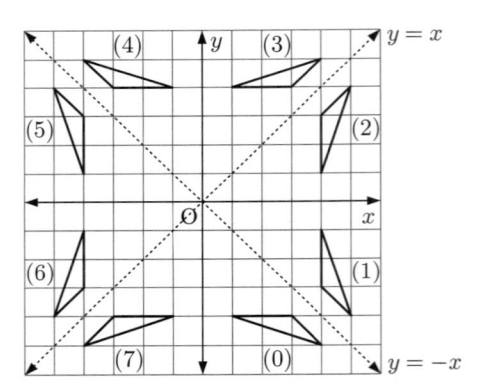

T_1: reflect in the line $y = -x$
T_2: rotate $90°$ anticlockwise about $O(0, 0)$
T_3: reflect in the x-axis
T_4: rotate $180°$ about $O(0, 0)$
T_5: reflect in the line $y = x$
T_6: rotate $90°$ clockwise about $O(0, 0)$
T_7: reflect in the y-axis.
Find the single transformation equivalent to:
  a T_5 then T_7 **b** T_1 then T_4
  c T_6 then T_6 **d** T_3 then T_2

TOPIC 6: MENSURATION

UNITS

You should be familiar with the following units:

- Length: mm, cm, m, km
- Area: mm², cm², m², ha, km²
- Volume: mm³, cm³, m³
- Capacity: ml, cl, l
- Mass: g, kg, t

To convert from **smaller** to **larger** units we **divide** by the conversion factor.

To convert from **larger** to **smaller** units we **multiply** by the conversion factor.

LENGTH

The **perimeter** of a figure is the measurement of the distance around its boundary.

For a **polygon** the perimeter is the sum of the lengths of all sides.

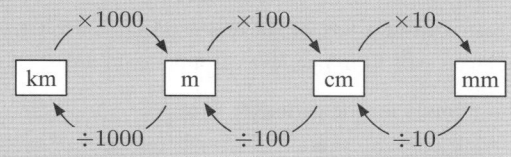

km →×1000→ m →×100→ cm →×10→ mm
km ←÷1000← m ←÷100← cm ←÷10← mm

AREA

The **area** of a figure is the amount of surface within its boundaries.

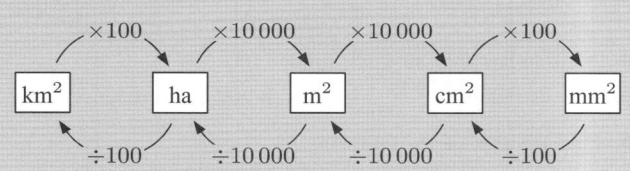

km² →×100→ ha →×10 000→ m² →×10 000→ cm² →×100→ mm²
km² ←÷100← ha ←÷10 000← m² ←÷10 000← cm² ←÷100← mm²

You should be able to use these formulae for area:

Rectangles

Area = length × width

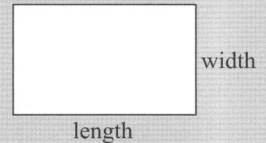

Triangles

Area = $\frac{1}{2}$ (base × height)

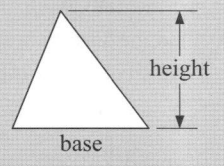

Parallelograms

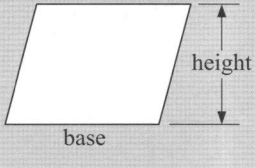

Area = base × height

Trapezia

Area = $\frac{1}{2}(a+b) \times h$

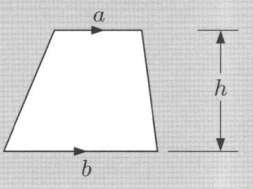

Circles and sectors

An **arc** is any continuous part of the circle. The length of an arc is called its **arclength**.

Every arc has a corresponding **sector**, which is the portion of the circle subtended by the same angle $\theta°$ as the arc.

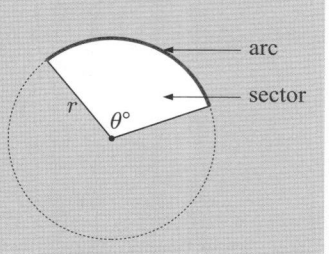

For a circle:
- Circumference $C = \pi d = 2\pi r$
- Area $A = \pi r^2$

For a sector of angle $\theta°$:
- Arclength $s = \left(\frac{\theta}{360}\right) \times 2\pi r$
- Area $A = \left(\frac{\theta}{360}\right) \times \pi r^2$

SURFACE AREA

Solids with plane faces

The **surface area** of a three dimensional figure with plane faces is the sum of the areas of the faces.

To assist in your calculations, you can draw a **net** of the solid, correctly labelling the dimensions.

Solids with curved surfaces

You should be able to use these formulae for surface area:

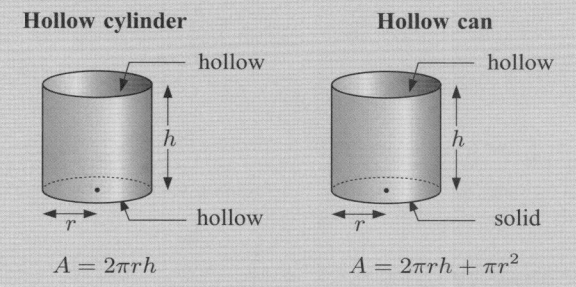

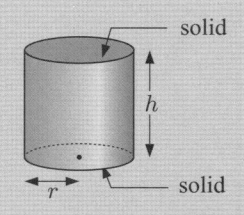

Hollow cylinder

$A = 2\pi rh$

Hollow can

$A = 2\pi rh + \pi r^2$

Solid cylinder

$A = 2\pi rh + 2\pi r^2$

Hollow cone

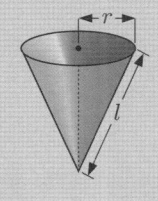

$A = \pi rl$

Solid cone

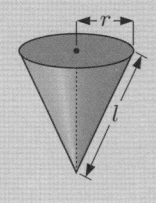

$A = \pi rl + \pi r^2$

Sphere

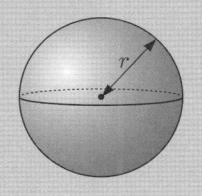

$A = 4\pi r^2$

VOLUME

The **volume** of a solid is the amount of space it occupies.

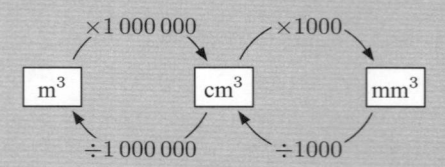

$$m^3 \xrightarrow{\times 1\,000\,000} cm^3 \xrightarrow{\times 1000} mm^3$$
$$m^3 \xleftarrow{\div 1\,000\,000} cm^3 \xleftarrow{\div 1000} mm^3$$

You should be able to use these formulae for volume:

Solids of uniform cross-section

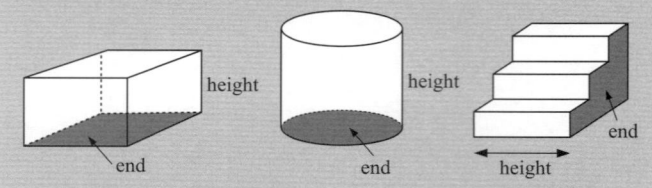

Volume of uniform solid = area of end × height

Pyramids and cones

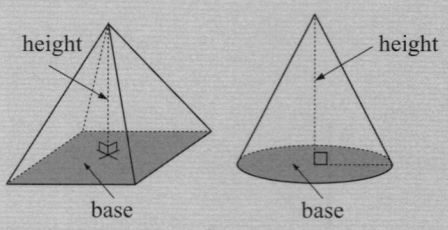

Volume of a pyramid or cone = $\frac{1}{3}$(area of base × height)

Spheres

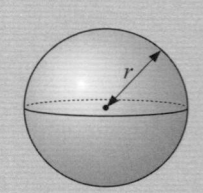

Volume of a sphere = $\frac{4}{3}\pi r^3$

You can find the volumes of compound solids by separating the solid into sections like those above.

CAPACITY

The **capacity** of a container is the quantity of fluid or gas required to fill it.

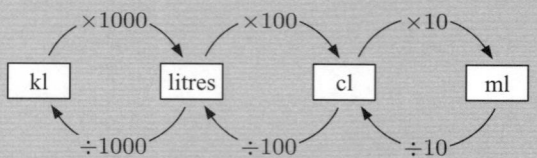

$$kl \xrightarrow{\times 1000} litres \xrightarrow{\times 100} cl \xrightarrow{\times 10} ml$$
$$kl \xleftarrow{\div 1000} litres \xleftarrow{\div 100} cl \xleftarrow{\div 10} ml$$

Connecting volume to capacity

$$1\,ml \equiv 1\,cm^3$$
$$1\,litre \equiv 1000\,cm^3$$
$$1\,kl = 1000\,litres \equiv 1\,m^3$$

MASS

The **mass** of an object is the amount of matter in it.

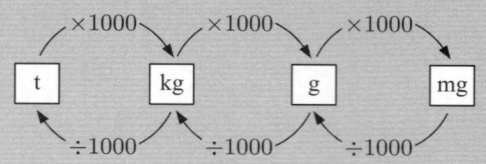

$$t \xrightarrow{\times 1000} kg \xrightarrow{\times 1000} g \xrightarrow{\times 1000} mg$$
$$t \xleftarrow{\div 1000} kg \xleftarrow{\div 1000} g \xleftarrow{\div 1000} mg$$

SKILL PRACTICE

1 Convert:
 a 72 mm to cm
 b 5.8 m to mm
 c 9.75 km to m
 d 28 000 000 cm to km.

2 Kevin counts the light poles on the footpath as he walks to school. Kevin walks 2.4 km, and counts 80 light poles. How far is it between each light pole?

3 Find the perimeter of:
 a (triangle with sides 15 cm, 12 cm)
 b (parallelogram 2 m, 3.5 m)
 c (L-shape with 1.5 m, 2.5 m)

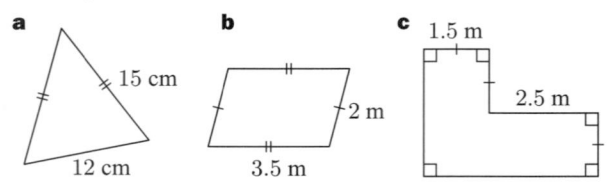

4 Convert:
 a 44 mm² to cm²
 b 0.059 ha to cm²
 c 21.85 ha to km²
 d 0.000 006 2 km² to mm²
 e 360 m² to cm²
 f 39 500 m² to ha.

5 A rectangle is 3.2 m by 2.4 m and has the same perimeter as a square. Find the length of the sides of the square.

6 The base area of a box of stickers is 85 cm². How many of these boxes will fit in one layer of a pallet of area 1.36 m²?

7 Find a formula for the perimeter P of:
 a (square side z)
 b (triangle sides a, b)
 c (T-shape with p, q)

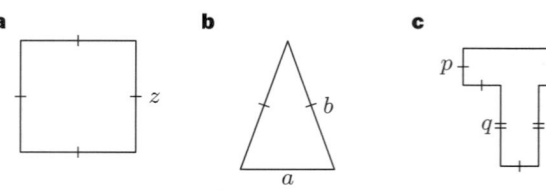

8 A circle has area 36.4 m². Find:
 a its radius
 b its circumference.

9 Find the area of the following:
 a (compound shape 3 cm, 5 cm, 8 cm)
 b (parallelogram 9 m, 6 m)

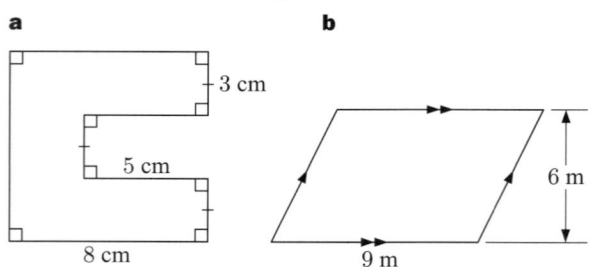

10 Find the surface area of:
 a a cube with sides 16 cm
 b a cuboid 36 mm × 48 mm × 21 mm.

11 Convert:
 a 3.71 litres into cl
 b 58 215 ml into litres.

12 Calculate the length of guard rail needed to construct a safety fence for the following viewing platform:

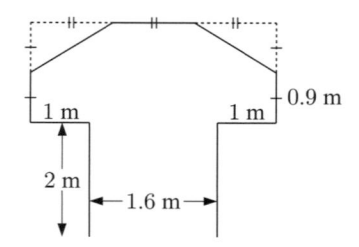

(diagram with 1 m, 1 m, 0.9 m, 2 m, 1.6 m)

13 Find the area of a kite whose diagonals have lengths 40 cm and 70 cm.

14 Find the surface area of:

   a   **b**

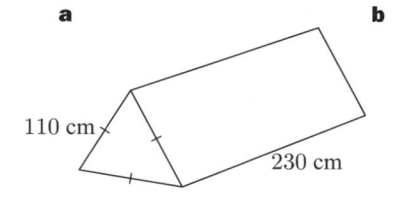

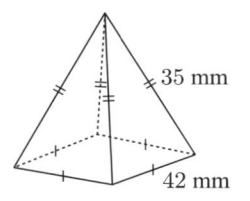

15 Find the perimeter and area of the following figures:

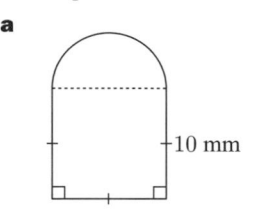

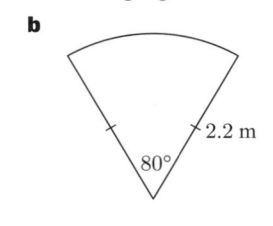

   a   **b**

16 Convert the following:

   a 7.25 m^3 to cm^3

   b $2\,900\,000\,000 \text{ mm}^3$ to m^3

   c 2500 cm^3 to mm^3.

17 A chef uses 75 ml of milk in each serve of mashed potatoes. He makes an average of 235 serves each week. How many litres of milk does he use?

18 Adrian's new garage has the dimensions shown. Find the surface area of sheet metal required for the walls and roof.

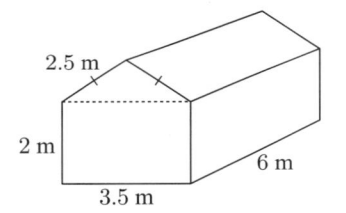

19 A sector has radius 4 cm and angle 250°. Find its area.

20

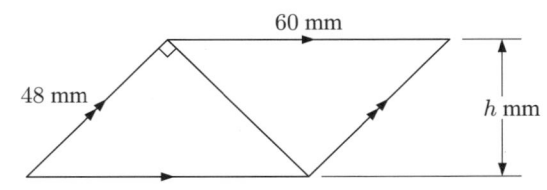

   a Find the area of the parallelogram.   **b** Find h.

21 Find the surface area of these solids:

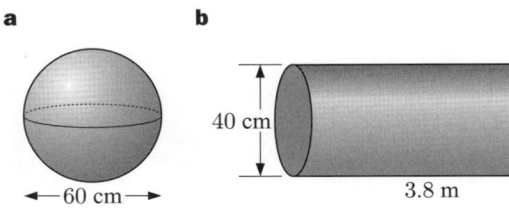

   a   **b**

22 A television cabinet has the dimensions shown. Find:

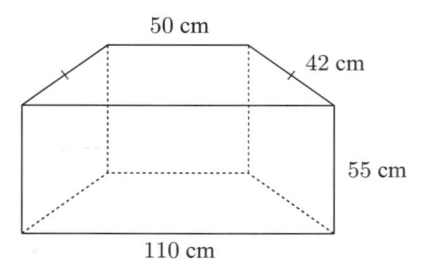

   a the area of its top

   b the total surface area of its four sides.

23 40 mm^3 of copper is required to make a single resistor. How many resistors can be made with 1000 cm^3 of copper?

24 The engine of a 500 cc motorbike holds 500 cm^3 of fuel-air mixture. Express this quantity in litres.

25 Find the perimeter and area of this figure:

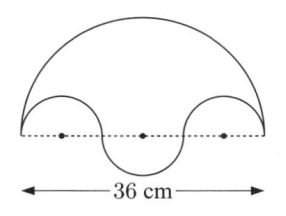

26 When full, a blow-up beach ball has diameter 36 cm. Find the surface area of rubber needed to make 200 of these balls.

27 Find the volume of the following:

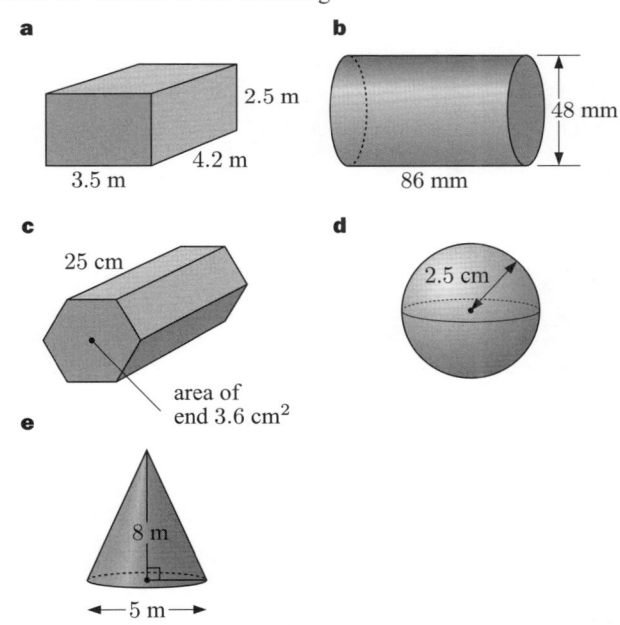

   a   **b**

   c   **d**

   e

28 Find the outer surface area of:

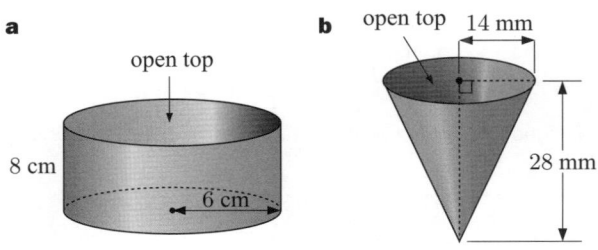

   a   **b**

29 A circular pie with radius 8.5 cm is served on a square plate with sides 21 cm long. What proportion of the area of the plate does the pie cover?

30 A cylindrical drinking flask has radius 3.42 cm and height 16.33 cm. Find its capacity.

31 How many cylindrical cookies with diameter 5 cm and thickness 1 cm could be made from a rectangular block of dough 20 cm × 15 cm × 8 cm?

32 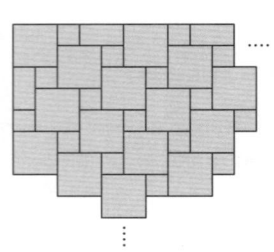 Three sizes of tile are used to form the 3.25 m × 2.25 m floor of a bathroom using the pattern shown. The large tiles are 10 cm × 10 cm. What proportion of the area is covered by the smallest tiles?

33 Pauline has a wooden block with the dimensions shown. She paints a 1 cm wide border around the edge of every face. Find:

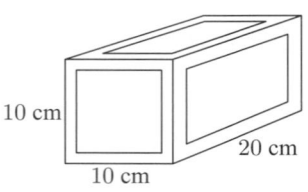

   a the total surface area of the block
   b the painted area
   c the unpainted area.

34 Find a formula for the area A of the following regions:

   a 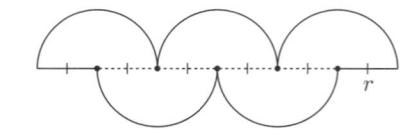
   b 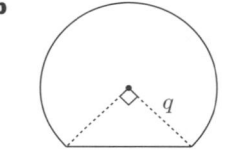
   c 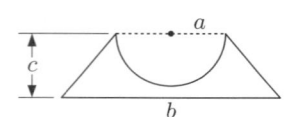
   d

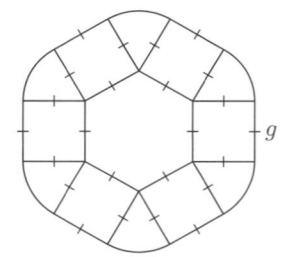

35 A solid cone has diameter 15 mm and slant height 34 mm. Find its surface area.

36 Emma has just bought 60 timber posts to help build a fence. Each post is a cylinder 1.8 m long with diameter 16 cm. The total mass of Emma's posts is 1.08 tonnes.
Find: **a** the mass of each post in kilograms
       b the volume of each post in m^3

37 Find a formula for the surface area A of the following solids:

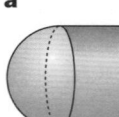

   a 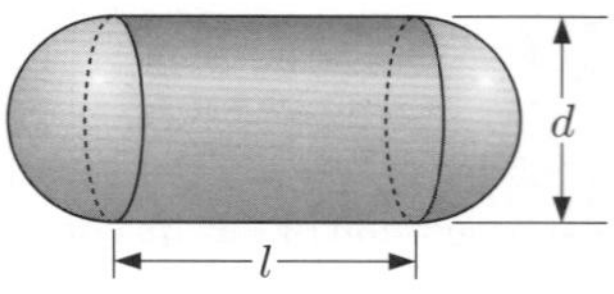
   b

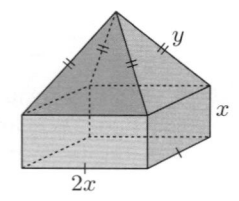

38 Eliza has a bucket with the dimensions shown. She fills it with water, but there is a hole in the bucket, so the water drips out at a rate of 1.2 ml/min. How much water remains in the bucket when Eliza returns 3 hours later?

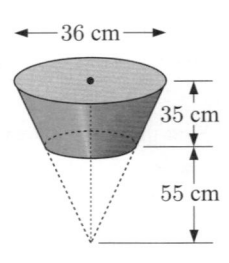

39 Find formulae for the volume V of the following objects:

   a 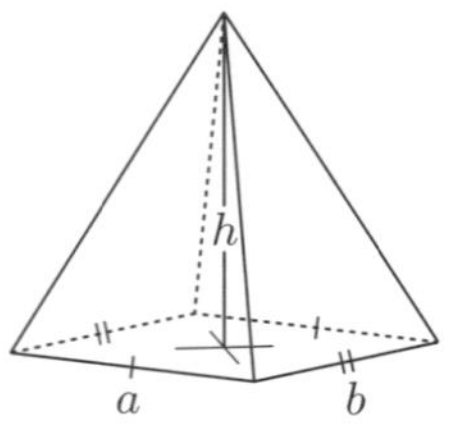
   b 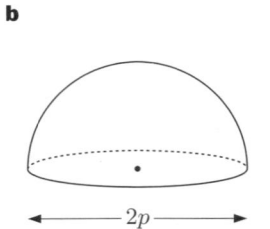
   c

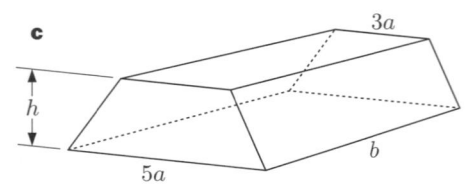

40 Des buys a 500 g wedge of his favourite cheese. The wedge is a right angle and is 6.1 cm high. Its volume is 460 cm^3.
Find the radius of the wedge.

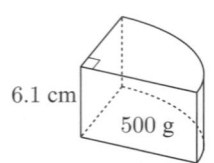

41 A concrete bench for a bus stop is made with the dimensions shown. Show that the volume of concrete used is given by the formula $V = a^2 l(\frac{\pi}{8} + 8)$.

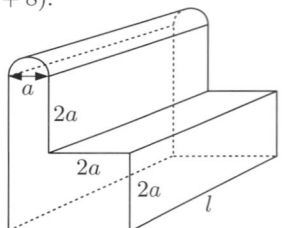

42 A metal door handle is formed from three cylindrical pieces. The handles are 4 cm deep and have radius 3 cm. The shaft in the middle has length 12 cm and radius 1.5 cm.

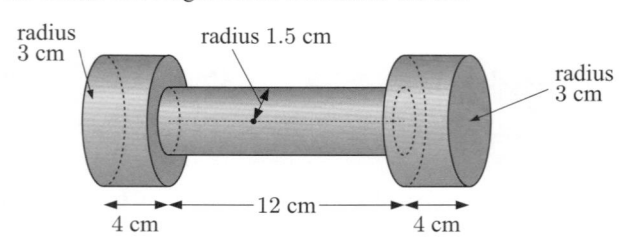

Find the total volume of the door handle.

43 **a** 55 litres of water is added to the cylindrical aquarium shown.
How far from the top does the water rise?

    b Glass marbles of diameter 12 mm are carefully added to the aquarium. How many marbles can be added without causing the water to overflow?

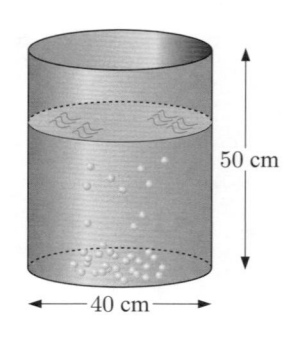

Cambridge IGCSE International Mathematics (0607) Extended   Exam Preparation & Practice Guide

TOPIC 7: COORDINATE GEOMETRY

THE NUMBER PLANE

The position of any point P can be written as the **ordered pair** (x, y), where:

x is the **x-coordinate** or **horizontal step** from the origin O

y is the **y-coordinate** or **vertical step** from the origin O.

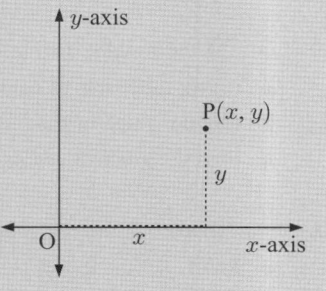

When plotting points, remember that the x-coordinate is always given first. This indicates the movement from the origin in the horizontal direction.

The number plane is divided into **four quadrants**:

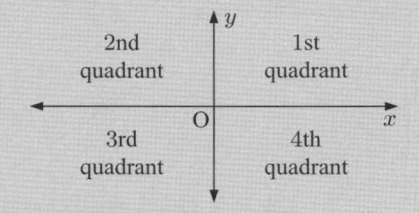

DISTANCE BETWEEN TWO POINTS

The distance between two points may be found using Pythagoras' theorem or using the **distance formula**.

If $A(x_1, y_1)$ and $B(x_2, y_2)$ are two points in a plane, the distance between these points is
$AB = \sqrt{(x_2 - x_1)^2 + (y_2 - y_1)^2}$.

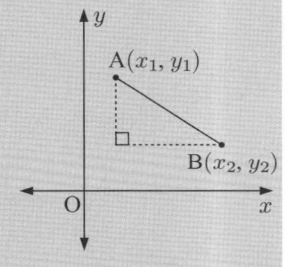

MIDPOINT OF A LINE SEGMENT

If $A(x_1, y_1)$ and $B(x_2, y_2)$ are two points then the **midpoint** M of AB has coordinates
$\left(\dfrac{x_1 + x_2}{2}, \dfrac{y_1 + y_2}{2}\right)$.

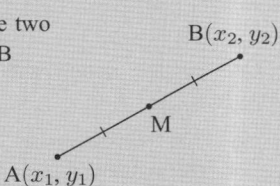

GRADIENT

The **gradient** of a line is a measure of its steepness.

Given two points $A(x_1, y_1)$ and $B(x_2, y_2)$, the **gradient** of AB is $\dfrac{y_2 - y_1}{x_2 - x_1}$.

The gradient of any **horizontal** line is **0**.
The gradient of any **vertical** line is **undefined**.

Two lines are **parallel** $\Leftrightarrow$ their gradients are **equal**.
Two lines are **perpendicular** $\Leftrightarrow$ their gradients are **negative reciprocals**.

Collinear points

Three or more points are **collinear** if they lie on the same straight line.

Points A, B and C are collinear if the gradient of AB equals the gradient of BC.

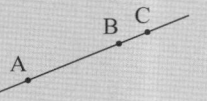

Using coordinate geometry

In problems where we use coordinate geometry to check geometrical facts:

- draw a clear labelled diagram of the situation
- for figures named ABCD, the labelling is in cyclic order.

STRAIGHT LINES

The **equation of a line** is an equation which connects the x and y values for every point on the line.

All **vertical** lines have equations of the form $x = a$.
All **horizontal** lines have equations of the form $y = b$.

The **x-intercept** of a line is the value of x where the line meets the x-axis.
The **y-intercept** of a line is the value of y where the line meets the y-axis.

The **gradient-intercept form** of the equation of a line with gradient m and y-intercept c, is $y = mx + c$.

The **general form** of the equation of a line is $ax + by = d$.

- For a line with gradient $\dfrac{p}{q}$, the general form of the line is $px - qy = d$.
- For a line with gradient $-\dfrac{p}{q}$, the general form of the line is $px + qy = d$.

We can find the equation of a line if we are given:

- its gradient and the coordinates of one point on the line
- the coordinates of two points on the line.

Graphing straight lines

Given the gradient-intercept form of a straight line, start at the y-intercept then use the gradient to find another point.

Given the general form of a straight line, find the x- and y-intercepts and draw the straight line through them.

LINEAR INEQUALITY REGIONS

Linear inequalities can be used to define regions of the Cartesian plane. To illustrate an inequality, we first shade out all unwanted points.

We then consider the boundaries:

- We use a solid boundary line ——— to indicate that the points on a boundary are wanted.
- We use a dashed boundary line ------ to indicate that the points on a boundary are unwanted.

Finally, we write $\mathcal{R}$ in the region left unshaded.

All points satisfying $ax + by < d$ lie on one side of the line $ax + by = d$ and all points satisfying $ax + by > d$ lie on the other side.

To find the region corresponding to an inequality, substitute into the inequality a point **not** on the boundary line.

- If a **true** statement results, then this point lies in the region we want.
- If a **false** statement results, then the required region is on the other side of the boundary line.

SKILL PRACTICE

1   **a**   Plot the points $\{(-2, -3), (1, 1), (5, 4), (2, 0)\}$ on a set of axes.

     b   What geometric shape do the points form? Give evidence to support your answer.

2   Find the gradient of each line segment:

   a   **b** 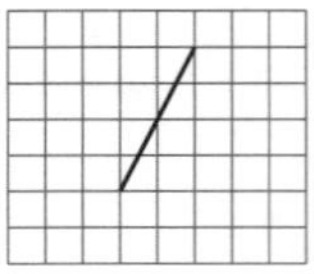   **c**

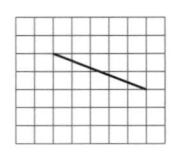

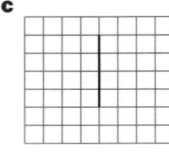

   d 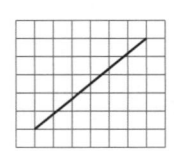   **e** 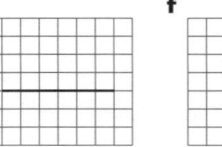   **f**

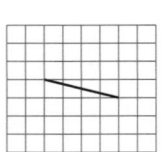

3   Find the distance between:

   a   A and B   **b**   B and C   **c**   A and C

   Hence classify triangle ABC.

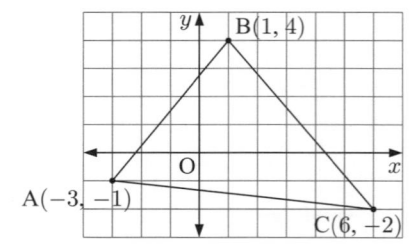

4   Find the equation of:

   a   the horizontal line 5 units above the x-axis

   b   the vertical line with x-intercept 3

   c   the line with gradient -2 and y-intercept 3

   d   the line with x-intercept 4 and y-intercept -2.

5   Classify triangle ABC:

   a $A(0, -3)$, $B(2, 1)$, $C(6, -1)$

   b $A(1, 1)$, $B(5, 4)$, $C(-2, -3)$.

6   Find the gradient of all lines perpendicular to a line with gradient:

   a $-\frac{1}{2}$ **b**   4   **c** -3

7

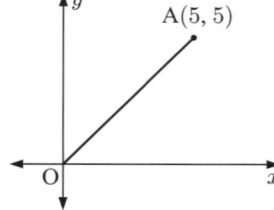

   a   Find the midpoint M of line segment AO.

   b   Find x such that $P(x, 2)$ is equidistant from A and O.

   c   Find the coordinates of point B on the line segment AO such that $AB = 3\sqrt{2}$ units.

8   Find the gradient of the line segment joining the following points:

   a $(1, 5)$ and $(3, -3)$ **b** $(1, 1)$ and $(-5, -9)$

   c $(4, 2)$ and $(15, 2)$ **d** $(5, 0)$ and $(5, -4)$.

9   Triangle ABC has vertices $A(1, 3)$, $B(4, 5)$, and $C(3, -1)$.

   a   Find the length of the line segment from A to the midpoint M of BC.

   b   Find the area of a circle drawn with diameter AM.

10   Consider the straight line $y = \frac{1}{2}x + \frac{3}{2}$.

   a   Construct a table of values for values of x from -3 to 3.

   b   Plot the graph of the line.

   c   Find the gradient and axes intercepts of the line.

11   Find t given that the line joining:

   a $R(-1, 3)$ to $S(-4, t)$ is parallel to a line with gradient -2

   b $A(t, 5)$ to $B(2, 3)$ is perpendicular to a line with gradient $\frac{2}{3}$.

12   Find the equations of these lines:

   a   **b**

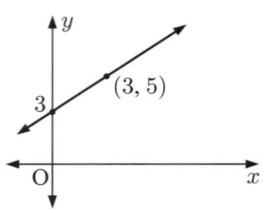

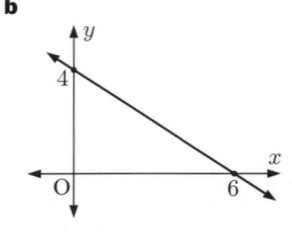

   c 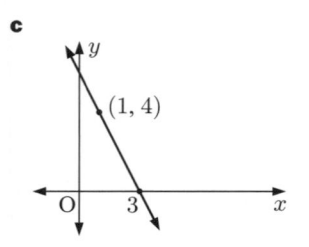   **d**

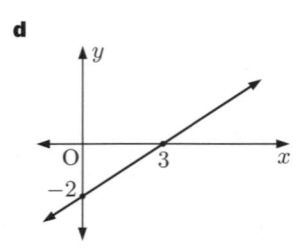

13   Do the points $D(-2, 5)$, $E(1, 6)$, and $F(-6, 3)$ lie in a straight line?

14   Consider the triangle with vertices $A(1, 2)$, $B(9, 3)$, and $M(3, 6)$.

   a   Find the coordinates of C such that M is the midpoint of line segment AC.

   b   Show that triangle ABC is isosceles.

15   Find the equation of the line:

   a   with gradient -1 and y-intercept 7

   b   with gradient 2 which passes through $(-2, -5)$

   c   which passes through $(2, -5)$ and $(-1, 3)$.

16   Find the equations connecting these variables:

   a   **b**

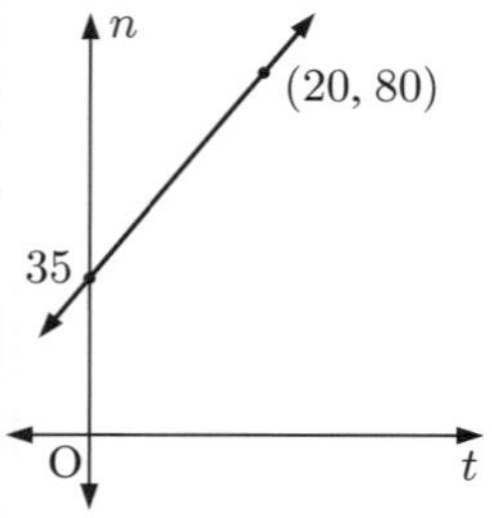

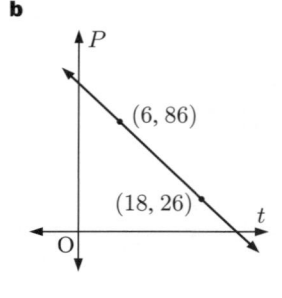

17 The graph alongside indicates the amounts charged by a water company for household water use.

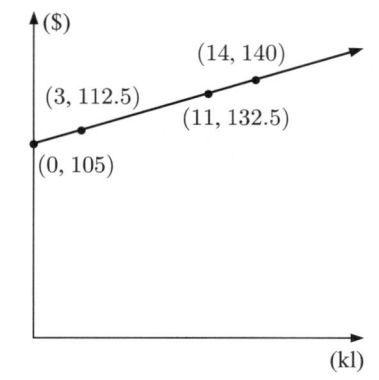

Points shown: (0, 105), (3, 112.5), (11, 132.5), (14, 140). Vertical axis ($), horizontal axis (kl).

  a What does the intercept on the vertical axis mean?

  b Find the gradient of the line, and interpret this value.

  c Determine the amount paid for using 7 kl of water.

  d If a household was billed $120, how much water did they use?

  e The water company is increasing the supply charge to $115, but decreasing the cost per kl to $1.40. Is a household using 7 kl of water going to be better or worse off?

18 Find the equation of the line through $(-1, 2)$ which is perpendicular to the line $2x - y = 5$.

19 Find z given that J$(z, -2)$, K$(-2, 1)$, and L$(3, -1)$ lie in a straight line.

20 Find a given that $(a, 4)$ lies on the line with equation $2x - 3y = a$.

21 Write inequalities to represent the following unshaded regions, $\mathcal{R}$:

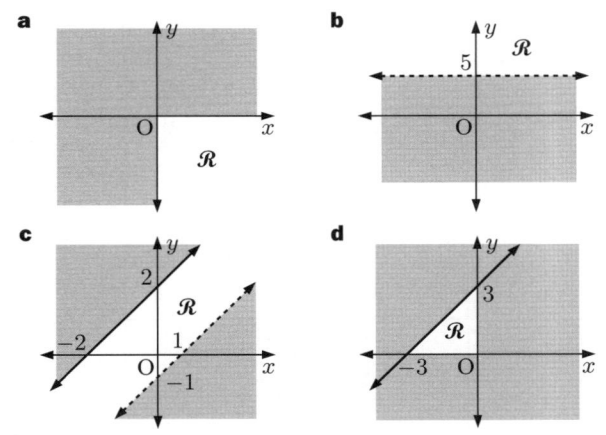

  a, **b**, **c**, **d** (graphs shown)

22 Convert:
  a $y = -\frac{1}{3}x + 2$ into general form
  b $2x + 3y = -4$ into gradient-intercept form

23 Find, in general form, the equations of:
  a line through $(-3, 7)$ and $(5, 13)$
  b line through $(\frac{1}{2}, \frac{9}{2})$ and $(2, 0)$

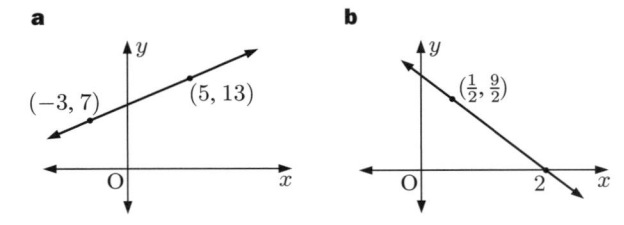

24 Consider the points A$(1, 1)$ and B$(7, 5)$.
  a Locate the midpoint M of line segment AB.
  b Find the gradient of AB.
  c Find the equation of the line through M which is perpendicular to AB.
  d Find the y-coordinate of the point C$(8, y)$ which is equidistant from A and B.
  e Show that C lies on the line through M which is perpendicular to AB.
  f Explain what geometrical fact you have demonstrated.

25 The graph alongside indicates the cost of sending parcels in the post using normal mail (line A) and express mail (line B). Points (5, 45) on B and (5, 20) on A.
  a Find and interpret the gradients of line A and line B.
  b How much more does it cost to send a 3 kg parcel using express mail than using normal mail?

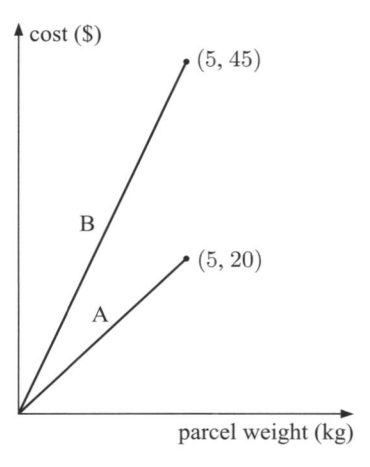

26 Find the equation of a line which is:
  a parallel to $3x - y = 5$ and passes through O$(0, 0)$
  b perpendicular to $x + 4y = -3$ and passes through $(3, 2)$.

27 Graph the regions defined by:
  a $3 < x \leqslant 5$ **b** $3x - y \geqslant 6$
  c $4x + 5y < 0$ **d** $x < 0$ and $y \leqslant 2$
  e $x + y > 0$ and $x \leqslant 4$
  f $x \geqslant 0$, $y < 2$, and $y > x$.

28 Two straight lines have equations $y = 3x - 7$ and $2x + ay = -1$. Find the value of a if the lines are:
  a parallel **b** perpendicular.

29 Graph the line with equation:
  a $y = 3x + 2$ **b** $4x - y = 8$
  c $y = -\frac{1}{2}x + 6$ **d** $x + 3y = 9$

30 Consider the points A$(1, 3)$, B$(4, 4)$, and C$(6, -2)$.
  a Find the length of:
     i OA **ii** AB **iii** BC **iv** OC.
  b What can be said about triangles OAC and BAC? Give reasons for your answer.
  c What sort of quadrilateral is OABC?
  d Find the gradient of:
     i OB **ii** AC
  e What property have you shown in **d**?
  f Find the midpoint of OB. Comment on the significance of this point.

31 **a** Graph the line with equation $y = \frac{1}{3}x + 4$.
  b Reflect $y = \frac{1}{3}x + 4$ in the y-axis and find the equation of the reflected line.
  c Reflect $y = \frac{1}{3}x + 4$ in the x-axis and find the equation of the reflected line.
  d Rotate $y = \frac{1}{3}x + 4$ through $90°$ clockwise about the origin. Find the equation of the rotated line.

32 Write down a set of inequalities to represent the unshaded region $\mathcal{R}$:

a **b**

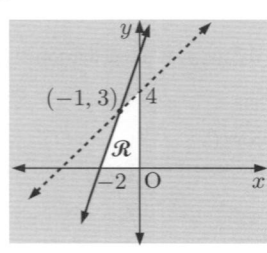

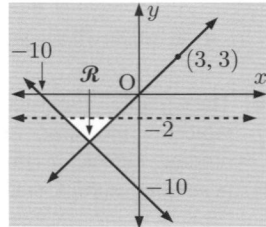

33 **a** Plot the points $A(1, 4)$, $B(5, 8)$, and $C(9, 9)$.
  b Classify quadrilateral OABC.
  c Find the equation of the line of symmetry of OABC.

34 The vertices of quadrilateral ABCD are $A(3, 0)$, $B(0, 4)$, $C(4, 7)$, and $D(7, 3)$.
  a Draw the quadrilateral on a set of axes.
  b Classify quadrilateral ABCD, giving reasons for your answer.
  c Find the equations of the lines of symmetry of ABCD.

35 Consider the region $\mathcal{R}$ defined by $x \geqslant 0$, $y \geqslant 2$, and $2x + y \leqslant 10$.
  a Graph the region $\mathcal{R}$ on grid paper.
  b How many points in $\mathcal{R}$ have integer coordinates?
  c How many of these points obey the rule $x + y = 7$?

36 Trapezium ABCD has vertices $A(2, 1)$, $B(6, 3)$, $C(7, 5)$, and $D(1, k)$. Point $E(4, -1)$ lies outside the trapezium.
  a Find k such that AB $\parallel$ CD.
  b Show that: **i** A, D, and E are collinear
      ii B, C, and E are collinear.
  c Find the length of
      i AE   **ii** DE   **iii** BE   **iv** CE
  d What can you conclude about triangles ABE and DCE? Explain your answer.

37 **a** Graph the region $\mathcal{R}$ for which $x \geqslant 1$, $y \geqslant 0$, $x + y \geqslant 4$, $3y - 2x \geqslant 2$ and $3x + 4y \leqslant 31$.
  b Find all points in $\mathcal{R}$ with integer coordinates, such that $2y - x = 7$.

TOPIC 8: TRIGONOMETRY

RIGHT ANGLED TRIANGLES

For the right angled triangle with angle θ, we label: 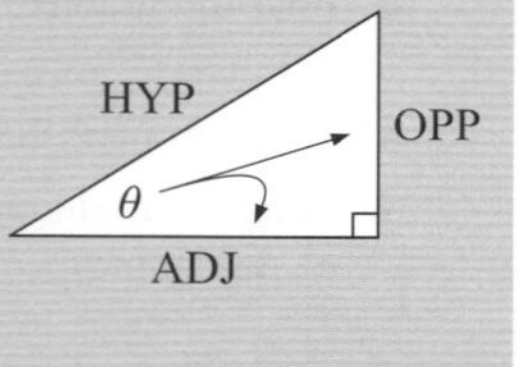
- the longest side or **hypotenuse** as **HYP**
- the side **opposite** θ as **OPP**
- the side **adjacent** to θ as **ADJ**.

The **trigonometric ratios** are:
- sine   or $\sin \theta = \dfrac{\text{OPP}}{\text{HYP}}$
- cosine   or $\cos \theta = \dfrac{\text{ADJ}}{\text{HYP}}$
- tangent   or $\tan \theta = \dfrac{\text{OPP}}{\text{ADJ}} = \dfrac{\sin \theta}{\cos \theta}$

You should be able to use the **inverse trigonometric ratios** $\sin^{-1}$, $\cos^{-1}$, and $\tan^{-1}$, to find angles.

You should be able to use the trigonometric ratios to find unknown side lengths and angles of right angled triangles. For problem solving questions you should remember to:
- draw a clearly labelled diagram of the situation
- explain, where appropriate, why you can apply right angled triangle trigonometry, for example in:
  ▸ isosceles and right angled triangles
  ▸ properties of circles and tangents
  ▸ angles of elevation and depression

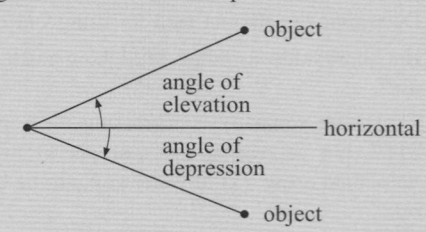

  ▸ true bearings problems where angles are measured clockwise from true north.
- answer the question in words.

THE UNIT CIRCLE

The **unit circle** is the circle with centre $O(0, 0)$ and radius 1 unit.

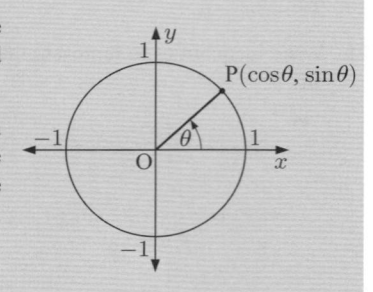

If point P is located on the unit circle at an angle θ measured anti-clockwise from the positive x-axis, then P has coordinates $(\cos \theta, \sin \theta)$.

You should memorise these trigonometric ratios for important angles:

θ	$\cos \theta$	$\sin \theta$	$\tan \theta$
$0°$	1	0	0
$30°$	$\dfrac{\sqrt{3}}{2}$	$\dfrac{1}{2}$	$\dfrac{1}{\sqrt{3}}$
$45°$	$\dfrac{1}{\sqrt{2}}$	$\dfrac{1}{\sqrt{2}}$	1
$60°$	$\dfrac{1}{2}$	$\dfrac{\sqrt{3}}{2}$	$\sqrt{3}$
$90°$	0	1	undefined

You can use the symmetry of the unit circle to find the trigonometric ratios for angles that are multiples of $30°$ and $45°$.

Three-dimensional problems

For three-dimensional objects we can **project** the image of a line onto a plane.

The projection of AB onto the base plane is AM.

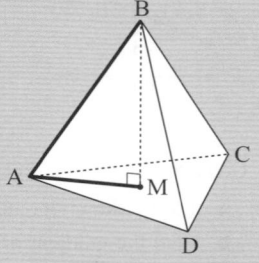

The **angle between a line and a plane** is the angle between the line and its projection onto the plane.

To find the **angle between two planes**:

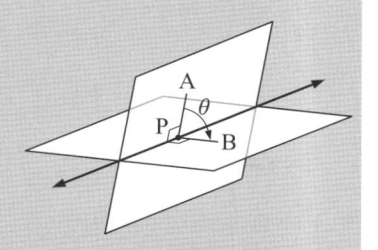

- choose a point P on the line where the planes intersect
- choose a point A in one plane so that AP is perpendicular to the line
- choose a point B in the other plane so that BP is perpendicular to the line, and so that $\widehat{APB} \leqslant 90°$
- the angle between the two planes is $\theta = \widehat{APB}$.

AREA OF A TRIANGLE USING SINE

The area of a triangle is a half of the product of two sides and the sine of the included angle between them.

$$\text{area} = \tfrac{1}{2}ab \sin C$$

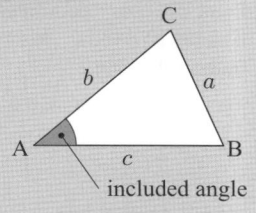

THE SINE RULE

In any triangle ABC with sides a, b, and c units, and opposite angles A, B, and C respectively,

$$\frac{\sin A}{a} = \frac{\sin B}{b} = \frac{\sin C}{c}$$

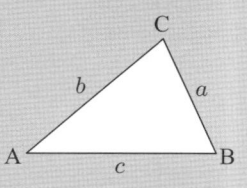

We use the sine rule when we are given:

- **two angles** and **a side**
- **two sides** and a **non-included angle**.

The **ambiguous case** may occur when we are given two sides and one angle, where the angle is opposite the shorter side.

It occurs because an equation of the form $\sin \theta = b$ produces answers of the form $\theta = \sin^{-1} b$ or $(180° - \sin^{-1} b)$.

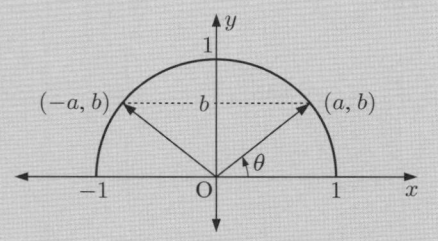

Sometimes the supplied information allows us to eliminate one of the answers.

THE COSINE RULE

In any triangle ABC with sides a, b, and c units, and opposite angles A, B, and C respectively,

$a^2 = b^2 + c^2 - 2bc \cos A$
$b^2 = a^2 + c^2 - 2ac \cos B$
$c^2 = a^2 + b^2 - 2ab \cos C$

$\cos A = \dfrac{b^2 + c^2 - a^2}{2bc}$

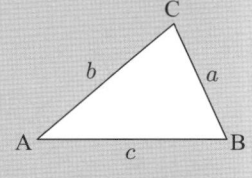

$\cos B = \dfrac{a^2 + c^2 - b^2}{2ac}$ $\qquad \cos C = \dfrac{a^2 + b^2 - c^2}{2ab}$

We use the cosine rule when given:

- **two sides** and the **included angle**
- **three sides**.

TRIGONOMETRIC GRAPHS

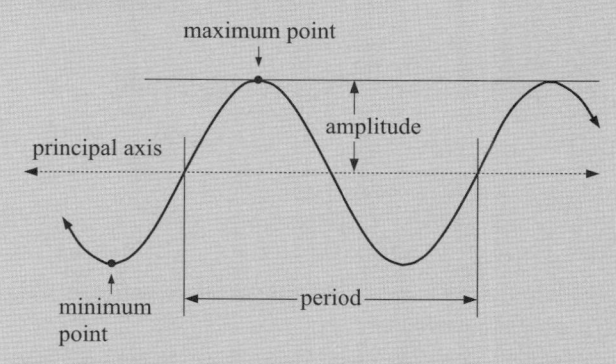

- A **periodic function** is one which repeats itself over and over in a horizontal direction.
- The **period** of a periodic function is the length of one repetition or cycle.
- The graph oscillates about the **principal axis** or **mean line**.
- A **maximum point** occurs at the top of a crest.
- A **minimum point** occurs at the bottom of a trough.
- The **amplitude** is the vertical distance between a maximum or minimum point and the principal axis.

The graph of $y = \sin x$:

- is continuous
- has range $\{y \mid -1 \leqslant y \leqslant 1, y \in \mathbb{R}\}$
- passes through the origin and continues indefinitely in both directions
- has amplitude 1
- has period 360°
- has both line and rotational symmetry.

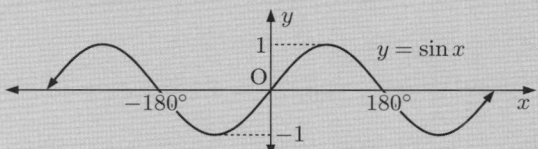

The graph of $y = \cos x$:

- is continuous
- has range $\{y \mid -1 \leqslant y \leqslant 1, y \in \mathbb{R}\}$
- has y-intercept 1
- has amplitude 1
- has period 360°
- has exactly the same shape as the sine graph, but is translated 90° to the left, or with vector $\begin{pmatrix} -90° \\ 0 \end{pmatrix}$
- has both line and rotational symmetry.

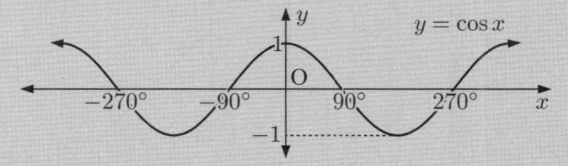

The graph of $y = \tan x$:

- is not continuous at $\pm 90°$, $\pm 270°$, $\pm 450°$, ...
- has range $\{y \mid y \in \mathbb{R}\}$
- passes through the origin
- has period 180°
- has rotational symmetry but not line symmetry.

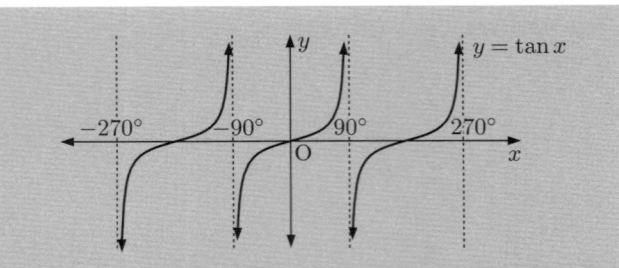

The graphs of $y = a\sin(bx)$ and $y = a\cos(bx)$

- a provides a stretch with invariant x-axis and scale factor a. The amplitude $= |a|$.
- b provides a stretch with invariant y-axis and scale factor $\frac{1}{b}$. The period $= \frac{360°}{b}$.

SKILL PRACTICE

1 For the triangle given, name the:
  a hypotenuse
  b side opposite α
  c side adjacent to α.

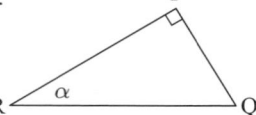

2 For the triangle given, find:
  a $\sin \theta$
  b $\cos \theta$
  c $\tan \theta$

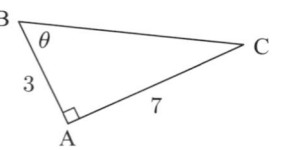

3 Find x, correct to 3 significant figures:
  a
  b

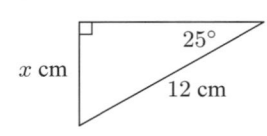

4 Without using a calculator, find the exact values of:
  a $\sin 135°$
  b $\tan 315°$
  c $\cos 240°$

5 Find x, correct to 2 decimal places:
  a
  b

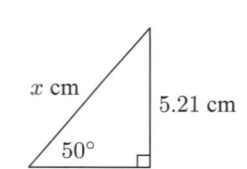

6 Find the length of metal required to make the bracket shown:

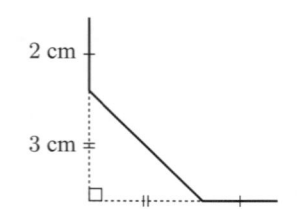

7 Find the area of:
  a
  b
  c

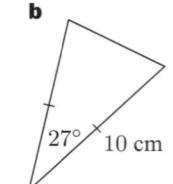

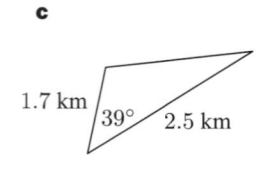

8 Find θ, correct to 1 decimal place:
  a
  b

9 Find the area of this trapezium:

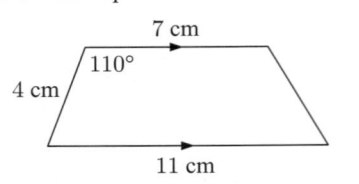

10 Point P has coordinates (0.5299, 0.8480). Without finding θ, state:
  a $\cos \theta$
  b $\sin \theta$
  c $\tan \theta$

11 If triangle PQR has area 175 cm^2, find the value of x.

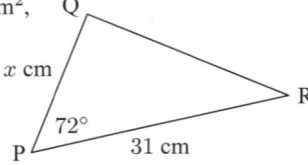

12 Sketch $y = \sin x$ for $0° \leqslant x \leqslant 360°$. Use your graph and your calculator to solve the following equations on the given domain. Give your answers correct to the nearest degree.
  a $\sin x = 0.6$
  b $\sin x = -0.5$

13 Find the exact value of the unknown:
  a
  b
  c

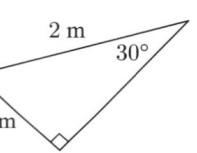

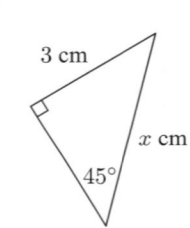

14 Find the value of x:
  a
  b
  c

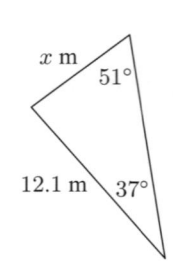

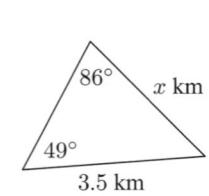

15 A hovercraft travels on a bearing of $130°$ until it is 23 km east of its starting point.
  a How far has the hovercraft travelled?
  b How far south is the hovercraft from its starting point?

16 In triangle ABC, find the measure of side b if $A = 67°$, $C = 31°$, and $c = 2.7$ km.

17 Find the unknowns, giving answers correct to 3 significant figures:

  a

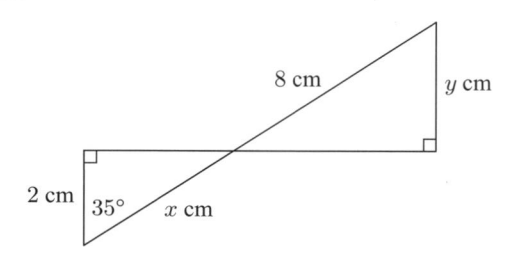

  b

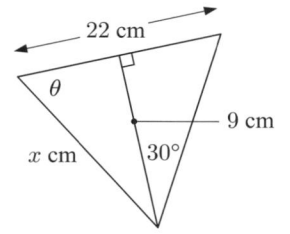

18 A, B and C are the checkpoints of an orienteering course. Find the bearing of:
  a B from A
  b C from B
  c B from C
  d C from A
  e A from B
  f A from C.

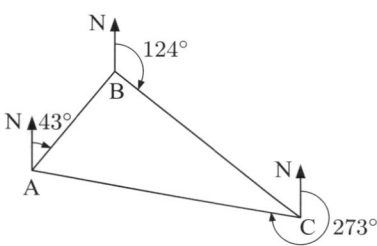

19 Find the value of x in:
  a 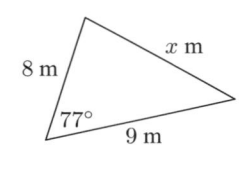
  b 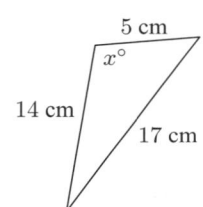
  c

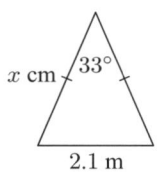

20 Show that $\tan 60° \sin 60° + \cos 60° = 2$.

21 Sketch $y = \cos x$ for $0° \leqslant x \leqslant 360°$. Use your graph and your calculator to solve the following equations on the given domain. Give your answers correct to the nearest degree.
  a $\cos x = -0.3$
  b $\cos x = 0.4$

22 Tangents are drawn from an external point P to a circle of radius 5 cm. Their points of contact are A and B. $A\hat{O}B = 150°$.
  a Explain why AP = BP.
  b Find the area of quadrilateral AOBP.

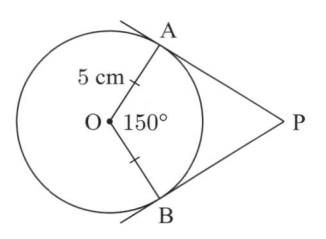

23 Find the angle between the given line segments and the base plane EFGH:
  a AH
  b CE

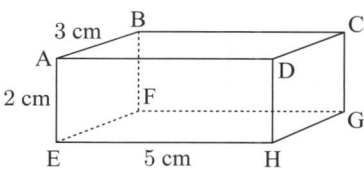

24 Find the measure of all angles of:

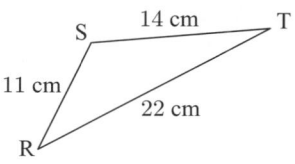

25 Robert hikes on a 110° course for 4.6 km. He then turns 90° to the right and hikes a further 3.9 km. Find the distance and bearing of Robert from his starting point.

26 A cat sits on the floor of a barn, 3 m from a wall. She patiently watches a mouse on a shelf attached to the wall. The cat's eyes are 20 cm above the floor, and the angle of elevation to the mouse is 55°. 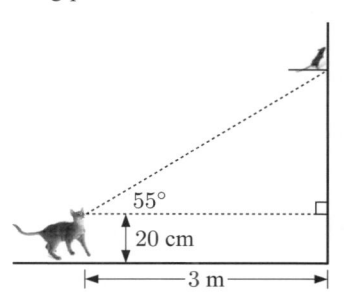
  a Find the angle of depression from the mouse to the cat.
  b How high is the shelf above the floor?

27 Point P is located on a unit circle at an anti-clockwise angle of 120° from the positive x-axis. N is located on the x-axis so that $P\hat{N}O = 90°$. 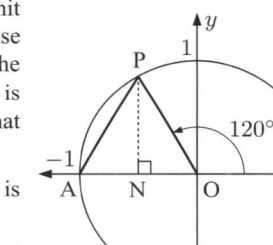
  a Explain why $\triangle APO$ is equilateral.
  b Hence state the exact length of:
    i ON
    ii PN
  c State the exact coordinates of P.
  d Use your answer to **c** to state:
    i $\cos 120°$
    ii $\sin 120°$
    iii $\tan 120°$.

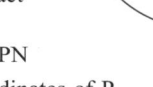

28 An aeroplane flies from Adelaide on a bearing of 221° for 7 km. It then changes course and flies for 230 km on a bearing of 073°. Find:
  a the distance of the aeroplane from Adelaide
  b the bearing of the aeroplane from Adelaide.

29 Sketch $y = \tan x$ for $0° \leqslant x \leqslant 360°$. Use your graph and your calculator to solve the following equations on the given domain. Give your answers to the nearest degree.
  a $\tan x = -0.4$
  b $\tan x = 3$

30 Find:
  a AC
  b CG
  c CE
  d $E\hat{C}G$

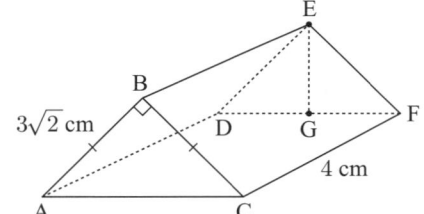

31 Triangle FGH has $F\hat{G}H = \theta$. FG = 3.1 m, GH = 4.5 m, and the area of the triangle is 5 m². Find the possible values of θ.

32 Sarah has just bought some farmland, and wishes to put a fence around the perimeter. It is irregularly shaped, so she hires a surveyor, who provides the radial survey alongside.
Find the perimeter of Sarah's property.

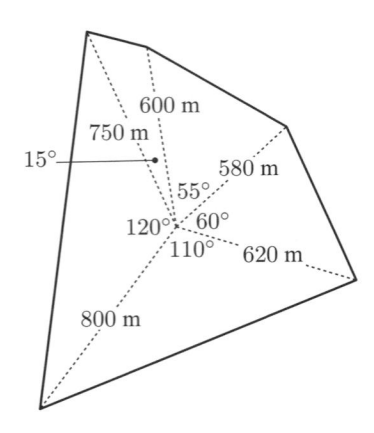

33 Sketch $y = \sin x$ and $y = \tan x$ on the same set of axes, for $-180° \leqslant x \leqslant 180°$. Explain their points of intersection using algebra.

34 Find the angle between the given line segments and the base of the cone:
  a AP
  b AB

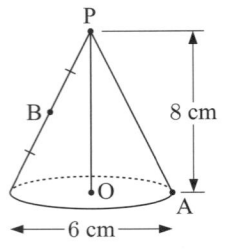

35 Find, correct to 3 significant figures, the area of:
  a
  b

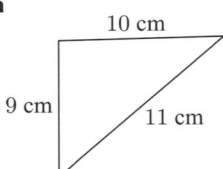

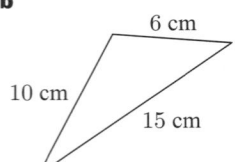

36 Draw free-hand sketches of the following for $0° \leqslant x \leqslant 720°$:
  a $y = \sin x$ and $y = \frac{1}{2}\sin x$
  b $y = \cos x$ and $y = 3\cos x$

37 A commentator at Old Trafford is 44 m above the field of play. The angle of depression to the ball is 34° and the angle of depression to the goal mouth is 26°. The commentator, the ball, and the goal mouth form a right angled triangle as shown.

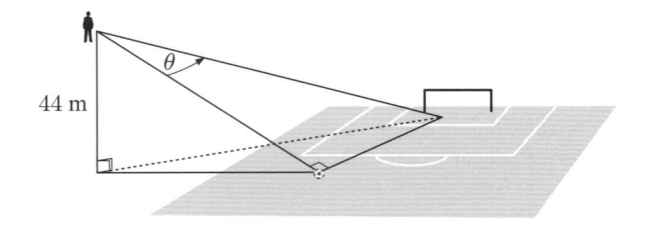

  a Find the distance from:
    i the commentator to the ball
    ii the commentator to the goal mouth
    iii the ball to the goal mouth.
  b Find the commentator's viewing angle θ between the ball and the goal mouth.

38 a On the same set of axes, for $0° \leqslant x \leqslant 720°$, sketch $y = \cos x$ and $y = \tan x$.
  b Explain how the asymptotes of $y = \tan x$ relate to the graph of $y = \cos x$.

39

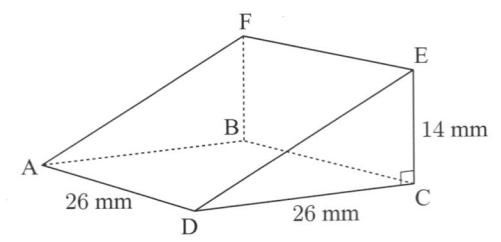

Find the angle between planes:
  a ABCD and ADEF
  b BED and ABCD.

40 Draw free-hand sketches of the following for $0° \leqslant x \leqslant 720°$:
  a $y = \cos x$ and $y = \cos(2x)$
  b $y = \sin x$ and $y = \sin(3x)$

41 Ali is constructing a housing for his car speakers. The ends of the housing have the shape shown:
Find the measure of angle β.

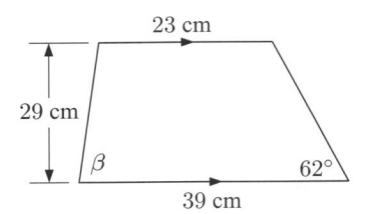

TOPIC 9: SETS

A **set** is a collection of objects or things.

The **elements** of a set are the objects or members which make up the set.

If x is in the set A we write $x \in A$.

The number of elements in set A is written $n(A)$.

Two sets A and B are **equal** if their elements are the same.

A is a **subset** of B if every element of A is also an element of B. We write $A \subseteq B$.

A is a **proper subset** of B if every element of A is also an element of B, but $A \neq B$. We write $A \subset B$.

The **universal set** U contains all elements under consideration.

The **empty set** $\varnothing$ or $\{\}$ has no elements.

The **complement** of A is the set of all elements of U which are *not* elements of A. We denote the complement A'.

You should be able to use set notation when dealing with the **special number sets** $\mathbb{N}$, $\mathbb{Z}$, $\mathbb{Q}$, and $\mathbb{R}$.

You should be able to use **inequalities** to describe sets on the number line.

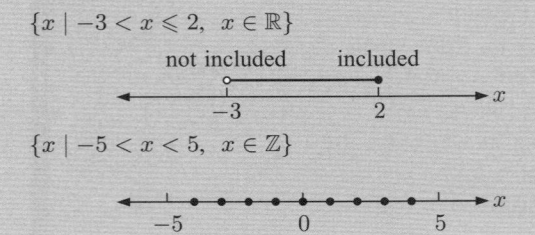

$\{x \mid -3 < x \leqslant 2, \; x \in \mathbb{R}\}$

$\{x \mid -5 < x < 5, \; x \in \mathbb{Z}\}$

VENN DIAGRAMS

A **Venn diagram** consists of a universal set U represented by a rectangle, and sets within it that are generally represented by circles.

The **union** of sets A and B, $A \cup B$, contains all elements belonging to A or B, or both A and B.

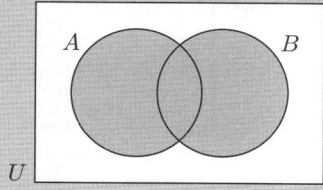

The **intersection** of sets A and B, $A \cap B$, contains all elements belonging to both sets.

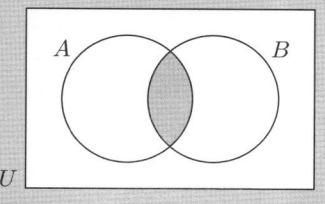

Two sets A and B are **disjoint** if $A \cap B = \varnothing$.

SKILL PRACTICE

1 $U = \{x \mid 1 \leqslant x \leqslant 10, \ x \in \mathbb{Z}\}$, $P = \{3, 4, 5, 6\}$, and $Q = \{4, 6, 8, 10\}$.
   a Find: **i** $n(U)$ **ii** $n(P)$ **iii** $n(Q)$
   b List the elements of P'.
   c True or false? **i** $P \subseteq Q$ **ii** $Q \subset U$
       iii $7 \in P$ **iv** $9 \in Q'$
   d Display P and Q on a Venn diagram, clearly showing each element of U.

2 True or false?
   a $\sqrt{5} \in \mathbb{Z}$ **b** $(\sqrt{5})^2 \in \mathbb{Z}$
   c $\sqrt{5} \in \mathbb{Q}$ **d** $\sqrt{5} + 1 \in \mathbb{R}$

3 a Sketch $\{x \mid -2 < x \leqslant 5, \ x \in \mathbb{Z}\}$ on a number line.
   b Is the set finite or infinite?

4 Suppose $A = \{$composite numbers between 10 and 20 inclusive$\}$ and $B = \{$even numbers between 10 and 20 inclusive$\}$.
   a List the elements of A and B.
   b Find: **i** $n(A)$ **ii** $n(B)$
   c True or false? **i** $A \subset B$ **ii** $A \subseteq B$

5 Illustrate using a Venn diagram:
   a $U = \{$quadrilaterals$\}$, $R = \{$rectangles$\}$, $S = \{$squares$\}$
   b $U = \{$quadrilaterals$\}$, $K = \{$kites$\}$, $T = \{$trapezia$\}$.

6 True or false?
   a $\mathbb{Z} \subset \mathbb{N}$ **b** $\mathbb{Z} \subset \mathbb{Q}$ **c** $\mathbb{R} \subseteq \mathbb{Q}$

7 a Sketch $\{x \mid 3 \leqslant x < 9, \ x \in \mathbb{R}\}$ on a number line.
   b Is the set finite or infinite?

8 Use a Venn diagram to illustrate the set of rational numbers that are not integers.

9 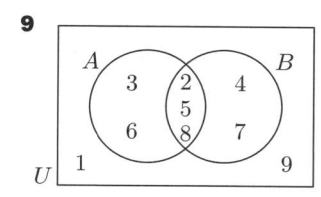
   a List the elements of:
      i A **ii** $A \cap B$
      iii $A' \cap B$
   b Find:
      i $n(A \cup B)$
      ii $n(A \cap B')$
      iii $n(A' \cup B')$

10 On separate Venn diagrams, shade the regions representing:
   a $A \cap B'$ **b** $A' \cup B$

11 Describe these sets using inequalities:
   a   **b** 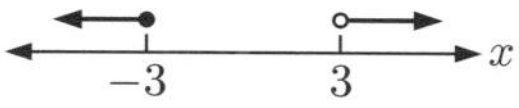
   c

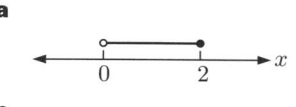

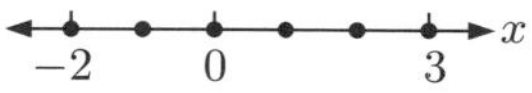

12 A survey was conducted to see which students liked pineapples (P) and durians (D). The results are shown in the Venn diagram. Determine the number of students:
   a in the survey
   b who like pineapples and durians
   c who like pineapples but not durians
   d who like pineapples or durians.

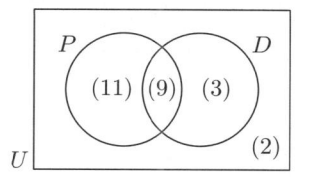

13 Use a Venn diagram to verify that $A \cap B = (A' \cup B')'$.

14 In the high school music society, 11 members like jazz (J), and 8 members like blues (B). There are 17 members in the society, of whom 4 like neither jazz nor blues.
   a Represent this information on a Venn diagram.
   b Determine the number of students who like:
      i both jazz and blues   **ii** jazz but not blues.

15 a Is $\{x \mid 0 \leqslant x \leqslant 1, \ x \in \mathbb{Q}\}$ finite or infinite?
   b Explain why you cannot illustrate this set on a number line.

16 Angela has 75 DVDs. She has watched 58 of them and her boyfriend Ben has watched 29 of them. 9 DVDs have been watched by neither of them. Find the number of DVDs which have been watched by:
   a both Angela and Ben   **b** Ben but not Angela.

17 Describe, in words and in symbols, the shaded region of:
   a   **b**

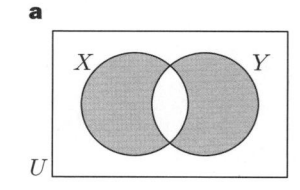

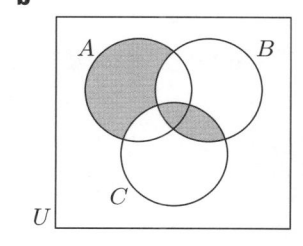

18 A survey is conducted at school to determine which places the students like spending their lunch times. 31 like the canteen (C), 19 like the library (L), and 20 like the student common room (R). 5 like all three places, while 12 like none of them. 12 students like both the canteen and the library, 9 like both the canteen and the common room, and 6 like both the library and the common room. How many students were surveyed?

TOPIC 10: PROBABILITY

Any event can be assigned a **probability** between 0 and 1 inclusive, which indicates the chance or likelihood of that event happening.

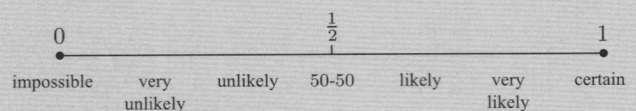

```
0               1/2              1
●───────────────●────────────────●
impossible  very   unlikely  50-50  likely  very   certain
            unlikely                        likely
```

If A is an event with probability $P(A)$ then $0 \leqslant P(A) \leqslant 1$.

EXPERIMENTAL PROBABILITY

In a probability experiment:

- The **number of trials** is the total number of times the experiment is repeated.
- The **outcomes** are the different results possible for one trial of the experiment.
- The **frequency** of a particular outcome is the number of times that this outcome is observed.
- The **relative frequency** of an outcome is the frequency of that outcome divided by the total number of trials.

$$\text{relative frequency} = \frac{\text{frequency}}{\text{number of trials}}$$

The relative frequency of an event is an estimate of its probability. The larger the number of trials, the more confident we are that the experimental probability obtained is accurate.

We sometimes use **two-way tables** to display experimental data. You should know how to use these tables to estimate the probabilities of different events.

EXPECTATION

Suppose the probability of an event occurring is p. If the experiment is repeated n times, the **expectation** of the event, or number of times we expect it to occur, is np.

SAMPLE SPACE

A **sample space** is the set of all possible outcomes of an experiment.

You should be able to represent sample spaces by:

- listing them
- using a two-dimensional grid
- using a tree diagram
- using a Venn diagram.

THEORETICAL PROBABILITY

For experiments in which there are a number of equally likely outcomes, the **theoretical probability** of an event A is

$$P(A) = \frac{\text{number of ways } A \text{ can happen}}{\text{total number of possible outcomes}} = \frac{n(A)}{n(U)}.$$

COMPLEMENTARY EVENTS

Two events are **complementary** if exactly one of them *must* occur.

The **complement** of event A is A'.

$$P(A) + P(A') = 1$$

INDEPENDENT EVENTS

Two events are **independent** if the occurrence or non-occurrence of either one of them cannot affect the occurrence of the other.

Two events A and B are independent if

$$P(A \text{ and } B) = P(A) \times P(B).$$

DEPENDENT EVENTS

Two events are **dependent** if the occurrence of one event affects the occurrence of the other.

You should be able to use tree diagrams to help find probabilities for compound events which are independent (sampling with replacement) and dependent (sampling without replacement).

MUTUALLY EXCLUSIVE EVENTS

If two events have no common outcomes then they are **mutually exclusive** or **disjoint**.

If A and B are mutually exclusive then

$$P(A \text{ or } B) = P(A) + P(B).$$

SKILL PRACTICE

1 Describe what it means for an event to have probability:

  a $\frac{1}{2}$ **b** 0.95   **c** 30%

2 Use words to describe the probability that:

  a it will snow in Cairo tomorrow
  b you will be sick at least once this year.

3 Last year Express Transport recorded the following results for parcels it carried for its clients:

Description	Frequency
Delivered on time	2812
Delivered late	465
Returned to sender	38
Lost	11

  a How many parcels did the company carry last year?
  b Add a relative frequency column to the table.
  c Estimate the probability that Express Transport will:
    i lose a parcel
    ii deliver a parcel at some time.

4 A survey was conducted to determine public opinion on whether smoking should be allowed in restaurants, clubs, and sports venues. Survey participants were also asked if they were smokers. The results are summarised in the table shown.

	Allowed	Not Allowed
Smokers	86	19
Non-smokers	24	215

  a Copy and complete the table to include "totals".

b Estimate the probability that a randomly chosen person:
   i is a smoker
   ii wants smoking banned from restaurants, clubs, and sports venues.

c Amy is a non-smoker. Estimate the probability that she wants smoking allowed.

5 Which of these pairs of events are dependent? Give reasons for your answers.

   a The light globe in the garage is blown.
   The milk in my fridge is off.

   b Chocolate is on special at the supermarket.
   John has enough money to buy chocolate.

6 List the sample space for the order in which Ting, Mei, and Ivy can sit.

7 Wern has a circular spinner divided into 3 unequal regions. From 300 spins, the spinner finished on blue 132 times, on red 86 times, and on green for the remainder.

   a Estimate the proportion of the spinner that is:
      i red   **ii** green.

   b Henry is blue-green colour-blind. If he were to use this spinner, estimate the probability he would record a result of 'non-red'.

8 The Venn diagram shows the parts sung by men in a choir.

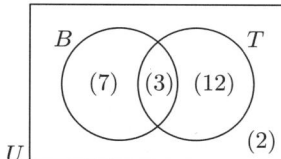

$B \equiv$ bass
$T \equiv$ tenor

Find the probability that a randomly chosen man in the choir sings:

   a bass   **b** tenor but not bass
   c tenor or bass **d** tenor, given he sings bass.

9 On average, Katie scores 23% of the points for her basketball team. If the team scores 76 points in their next match, how many would we expect Katie to have scored?

10 A cloth bag contains 6 coloured tokens, 4 red and 2 white. Two tokens are randomly selected without replacement.

   a Determine the probability of getting:
      i two red tokens   **ii** a red then a white token
      iii a red and a white token
      iv at least one white token.

   b The first token is drawn out but is kept hidden from view. The second token drawn is red. Find the probability that the first token was:
      i red   **ii** white.

11 250 people were surveyed on their use of dishwashing liquid. Some of their results are shown.

Brand	Frequency	Relative frequency
Sparkle	114	
Fresh		
Shine		0.18
Megawhite	23	
Total	250	

   a Copy and complete the table.

   b Which brand is the market leader?
   c Estimate the probability that a randomly selected home uses: **i** Megawhite   **ii** Fresh.

12 The six faces of a die are marked 1, 1, 1, 2, 2, and 3.
   a Find the probability that any roll will be a 2.
   b If the die is rolled 120 times, how many times would we expect a result of:
      i 3   **ii** 1   **iii** 4

13 A coin is tossed and a triangular spinner labelled P, Q, and R is spun. Illustrate the sample space of possible results on:
   a a 2-dimensional grid   **b** a tree diagram.

14 In a stable of 36 horses, 27 are thoroughbreds (T) and 16 are mares (M). 4 horses are neither thoroughbred nor a mare.

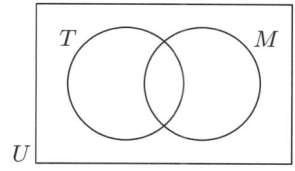

   a Copy and complete the Venn diagram to display the horses in the stable.

   b Find the probability that a randomly selected horse from the stable:
      i is a thoroughbred mare
      ii is a mare that is not thoroughbred
      iii is thoroughbred or a mare
      iv is thoroughbred, given it is a mare.

15 The table shown describes the origins of students at an international school in Hong Kong.

Origin	Boy	Girl
Africa	3	1
Asia	312	247
Australia or New Zealand	24	15
Europe	86	105
South America	0	2
North America	28	32

   a How many students are there in the school?
   b Find the probability that a randomly chosen student:
      i is a girl
      ii is a boy from Europe
      iii is a girl from the Americas

   c A student from Asia is chosen at random. What is the probability that the student is a boy?
   d A girl is chosen at random. What is the probability that she is from Australia or New Zealand?

16 Each time he bats, Danny hits a home run with probability 0.045. In a season he bats 62 times. How many home runs do you expect him to score?

17 When an ordinary die is rolled, the result is recorded as *low* or L, if it is a 1 or a 2.
Two dice are rolled simultaneously.

   a Construct a tree diagram to display the possible outcomes, including probabilities for each event.
   b Find the probability that:
      i both results are low   **ii** at least one result is not low.

18 List the possible outcomes when four coins are tossed simultaneously. Hence determine the probability of getting:
   a at most three heads
   b two heads, given there is at least one head
   c one tail, given there are at least two heads.

19 Oscar has 12 square tiles, 7 red, 3 blue, and 2 green. He puts them all in a bag and asks his sister Christa to draw one out at random. Oscar notes down the colour of Christa's tile, and she puts it back in the bag. Oscar then draws a tile from the bag himself.

  a Draw a tree diagram to display the possible outcomes.
  b Find the probability that:
   i Christa draws a red tile and Oscar draws a green tile
   ii two blue tiles are drawn
   iii a blue and a green tile are drawn
   iv two different colours are drawn
   v no green tiles are drawn
   vi exactly one red tile is drawn.

20 In this weekend's match, the probability that the home team will win is $\frac{2}{3}$. The probability that Ben will forget his lunch tomorrow is $\frac{1}{10}$.

  a Are these events dependent? Explain your answer.
  b Find the probability that:
   i the home team wins and Ben forgets his lunch
   ii the home team wins or Ben remembers his lunch
   iii the home team does not win, given that Ben forgets his lunch
   iv Ben remembers his lunch, given that the home team does not win.

21 Gerrit has 7 hens in his coop, 4 white and 3 brown. In the morning he opens the coop to let them out into the yard, and watches them carefully.

  a Determine the probability that:
   i the first two hens out are white
   ii the last hen out is brown.
  b The first hen out is brown. Determine the probability that the second hen out is white.
  c Determine the probability that the second and third hens have the same colour.

22 Kylie draws a card from an ordinary playing deck.
Which of the following pairs of events are mutually exclusive? Explain your answers.

  a A : The card is a 2. B : The card is a heart.
  b A : The card is a picture card. B : The card is a 7.

23 There are 100 tickets in a raffle of which Kevin has bought 5. The winners for two prizes are chosen at random, in succession, without replacement.

  a Draw a tree diagram to illustrate the possible outcomes.
  b Determine the probability that Kevin wins:
   i both prizes **ii** second prize **iii** only second prize.
  c Ping has bought 3 tickets in the same raffle. Determine the probability that:
   i both Kevin and Ping win prizes
   ii Ping wins a prize but Kevin does not.

24 a List the 6 different ways in which four friends A, B, C, and D can sit around a circular table. There is no 'head' of the table, so regard ABCD as the same as BCDA.

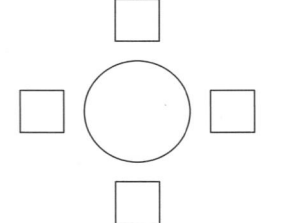

  b The four friends sit down at random. Find the probability that:
   i A sits opposite C
   ii D sits immediately to the left of B
   iii B sits between A and D.
  c A sits opposite D. Find the probability that C and B sit together.
  d B sits next to C. Find the probability that A and D sit together.
  e B sits immediately to the right of A. Find the probability that A and C sit opposite one another.

25 Ling is pondering which two games she will play at the local fair. She decides to toss a coin to decide.
If the coin lands heads, she pays £2 to spin a spinner. She will then have 30% chance of winning £5.
If the coin lands tails, she pays £1 to draw a ball from a bag. There are 9 white balls in the bag and one black one. If she draws the black ball she wins £8.

  a Draw a tree diagram to illustrate the possible outcomes.
  b Determine the probability that after the game, Ling has:
   i won £3   **ii** lost £1   **iii** won money.
  c Determine the expected result from Ling playing a game.

26 In a penalty shoot out, Adrian, Brad, Charles, Danny, and Edward each had a shot at goal. The table shows the probabilities of each player scoring a given penalty.
Find the probability that Adrian scored, given that at least four of the boys scored.

Adrian	0.75
Brad	0.85
Charles	0.80
Danny	0.90
Edward	0.85

TOPIC 11: STATISTICS

VARIABLES

Categorical variables
- describe a particular quality or characteristic
- can be divided into categories
- information collected is **categorical data**.

Quantitative variables
- **discrete variables**
  ▸ take exact number values
  ▸ often result from **counting**
  ▸ information collected is **discrete numerical data**.
- **continuous variables**
  ▸ take numerical values within a continuous range
  ▸ usually result from **measuring**
  ▸ information collected is **continuous numerical data**.

STATISTICAL GRAPHS

You should be able to display data using various graphs and charts, and use these to help analyse the data. In particular, we use:

- **bar charts**
  ▸ vertical or horizontal
  ▸ bars are separated for discrete data
  ▸ all bars same width

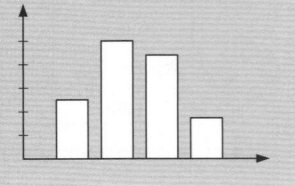

- pie charts
  - each sector angle is found as a fraction of 360°

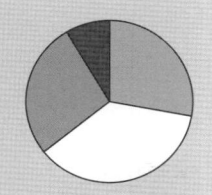

- scatter diagrams
  - used to plot pairs of points

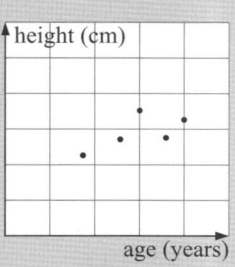

- line graphs
  - used only for continuous data
  - we plot pairs of points and join them with straight line segments

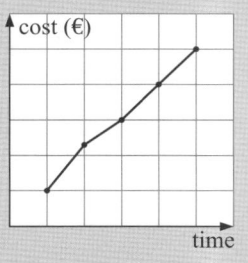

- stem-and-leaf plots
  - display all data values
  - require a key
  - may be unordered or ordered

stem	leaf
1	3 7
2	0 1 4 4 6 9
3	2 5 6 8
4	3 4

Key: 1 | 3 means 1.3 m

You should also be able to construct graphs to *compare* two sets of data. These include:

- side-by-side bar charts
- back-to-back bar charts
- back-to-back stem plots.

CENSUS OR SAMPLE

A **census** involves collecting data about *every* individual in a population.

A **sample** involves collecting data about a part of the population. It must be unbiased, and sufficiently large to reflect the characteristics of the whole population.

THE CENTRE OF A DISCRETE DATA SET

- Mean
  - $\overline{x} = \dfrac{\Sigma x}{n}$ where Σx is the sum of the data.

- Median
  - the middle value of an ordered data set
  - for n data values, it is the $\left(\dfrac{n+1}{2}\right)^{\text{th}}$ data value
  - for an **odd number** of data, the median is one of the data
  - for an **even number** of data, the median is the average of the two middle values.

- Mode
  - the most frequently occurring value in the data set
  - a data set is **bimodal** if it has two modes
  - if a data set has three or more modes, we do not use the mode as a measure of centre.

MEASURING THE SPREAD OF DISCRETE DATA

- **Range** = maximum data value − minimum data value
- **Lower Quartile (Q_1)**:
  - middle value of the **lower** half of the data
  - 25% of the data have values less than or equal to Q_1.
- **Upper Quartile (Q_3)**:
  - middle value of the **upper** half of the data
  - 75% of the data have values less than or equal to Q_3.
- **Interquartile Range (IQR)**:
  - is the range of the middle half of the data
  - IQR = $Q_3 - Q_1$

FREQUENCY TABLES

For frequency tables, the **mean** is calculated using $\overline{x} = \dfrac{\Sigma fx}{\Sigma f}$.

A 'product' column helps in finding Σfx.

The **median** and **mode** are found in the same way as previously.

GROUPED DISCRETE DATA

When data has been grouped, calculating the exact mean and median becomes impossible. However, we can *estimate* them:

Mean

- use the **midpoint** of each class to represent x
- mean = $\dfrac{\Sigma fx}{\Sigma f}$

Median

- create a **cumulative frequency graph** or **ogive**
- the median is the 50th percentile.

USING TECHNOLOGY

You should be able to use your graphics calculator to:

- enter a list of data
- obtain the descriptive statistics for the data set
- create statistical graphs.

CONTINUOUS DATA

Continuous data is placed into **class intervals** which are shown by **inequalities**.

To estimate the **mean** of continuous data, we use the same method as for grouped discrete data.

Histograms

Histograms are used when graphing continuous data. They are similar to bar charts, but there are no gaps between the columns.

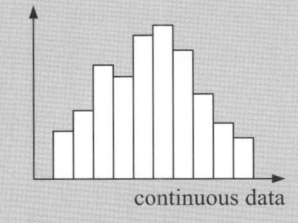

If the width of each class interval is the same:

- the histogram is a **frequency histogram**
- the height of each column is the frequency of the class
- the **modal class** is the class with the highest frequency.

If the class interval widths are not the same, then it is **not** a frequency histogram.

- the height of each column is the **frequency density** of the class
- frequency density = $\dfrac{\text{frequency}}{\text{class interval width}}$
- the **modal class** is the class with the highest frequency density.

Cumulative frequency

The **cumulative frequency** gives a running total of the scores up to a particular value.

We find cumulative frequencies using a frequency table, and graph this data on a **cumulative frequency curve**:

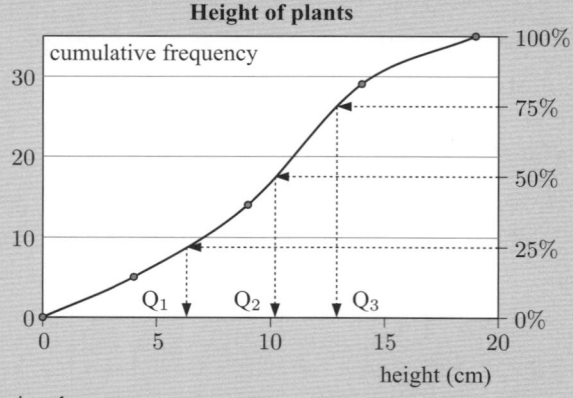

Height of plants

Notice that:

- $Q_1 = 25^{\text{th}}$ percentile,
- $Q_2 = 50^{\text{th}}$ percentile = the median,
- $Q_3 = 75^{\text{th}}$ percentile
- we draw lines on the graph to show the method
- **percentiles** can be easily read using the percentage scale.

CORRELATION

Correlation is a measure of the strength of the association between two variables.

Correlation may be described as:

- positive or negative (or no correlation)
- strong, moderate or weak
- linear or non-linear.

LINE OF BEST FIT

To draw a line of best fit by eye:

- find the mean point $(\overline{x}, \overline{y})$
- plot the mean point on the scatter diagram
- draw a straight line through the mean point so there are roughly the same number of points above the line as below it.

Linear regression is a formal method for finding a line which best fits a set of data. You should know how to find the equation of the line of best fit for a data set using your graphics calculator.

We can use the line of best fit to estimate the value of one variable given a value for the other.

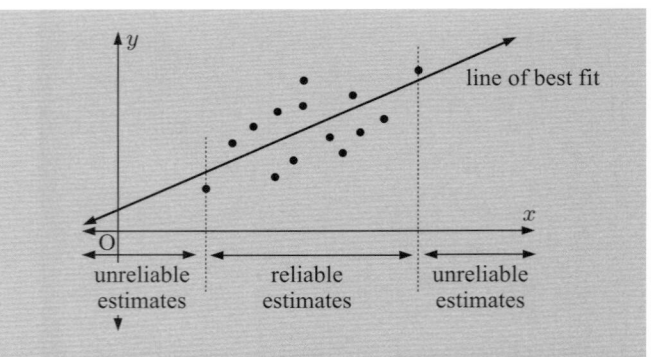

In general, it is reasonable to estimate values within the domain of the data, but unreliable to estimate values outside the domain.

SKILL PRACTICE

1 A gelateria sells four sizes of gelato. The sales for one evening are shown in the table below.

Size	small	medium	large	extra large
Number sold	37	25	18	11

Find the sector angles for each category, and hence construct a pie chart for the data.

2 Classify the following variables as categorical, discrete, or continuous:

   a the number of sweets in a 500 g packet
   b the weights of eggs produced at a farm
   c the breakfast cereal a person mainly eats
   d the points scored in a game by basketball players.

3 **a** Draw an ordered stem-and-leaf plot to display the ages of members in a croquet club:
   23, 26, 31, 35, 36, 44, 47, 48, 48, 51, 56, 58, 59, 62, 64, 64
   b What percentage of members are at least 50 years of age?

4 A pie chart was constructed from data on political preference collected from 200 people. The segment for "conservative" has angle $54°$. How many people gave the answer "conservative"?

5 State whether a census or a sample would be used for the following investigations:

   a The heights of 15 year old students at a school.
   b The average monthly water bill for households in Sydney.

6 Johann's times for each kilometre of a 5000 m race are shown in the table below:

Kilometre	1	2	3	4	5
Time (s)	176	185	183	181	174

   a Display the progress of Johann throughout the race using a straight line graph.
   b Explain why a line graph is more appropriate than a scatter diagram in this case.
   c Find the total time Johann took for the race. Express your answer in minutes and seconds.

7 35 football players were asked "How many goals did you score in your last game?" The following data was collected:

```
0 3 1 2 0 1 0 1 4 2 0 1
1 2 4 1 5 2 1 0 0 1 3 2
1 2 0 4 1 3 1 0 4 2 1
```

a What is the variable in this investigation?
b Is the data discrete or continuous? Explain your answer.
c Construct a vertical bar chart for the data.
d What percentage of the footballers surveyed scored:
  i less than 2 goals   ii 3 goals?

8 Discuss any possible bias in the following situations:
  a policemen in one police station are interviewed about problems with crime
  b only dog and cat breeders are interviewed about possible changes to pet registration laws
  c a survey where people must submit their answers on the internet.

9 The maximum daily temperatures in London and Paris during the same week are recorded below:

Day	Mon	Tues	Wed	Thu	Fri	Sat	Sun
London (°C)	15	17	16	14	13	15	16
Paris (°C)	18	20	20	19	16	18	21

  a Draw a side-by-side bar chart to illustrate the data.
  b In which city was the temperature higher?
  c Comment on any pattern in the data.

10 Find the a mean b median c mode for each of the following data sets:
  a 5 6 6 1 7 5 9 7 7 3
  b 0 0 0 1 2 2 2 2 3 3 3 4 4 4 4 5 5 6 7
  c 8.9 7.5 8.6 8.1 8.9 9.1 8.0 7.9 8.7 7.1
  d 80 85 81 82 79 83 86 81 79 85 86

11 For each of the scatter diagrams below, state:
  i whether there is a positive, negative, or no association between the variables
  ii whether the relationship between the variables appears to be linear
  iii the strength of the association (zero, weak, moderate, or strong).

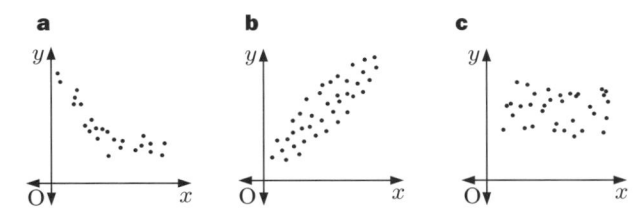

a
b
c

12 The mean of 18 scores is 27.5. What is the sum of the scores?

13 The bar graph below shows the delivery times of parcels sent overseas using two different shipping companies.

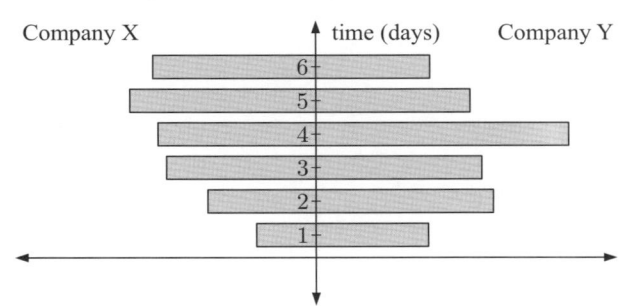

Company X   time (days)   Company Y

Which company would you recommend using? Explain your answer.

14 The purchase prices of ten cars are:
  £7000   £7000   £25 000   £28 000   £14 000
  £18 000   £13 000   £7000   £29 000   £47 000

  a Find the mean, median, and modal purchase prices.
  b Is the mode a suitable measure of centre for this data? Explain your answer.
  c Give a reason why the mean may not be the most suitable measure of centre for this data.

15 Find q, given that 8, 7, 8, 9, 12, q, q, and 6 have a mean of 7.

16 An experiment was conducted to determine whether the number of blueberries on a plant was affected by the use of a fertiliser. The results are summarised in the table below:

Blueberries on a plant

Without fertiliser	17	38	41	21	30	42
	40	25	19	42	37	33
	26	35	30	45	23	19
With fertiliser	44	39	26	51	45	35
	31	45	54	43	27	62
	58	28	47	31	60	40

  a Display the data on a back-to-back stem-and-leaf plot.
  b Determine the highest number of blueberries on a plant with and without fertiliser.
  c Comment on whether the fertiliser is effective.

17 The students in Year 9 at a school were asked how many sports they played. The results are shown opposite. Calculate the:
  a mode
  b median
  c mean
  d range of the data.

Number of sports	Frequency
0	10
1	13
2	25
3	15
4	2
Total	65

18 10 students were asked how many hours of revision they did for their Maths exam, and their final exam score. The results are given in the table below:

Student	A	B	C	D	E	F	G	H	I	J
Hours of revision	4	5	0	3	3	2	9	6	1	8
Exam score (%)	63	70	45	55	50	68	92	80	52	81

  a Draw a scatter diagram with the hours of revision on the horizontal axis.
  b Comment on the relationship between the hours of revision done and the exam score.

19 An inspector takes a random sample of 50 trees in a plantation and measures their height in metres. The results are shown in the table below:

Height (h m)	Frequency
$2.0 \leqslant h < 3.0$	2
$3.0 \leqslant h < 4.0$	8
$4.0 \leqslant h < 5.0$	11
$5.0 \leqslant h < 6.0$	10
$6.0 \leqslant h < 7.0$	12
$7.0 \leqslant h < 8.0$	7

  a Explain why *height* is a continuous variable.
  b What is the modal class?

c Estimate the mean height.
d If there are 5000 trees in the plantation, estimate the number which measure:
   i 5 metres or more
   ii between 3 and 5 metres.

20 A randomly selected sample of office workers was asked "How many work meetings did you attend in the past fortnight?". The results are illustrated in the graph below:

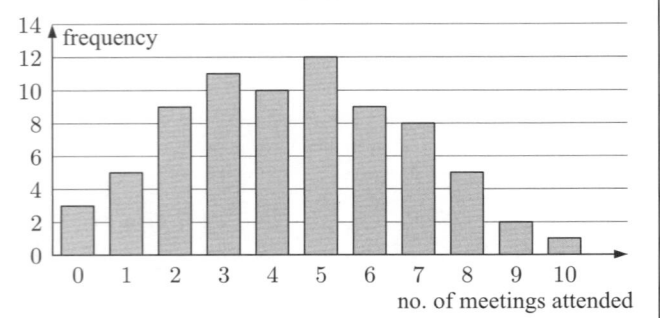

a How many workers were surveyed?
b How many workers attended between 2 and 5 meetings (inclusive)?
c What percentage of workers attended more than 6 meetings?

21 A sample of 25 measurements has a mean of 27.4 and another sample of 15 measurements has a mean of 22.6. Find the mean of all 40 measurements.

22 The following frequency table records the number of cyclists stopping for coffee at a local café over a 6 week period.

No. of cyclists	Frequency
2	1
3	1
4	2
5	3
6	5
7	8
8	12
9	7
10	3

a For this data, find the:
   i   mean
   ii median
   iii mode
   iv range.
b Construct a vertical bar chart for this data. Show the positions of the mean, median, and mode on the horizontal axis.
c Which measure of centre would be most suitable for this data set? Explain your answer.

23 A department store keeps a record of how many shoppers come into their store each day. The results are:

```
1161  963  721  750  1376  502  450  192
380   729  527  399  312   1201 480  1052
479   363  1481 123  799   521  367  925
157   1005 195  144  586   177  423  571
601   291  538  258  687   590  665  1185
```

a Construct a tally and frequency table using class intervals 0 - 199, 200 - 399, 400 - 599, ...
b Draw a vertical bar chart to display the data.
c On how many days did the store receive less than 400 visitors?
d What is the modal class for the data?

24 The mean, median, and mode of a set of six numbers are all 9. Two of the numbers are 7 and 11. If the largest number is twice the smallest number, find the smallest number.

25 The following table shows the number of hot drinks ordered at a café each month, along with the average daily temperature for the month.

Month	Jan	Feb	Mar	Apr	May	Jun
Temperature (°C)	25	22	18	15	13	10
Hot drinks ordered ('00)	26	15	41	36	52	60

Month	Jul	Aug	Sep	Oct	Nov	Dec
Temperature (°C)	12	15	20	22	25	28
Hot drinks ordered ('00)	43	43	34	27	20	5

Draw a scatter diagram with the independent variable *temperature* on the horizontal axis. Comment on the relationship between the two variables.

26 For the following data sets, make sure the data is ordered and then find:
   i   the median
   ii the upper and lower quartiles
   iii the range
   iv the interquartile range.
a 7, 7, 7, 8, 8, 8, 9, 9, 9, 10, 10, 10, 10, 11, 11, 11, 11, 11, 12, 13, 13
b 4, 2, 3, 2, 2, 1, 7, 0, 0, 1, 1, 1, 1, 5, 5, 2, 3, 6, 10
c 88.6, 87.9, 89.9, 89.7, 90.9, 92.7, 89.9, 88.1, 89.5

27 35 businesses in a city were surveyed to find out how many people they employed. The following data was collected:

```
68  61  50  52  42  66  41
57  32  71  74  75  77  20
19  61  8   65  21  63  58
47  49  65  36  70  44  37
31  28  31  53  56  11  67
```

a Construct a tally and frequency table for this data, using suitable class intervals.
b Draw a vertical bar chart to display the data.
c How many businesses employed less than 30 people?
d What percentage of businesses had at least 50 employees?
e Find the modal class for the data.

28 A sample of 120 university students were asked for their highest score on the refectory's arcade machine. The results were:

Highest score	0 - 1999	2000 - 3999	4000 - 5999	6000 - 7999
Frequency	25	32	48	15

a Estimate the mean high score for the group of students.
b What is the modal class?
c Estimate the median high score.

29 The weights of a zoo's colony of adult meerkats are:

Weight (w g)	Frequency
$600 \leqslant w < 650$	2
$650 \leqslant w < 700$	3
$700 \leqslant w < 750$	7
$750 \leqslant w < 800$	11
$800 \leqslant w < 850$	2

a What is the modal class?
b Estimate the mean weight.
c A larger colony of 50 meerkats is discovered in Namibia. Estimate the number of meerkats in this colony that weigh at least 800 g.
d Discuss whether your estimate might be reasonable.

30 For the data set given, find:
  a the minimum value
  b the maximum value
  c the median
  d the lower quartile
  e the upper quartile
  f the range
  g the interquartile range.

Stem	Leaf
10	0 1
11	0 0 1 3 5
12	2 3 4 5 6 6 7 7
13	0 1 2 2 9
14	2 2 2
15	0 4

11 | 2 represents 112

31 The histogram shows the times (in minutes) that trains into Birmingham New Street Station arrive late.

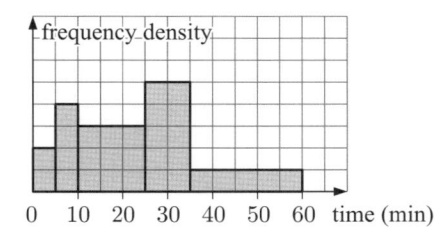

  a Given that 8 trains were between 5 and 10 minutes late, find the sample size used.
  b Find the modal class.
  c Estimate the mean of the data.

32 The mean weights of oranges (to the nearest gram) in a collection of crates from two citrus orchards in Spain are:

Orchard A:
242 230 251 210 235 240 228 246 255 208
246 235 229 240 233 201 241 233 244 234
231 241 236 237 227 244 236 222 245 228
244 205 230 224 248 215 258 239 221 231

Orchard B:
230 220 226 241 227 223 232 236 229 234
222 217 230 213 225 231 220 238 224 215
222 237 202 220 241 233 240 245 229 233
200 231 228 259 225 248 218 225 232 207

  a Construct a tally and frequency table for each set of data, using suitable class intervals.
  b Construct a frequency histogram for each data set.
  c Find the modal class of each data set.
  d Compare the distribution of each data set.
  e Find the range of each data set.
  f Which orchard produced heavier oranges? Explain your answer.

33 The table shows the *household income* and the *weekly food bill* (in £) for eight families.

Family	Household Income (£'000)	Weekly Food Bill (£)
A	45	172
B	22	91
C	38	140
D	36	96
E	23	109
F	40	243
G	28	162
H	55	271

  a Find the mean household income $\bar{x}$ and the mean weekly food bill $\bar{y}$.
  b Draw a scatter diagram for this data.
  c Describe the association between *household income* and the *weekly food bill*.
  d Plot $(\bar{x}, \bar{y})$ on the scatter diagram.
  e Draw a line of best fit through $(\bar{x}, \bar{y})$ on the scatter diagram.
  f Another household has an income of £70 000. Estimate the household's weekly food bill.
  g Comment on the reliability of your estimate in **f**.

34 The given graph estimates the weight of a crop of pumpkins.

Cumulative frequency curve of pumpkin weights

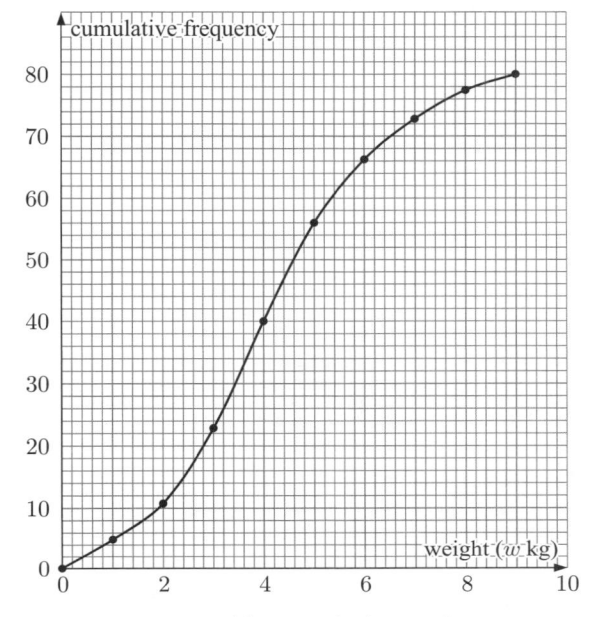

  a How many pumpkins were in the crop?
  b Estimate the median pumpkin weight.
  c Pumpkins heavier than 7.5 kg will be entered in the local fair. How many pumpkins will be entered?
  d What is the IQR for the pumpkin crop?

35 The students at a sporting academy were asked to bench press as much weight as they could. The results were recorded and tabled.

Weight (w kg)	Frequency
$0 \leqslant w < 50$	2
$50 \leqslant w < 75$	13
$75 \leqslant w < 100$	19
$100 \leqslant w < 110$	36
$110 \leqslant w < 150$	10

  a Draw a histogram of the data.
  b What is the modal class?
  c Use a graphics calculator to estimate the mean weight lifted.

36 Simon and Barney are throwing darts at a dart board. The points scored by each of them over 12 darts were:

  Simon: 50 15 1 60 57 40
          20 60 25 60 36 60

  Barney: 50 15 50 20 54 48
          50 51 48 48 16 57

  a Calculate the mean and median for each player.
  b Calculate the range and interquartile range for each player.
  c Which player is more consistent? Explain your answer.

37 The yield of apples depends on the number of cold days experienced by the tree. The following table shows the total yield of apples from an orchard over several years with different numbers of cold days.

Cold days (n)	32	55	67	28	43	50
Yield (Y tonnes)	41.1	26.8	16.2	45.3	38.6	32.5

  a Name the independent and dependent variables.
  b Draw a scatter diagram for the data.
  c Use technology to find the equation of the linear regression line.
  d Interpret the gradient and vertical intercept of this line.
  e Use the regression line to predict the yield from the orchard if the number of cold days was:
    i 40   **ii** 100.
  f Comment on the reliability of your answers in **e**.

38 The times taken for a group of students to complete a logic puzzle were recorded and the results tabled.

Time (t min)	Frequency
$0 \leqslant t < 1$	3
$1 \leqslant t < 2$	11
$2 \leqslant t < 3$	19
$3 \leqslant t < 4$	15
$4 \leqslant t < 5$	7
$5 \leqslant t < 6$	5

  a Construct a cumulative frequency curve for the data.
  b Estimate the IQR and explain what it means.
  c Estimate the 80th percentile and explain what it represents.
  d The fastest 10 students will be selected for an inter-school competition. Estimate the cut-off time for selection.

39 The average value of a rare bottle of wine sold at auction over the period from 1971 to 2005, is shown in dollars in the table below.

Year	1971	1980	1990	1995	2000	2003	2005
Value ($)	221	350	587	723	976	1062	1224

  a Let x be the number of years since 1971 and V be the value of the bottle of wine in dollars.
  Draw a scatter diagram of V against x.
  b Find a linear model for V in terms of x.
  c Comment on the association between the variables. Is the linear model appropriate? Why or why not?
  d How could you estimate the value of the bottle of wine in 2020?

PRACTICE EXAM 1

Paper 2:   45 minutes / 40 marks

1 Make y the subject of the formula: $z = \dfrac{2}{3y - 4}$.   [3]

2 Write an equivalent logarithmic statement for:
  a $3^6 = 729$ **b** $5\sqrt{5} = 5^{1.5}$ [4]

3 Find the exact value of the unknown in:
  a (right triangle with 60° angle, x m hypotenuse, 5 m adjacent)
  b (right triangle with 6 cm, $x°$, $3\sqrt{2}$ cm)   [4]

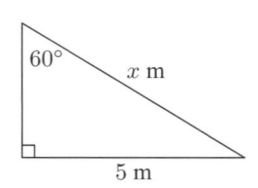

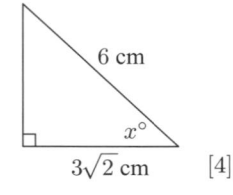

4 **a** Write down the next three terms of the sequence
    81, 80, 78, 75, ...   [1]
  b Write down a formula for the nth term of the sequence.   [2]
  c What is the first negative term of the sequence, and where does it occur?   [2]

5 Calculate the simple interest on a £1800 loan at 6% for 40 months.   [2]

6 Given $g(x) = \dfrac{2-x}{x^3}$,   find:
  a $g(-1)$ **b** $g(3)$ **c** $g(-5)$ [3]

7 Show that $0.\overline{12}$ is a rational number.   [2]

8 Write these sets in interval notation:
  a (number line with closed dots at 0,2,4,6,8)
  b (number line with open circle at −1, closed dot at 2, arrows both ways)   [2]

9 Find:
  a the area of: (triangle with sides 12 cm, 18 cm and included angle 30°)
  b **i** the measure of angle A in $\triangle ABC$ if $a = \sqrt{61}$ cm, $b = 9$ cm and $c = 4$ cm.
    ii Hence find the exact area of the triangle.   [8]

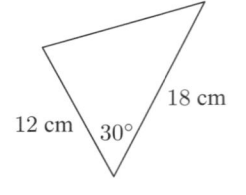

10 Match each scatter diagram to its description:
  a strong, negative, linear correlation   [1]
  b moderate, non-linear correlation   [1]
  c no correlation   [1]

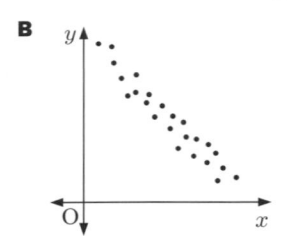

  A (scatter plot — scattered points)
  B (scatter plot — downward linear trend)
  C (scatter plot — upward curved trend)

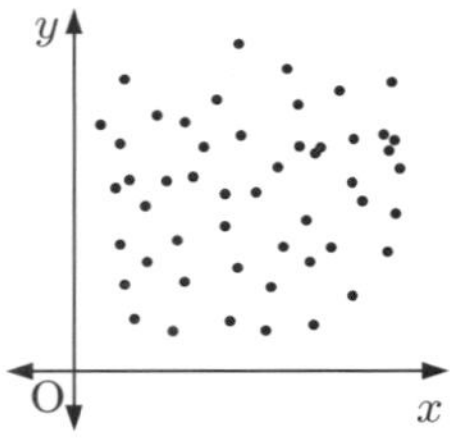

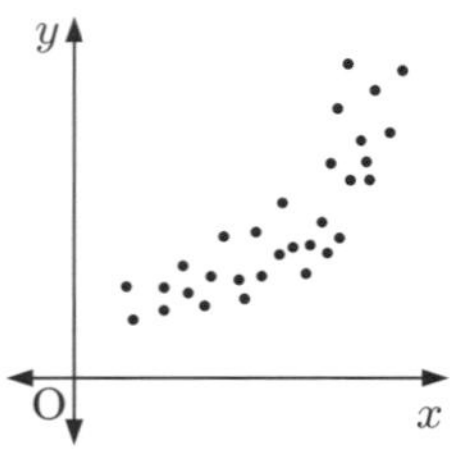

11 Write in component form:
  a $\overrightarrow{XZ}$ [1] 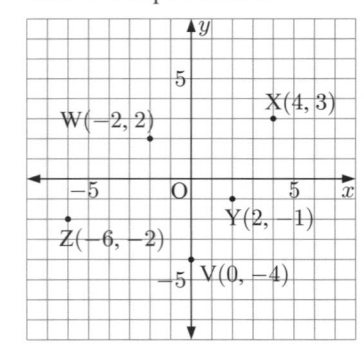
  b $\overrightarrow{WY}$ [1]
  c $\overrightarrow{VY}$ [1]
  d $\overrightarrow{WZ}$ [1]

  (grid showing points W(−2, 2), X(4, 3), Y(2, −1), Z(−6, −2), V(0, −4))

Paper 4: 2 hours 15 minutes / 120 marks

1 Triangle PQR has vertices P(2, 2), Q(3, 5), and R(5, 3).
   a Draw triangle PQR on a set of axes. Label this triangle A. [2]
   b Rotate triangle A 180° about O. Label this image B. [2]
   c Reflect image B in the x-axis. Label this image C. [2]
   d What single transformation maps A onto C? [1]

2 Consider the function $f(x) = 3x^2 - 6x + 1$.
   a Find $f(3)$. [2]
   b Find the axes intercepts of the function. [3]
   c Find the axis of symmetry of the function, and hence find its vertex. [3]
   d Use **c** to write $f(x)$ in the form $a(x-h)^2 + k$. [1]
   e Suppose $g(x)$ is obtained by translating $f(x)$ through $\binom{1}{2}$. Find $g(x)$, giving your answer in the form $g(x) = ax^2 + bx + c$. [3]

3 The height of a solid cylinder is three times its radius r.
   a Write down an expression for the volume of the cylinder, in terms of r. [2]
   b Find r given that the cylinder has volume 192π cm^3. [3]
   c Find the surface area of the cylinder correct to the nearest cm^2. [4]
   d Suppose this cylinder was made of pure lead. Each cubic centimetre of lead weighs 11.37 g. Find the mass of the cylinder. [2]
   e The cylinder is melted down and moulded into a sphere. Find the radius of the sphere. [3]

4 a Graph the region $\mathcal{R}$ for which $x \geqslant 0$, $y \geqslant 1$, $2x + y \geqslant 3$, and $2x + 3y \leqslant 15$. [4]
   b Find all points in $\mathcal{R}$ with integer coordinates, such that $x + 2y = 4$. [3]

5 A school superintendent has a theory that the number of detentions given in a school per student is proportional to the percentage of female teachers on its staff. He has collected the following data from the schools in his district.

No. of female teachers (x)	60	30	2	10	56
No. of detentions per student per year (y)	0.72	0.36	0.03	0.05	0.70
No. of female teachers (x)	90	35	40	84	24
No. of detentions per student per year (y)	1.08	0.36	0.48	0.95	0.28

   a Draw a scatter diagram for the data. [3]
   b Comment on the correlation between the two variables. [2]
   c Is the superintendent's theory correct? Explain your answer. [2]
   d Use technology to determine the equation of the line of best fit. Plot this on the graph. [2]
   e Interpret the gradient of the line of best fit. [1]

   f Estimate the number of detentions per student given in a year in a school with:
      i 38 female teachers   **ii** 115 female teachers.
   Comment on whether these predictions are reasonable. [4]

6 Clinton is in a kayak race. His progress through the course is shown on the travel graph below:

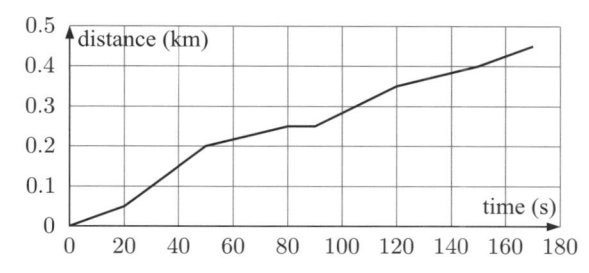

   a How long is the course? [1]
   b **i** When is Clinton travelling fastest?
      ii What is his speed at this time? [3]
   c During the race, Clinton's kayak becomes stuck on a rock. When does this happen, and for how long is he stuck? [2]
   d What is Clinton's average speed in:
      i km/h   **ii** m/s? [2]
   e Clinton's main rival Michel paddles at an average speed of 10.1 km/h, but gets stuck at the same rock for 10 seconds. Who finished in the quicker time? [2]

7 Triangle ABC is isosceles with AC = BC.
   BC is produced to E and AC is produced to D such that CE = CD.
   Prove that: 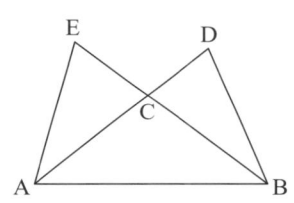
   a AE = BD [4]
   b triangles ABE and ADB are congruent [4]
   c ABDE is a cyclic quadrilateral. [3]

8 Toby recorded the length of time he spent each day surfing the internet. He summarised the results in the table below:

Surfing time (mins)	Frequency
$0 \leqslant t < 15$	3
$15 \leqslant t < 30$	6
$30 \leqslant t < 45$	8
$45 \leqslant t < 60$	5
$60 \leqslant t < 75$	4
$75 \leqslant t < 90$	2

   a For how many days did Toby record data? [1]
   b Construct a histogram to display the data. [4]
   c State the modal class for the data. [1]
   d Use technology to estimate the mean length of time Toby spends surfing the internet each day. [1]
   e Construct a cumulative frequency graph for the data. [4]
   f Toby decides to cut back on his internet surfing. He decides that next month he will 'fine' himself £5 for every day he spends more than 50 minutes online. If there are 30 days in the month, estimate how much money Toby will pay in 'fines'. [3]

9 a Suppose $f(x) = 5 \times 4^{-x}$.

  i Evaluate $f(-\frac{3}{2})$.
  ii Show that $f(x-1) = 4f(x)$.
  iii State the domain and range of $f(x)$.
  iv Write down the equation of the asymptote to $f(x)$.
  [5]

  b Given $f(x) = \dfrac{2x-1}{x+1}$, find $f^{-1}(x)$. Hence graph $f(x)$ and $f^{-1}(x)$ on the same set of axes. [5]

10 Triangle PQR has vertices P(−1, −5), Q(6, −1), and R(2, 1).

  a Show that triangle PQR is right angled, and state the vertex where the right angle occurs. [5]
  b Find the distance from P to X, the midpoint of QR. [3]
  c Point Y is located such that PXQY is a parallelogram. Find the coordinates of Y. [3]

11 a Use the method of elimination to solve simultaneously:
$$x - 4y = -12$$
$$x + 2y = 12$$
[4]

  b Use your graphics calculator to solve simultaneously:
$$x - 4y = -12$$
$$|x| + 2y = 12$$
[2]

12 Gemma and Jo have a total of €28.60 between them. Gemma has three times as much money as Jo. How much money does each have? [4]

Paper 6: 1 hour 30 minutes / 40 marks

A. Investigation: The Hilbert curve   25 marks

The Hilbert curve was first described in 1891. It is used in compressing and processing digital images.

The starting image of the Hilbert curve (called the *first order* Hilbert curve) is a 'cup', which is three sides of a square as shown.

To get the second order Hilbert curve, we start with a square of the original size, and place a smaller 'cup' in each corner:

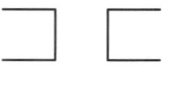

We then draw connecting line segments as shown:

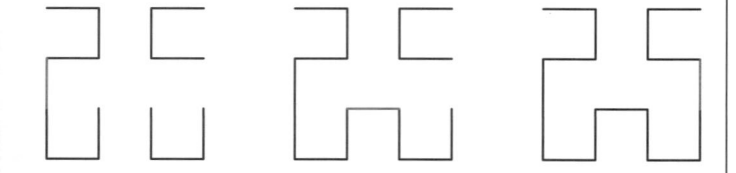

To get to the third order Hilbert curve, we use a similar process using the new shape:

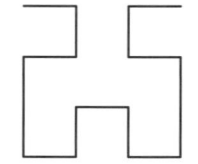

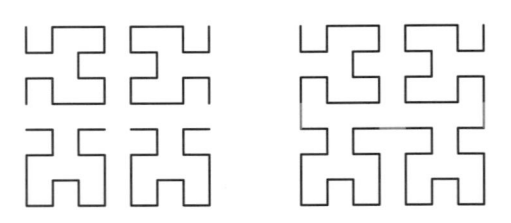

The fourth and fifth order Hilbert curves are shown below, enlarged so you can see the detail:

Rules

The next Hilbert curve is obtained from the current one by:

- shrinking the current curve
- putting a 'copy' in each corner of the square, and rotating it by some multiple of 90 degrees, and
- connecting them up with line segments.

Part I

In this part, we will consider the effect of successive iterations on the line segments of the curve.

1 Use the diagrams given to fill in the following table:

Hilbert curve order	Length of line segments
1	1
2	$\frac{1}{3}$
3	
4	
5	

2 Consider the denominator of each of the fractions in **1**. Describe what is happening to the denominator in words.
  Hint: Consider powers of 2.

3 Suggest a formula for the length of line segments in the nth order Hilbert curve. Hence find the length of the line segments in the:

   a sixth   **b** seventh iteration.

4 As n gets very large, what happens to the length of the line segments in the curve?

Part II

In this part we consider the *number* of line segments needed to make each iteration of the Hilbert curve. Remember that all of the line segments in the curve have the same length.

1 Consider the first order Hilbert curve. How many line segments are needed to make it?

2 Now consider the second order Hilbert curve in **Part I** above. How many line segments are needed to make it?

3 Let u_n be the number of line segments in the nth Hilbert curve. Using the list of rules for generating the next Hilbert curve, write a formula for u_{n+1} in terms of u_n.

4 Complete the table below:

Hilbert curve order	Number of line segments
1	
2	
3	
4	
5	

5 a Using the table above, write a formula for u_n in terms of n only.

   b Hence, find the number of line segments needed for:

   i u_6 **ii** u_{10} **iii** u_{15}.

6 Consider your answer to **5 a**. By replacing n with the expression $n+1$, show that this formula is equivalent to the one you found in **3**.

7 As n gets very large, what happens to the number of line segments needed?

Part III

Now, we will consider the effect that successive iterations have on the *total* length of the Hilbert curve.

1 Copy your answers from **Part I** and **Part II** into the table below, and hence fill out the *Total length* column.

Hilbert curve order	Length of each line segment	Number of line segments	Total length
1	1	3	$1 \times 3 = 3$
2	$\frac{1}{3}$		
3			
4			
5			

2 Suggest a formula for L_n, the total length of the nth order Hilbert curve. Hence find:

   a L_6 **b** L_7 **c** L_{20}

3 a As n gets larger, what happens to L_n?

   b Use your answers to **Part I**, question **4** and **Part II**, question **7** to explain your answer to **a**.

B. Modelling: Hard disk storage   15 marks

Ada, a computer science student, is investigating how the storage capacity of computer hard drives has increased over the past thirty years. She has collected the following data:

Year	1980	1984	1988	1995	1997
Storage (S GB)	0.005	0.015	0.04	1.2	4.3

Year	2000	2005	2007	2010
Storage (S GB)	17.3	140	320	1000

1 Let t be the number of years since 1980. Draw a scatter diagram of S against t.

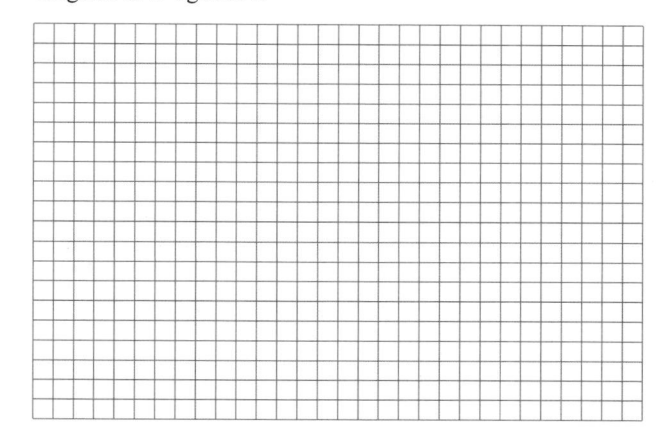

2 The storage capacity S fits a model of the form $S \approx 0.00276 \times 1.534^t$, where t is the number of years after 1980.

   a Find the approximate size of a hard drive in:

   i 1987   **ii** 1992.

   b What is the percentage increase in storage capacity between 1987 and 1992?

   c Will hard drive capacity increase by the same percentage every five years? Explain your answer.

3 During a major cleanout, Ada finds an old 6 GB hard drive without a date of manufacture.

   a Use the model to estimate when the hard drive was made.

   b Comment on the reliability of your prediction.

4 A study has predicted that in 2020, hard drives will reach 14 terabytes (14 000 GB) in size.

   a Use the model to predict the storage capacity of a hard drive in 2020.

   b Explain why your answer might be different from the study's prediction.

PRACTICE EXAM 2

Paper 2:   45 minutes / 40 marks

1 Simplify:

   a $-(-1)^3$ [1]

   b $\sqrt{28} - \sqrt{7}$ [2]

   c $\log 9 + \log 15 - \log 5$ [2]

2 Consider the sequence $-1, 2, -4, 8, -16, \ldots$.

   a Find the next two terms of the sequence.   [2]

   b Write down a formula for the nth term of the sequence.   [3]

3 Match each graph to its description:

  a y is inversely proportional to x [1]

  b y is proportional to x^2 [1]

  c y is directly proportional to x. [1]

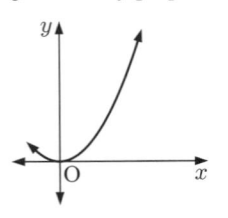

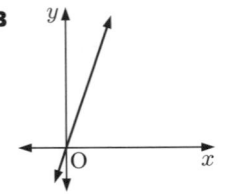

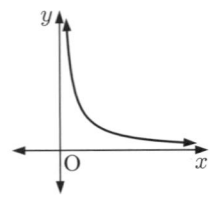

  A, **B**, **C** (graphs)

4 Make j the subject of the formula: $m = \dfrac{2-j}{3s}$ [3]

5 The given data shows the duration of some new movies. Find the:

  a maximum movie length

  b median movie length

  c modal movie length

  d percentage of movies shorter than 90 minutes. [5]

Length of movies (min)

```
 7 | 4
 8 | 6 8 9 9
 9 | 1 2 3 3 3 5 7
10 | 3 4 6 6 7
11 | 0 0 1
```

Key: 10 | 3 means 103 minutes

6 Find: **a** $36^{\frac{3}{2}}$ **b** 5^{-3} [2]

7 Solve the simultaneous equations:
$$2x + 5y = 11$$
$$3x - 2y = -12$$ [4]

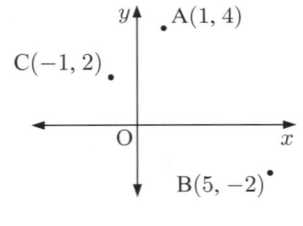

8 A(1, 4), C(−1, 2), B(5, −2)

  a Find $\overrightarrow{AB}$ in component form.

  b Write the coordinates of the image of C(−1, 2) translated through $\overrightarrow{AB}$.

  c Find the midpoint of the line AB. [3]

9 Write as a single fraction, simplifying as far as possible:
$$\dfrac{3}{x(x-1)} + \dfrac{x-4}{x-1}$$ [3]

10 Find a, leaving your answer in surd form: (triangle with sides 5 cm, 4 cm, a cm and 60° angle between the 5 cm and 4 cm sides) [4]

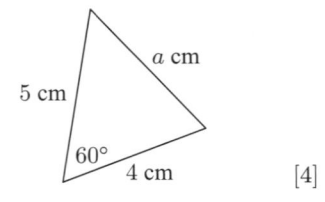

11 $\frac{2}{5}$ of all passengers at Exeter International Airport catch domestic flights, and $\frac{3}{4}$ of all passengers pay by credit card. Find the probability that a randomly selected passenger:

  a did not pay by credit card [1]

  b is catching a domestic flight and that they paid by credit card. [2]

Paper 4: 2 hours 15 minutes / 120 marks

1 The slant height of a cone is 2 cm more than its radius r cm.

  a Write down an expression for the total surface area of the cone, in terms of r. [2]

  b Find r given the cone has surface area 220π cm^2. [4]

  c Find the volume of the cone, correct to 3 significant figures. [3]

2 Peter recorded the times he took to drive to work each day for an eight week period. He summarised the results in the table alongside:

Drive time (mins)	Frequency
$20 \leqslant d < 25$	6
$25 \leqslant d < 28$	10
$28 \leqslant d < 30$	12
$30 \leqslant d < 32$	7
$32 \leqslant d < 36$	5

  a Construct a histogram to illustrate the data, showing frequency density on the vertical axis. [4]

  b State the modal class for the data. [1]

  c Estimate the mean time it takes Peter to get to work. [1]

  d Construct a cumulative frequency graph for the data. [4]

  e Estimate the time Peter should allow to get to work, to give him an 80% chance of arriving on time. [1]

3 Dale invests £2500 in an account at 7.3% p.a. compound interest, calculated annually.

  a Find the value of Dale's investment after:

  i 2 years **ii** 7 years. [4]

  b Write down an equation for the value V of the account after n years. [2]

  c Graph V against n. [2]

  d Use logarithms to determine when Dale's account is worth £8900. [3]

4 Consider the point A(2, 4) relative to the origin O(0, 0).

  a Find the length OA. [2]

  b Find the equation of the line OA. [2]

  c Find the point A′, which is the image of A when reflected in the y-axis. [1]

  d Find the equation of the line OA′. [2]

  e Find the area of triangle OAA′. [2]

  f Find the angle through which A′ would need to be rotated about O to give A. [3]

5 Consider the function $f(x) = \dfrac{3}{x-2}$.

  a Find $f(4)$. [1]

  b Solve $f(x) = 6$. [1]

  c State the equations of any asymptotes of $y = f(x)$. [2]

  d Find $f^{-1}(x)$. [2]

  e Graph $y = f(x)$ and $y = f^{-1}(x)$ on the same set of axes. [2]

  f State the equation of the function obtained when $f(x)$ is translated through $\begin{pmatrix} 4 \\ -1 \end{pmatrix}$. [2]

6 Solve for x:
  a $4x - 5 \leq 3(x+2)$ [2]
  b $|x - 4| < -2x^2 + 5x + 3$ [2]

7 C lies on a circle with radius 1 as shown.
  a Find k. [2]
  b Find the gradient of:
    i AC
    ii CB [4]
  c Use **b** to show that angle ACB is a right angle. [2]

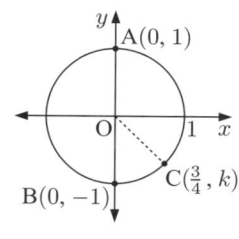

A(0, 1), B(0, −1), C($\frac{3}{4}$, k)

8 Four years ago Emma started an agricultural business. Its profits in the first 5 years are shown in the table below:

Year (n)	1	2	3	4	5
Profit P_n (€ '000s)	−5	−2.8	−0.2	2.8	6.2

  a Find Emma's total profit for the first 4 years. [1]
  b Use the method of differences to find a model for the general profit term P_n. [5]
  c Predict the profit for the 6th and 7th years of Emma's business. [2]
  d Comment on the reliability of your predictions in **c**. [1]

9 In the diagram shown, N divides $\overrightarrow{BC}$ in the ratio 1 : 2, and M divides $\overrightarrow{AC}$ in the ratio 1 : 2.

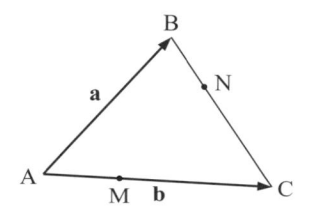

  a Write a vector expression for:
    i $\overrightarrow{AM}$ **ii** $\overrightarrow{BN}$ **iii** $\overrightarrow{MN}$ [4]
  b Describe quadrilateral ABNM. [1]
  c If △MNC has area 40 cm^2, find the area of ABNM. [4]

10 The mid-year and final examination results for a class of 16 students were as follows. The teacher unfortunately lost student P's final exam before it was marked.

Student	A	B	C	D	E	F	G	H
Mid-year exam	51	43	75	96	61	79	68	73
Final exam	56	42	86	93	58	83	64	81
Student	I	J	K	L	M	N	O	P
Mid-year exam	74	90	81	57	66	68	76	83
Final exam	77	94	85	61	72	67	80	?

  a Calculate the mean, median, and interquartile range for each set of exam scores. [4]
  b Hence compare the class results for the two exams. [2]
  c Plot the results for the individual students (except P) on a scatter diagram, with the mid-year exam score on the x-axis and the final exam score on the y-axis. [3]
  d Comment on the correlation between the variables. [1]
  e Find the equation of the linear regression line. [2]
  f Suggest what mark student P should be given for the final exam. Explain your reasoning. [2]

11 Two concentric circles have centre O. PA and PB are tangents to the inner circle at X and Y respectively.

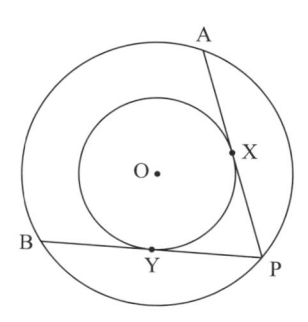

Show that:
  a PX = PY [1]
  b PX = XA [2]
  c angle BAX = angle ABY [3]
  d angle OXY = angle OYX [1]
  e AXYB is a cyclic quadrilateral. [3]

12 Marvin has 20 discs identical in shape. Each of the numbers 1, 2, 3, ..., 20 is written on a disc. All of the discs are then placed in a bag.

  a List the following sets of numbers in Marvin's bag:
    $P = \{$prime numbers$\}$
    $C = \{$composite numbers$\}$
    $M = \{$multiples of 5$\}$. [3]
  b Which of the sets in **a** are mutually exclusive? Explain your answer. [2]
  c Illustrate the sets P, C, and M on a Venn diagram, listing every element in its appropriate region. [3]
  d If two discs are drawn from the bag one after the other, without replacement, find the probability that:
    i the first number is composite and the second number is prime [2]
    ii both are composite multiples of 5 [2]
    iii the second number is prime, given that it is a multiple of 5. [1]

Paper 6:   1 hour 30 minutes / 40 marks

A. Investigation: Freya Sums   25 marks

Part I

The **Freya sum** of a 2×2 grid of numbers is the sum of those numbers.

1 Find the Freya sum of: $\begin{array}{|cc|} \hline 1 & 1 \\ 1 & 1 \\ \hline \end{array}$

2 For an $n \times n$ grid, we find the Freya sums of 2×2 boxes inside the grid, putting the answers in an $(n-1) \times (n-1)$ grid. We repeat this process until we have only one number left.

Use the following procedure to find the Freya sum of this grid:

$\begin{array}{|ccc|} \hline 1 & 1 & 1 \\ 1 & 1 & 1 \\ 1 & 1 & 1 \\ \hline \end{array}$

  a Draw a 2×2 grid.
    In the top left-hand corner of your grid, write the Freya sum of the smaller box:
    $\begin{array}{|ccc|} \hline 1 & 1 & 1 \\ 1 & 1 & 1 \\ 1 & 1 & 1 \\ \hline \end{array}$

In the top right-hand corner, write the Freya sum of this smaller box:

$$\begin{array}{|ccc|} 1 & 1 & 1 \\ 1 & 1 & 1 \\ 1 & 1 & 1 \end{array}$$

In the bottom left-hand corner, write the Freya sum of this smaller box:

$$\begin{array}{|ccc|} 1 & 1 & 1 \\ 1 & 1 & 1 \\ 1 & 1 & 1 \end{array}$$

Complete the last entry of your grid.

b You should now have a completed 2×2 grid. Find its Freya sum. This final answer is the Freya sum of the 3×3 grid:

$$\begin{array}{|ccc|} 1 & 1 & 1 \\ 1 & 1 & 1 \\ 1 & 1 & 1 \end{array}$$

3 We now find the Freya sum of:

$$\begin{array}{|cccc|} 1 & 1 & 1 & 1 \\ 1 & 1 & 1 & 1 \\ 1 & 1 & 1 & 1 \\ 1 & 1 & 1 & 1 \end{array}$$

a What size grid will the first step produce?

b In the top left-hand corner, place the Freya sum of this smaller box:

$$\begin{array}{|cccc|} 1 & 1 & 1 & 1 \\ 1 & 1 & 1 & 1 \\ 1 & 1 & 1 & 1 \\ 1 & 1 & 1 & 1 \end{array}$$

Place the Freya sum of this smaller box in the correct place:

$$\begin{array}{|cccc|} 1 & 1 & 1 & 1 \\ 1 & 1 & 1 & 1 \\ 1 & 1 & 1 & 1 \\ 1 & 1 & 1 & 1 \end{array}$$

Fill in the rest of your first-step grid.

c Repeat the process to find the Freya sum of the 4×4 grid of 1s.

4 What is the Freya sum of a 5×5 grid of 1s?

5 Suggest a formula for the Freya sum of an $n \times n$ grid of 1s.

6 What is the Freya sum of a 17×17 grid of 1s?

Part II

In this part we investigate the contribution to the final Freya sum, made by the individual numbers in the original grid. To do so we examine grids with all zeros except for a single '1' entry.

1 Find the Freya sums of:

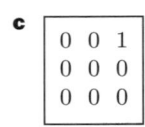

a
$$\begin{array}{|ccc|} 1 & 0 & 0 \\ 0 & 0 & 0 \\ 0 & 0 & 0 \end{array}$$

b
$$\begin{array}{|ccc|} 0 & 1 & 0 \\ 0 & 0 & 0 \\ 0 & 0 & 0 \end{array}$$

c
$$\begin{array}{|ccc|} 0 & 0 & 1 \\ 0 & 0 & 0 \\ 0 & 0 & 0 \end{array}$$

d
$$\begin{array}{|ccc|} 0 & 0 & 0 \\ 1 & 0 & 0 \\ 0 & 0 & 0 \end{array}$$

e
$$\begin{array}{|ccc|} 0 & 0 & 0 \\ 0 & 1 & 0 \\ 0 & 0 & 0 \end{array}$$

2 Explain why some of the answers in **1** equal others.

3 Complete this 3×3 **base grid** with the results from **1**, and fill in the remaining places.

$$\begin{array}{|ccc|} a & b & c \\ d & e & .... \\ .... & .... & .... \end{array}$$

4 Find the Freya sums of:

a
$$\begin{array}{|ccc|} 1 & 0 & 0 \\ 0 & 0 & 1 \\ 0 & 0 & 0 \end{array}$$

b
$$\begin{array}{|ccc|} 0 & 1 & 0 \\ 0 & 0 & 0 \\ 0 & 0 & 1 \end{array}$$

c
$$\begin{array}{|ccc|} 0 & 1 & 0 \\ 0 & 1 & 0 \\ 1 & 0 & 0 \end{array}$$

5 Use the base grid created in **3** to explain your results to **4**.

6 Find the 4×4 base grid by considering the following grids, along with any others that may be necessary.

$$\begin{array}{|cccc|} 1 & 0 & 0 & 0 \\ 0 & 0 & 0 & 0 \\ 0 & 0 & 0 & 0 \\ 0 & 0 & 0 & 0 \end{array} \quad \begin{array}{|cccc|} 0 & 1 & 0 & 0 \\ 0 & 0 & 0 & 0 \\ 0 & 0 & 0 & 0 \\ 0 & 0 & 0 & 0 \end{array} \quad \begin{array}{|cccc|} 0 & 0 & 0 & 0 \\ 0 & 1 & 0 & 0 \\ 0 & 0 & 0 & 0 \\ 0 & 0 & 0 & 0 \end{array}$$

7 Use your base grid from **6** to find the Freya sum of:

$$\begin{array}{|cccc|} 0 & 0 & 0 & 1 \\ 0 & 1 & 1 & 0 \\ 1 & 0 & 1 & 0 \\ 0 & 1 & 0 & 0 \end{array}$$

Part III

In this part we find the Freya sums of grids with entries other than '0' or '1'.

1 Calculate, without using the 3×3 base grid, the Freya sum of:

$$\begin{array}{|ccc|} 1 & 2 & 3 \\ 4 & 5 & 6 \\ 7 & 8 & 9 \end{array}$$

2 Use your 3×3 base grid to check your answer to **1**.

3 Rearrange the numbers 1 to 9 inside a 3×3 grid to make the highest possible Freya sum. What is that sum?

4 Use your 4×4 base grid to find the Freya sum of:

$$\begin{array}{|cccc|} 1 & 2 & 3 & 4 \\ 5 & 6 & 7 & 8 \\ 9 & 10 & 11 & 12 \\ 13 & 14 & 15 & 16 \end{array}$$

5 The Freya sum of an $n \times n$ grid with entries $1, 2, 3, ..., n^2$ (in order), is $2^{2n-3}(1+n^2)$.

a Verify this formula for $n = 1, 2, 3$ and 4.

b Find the Freya sum of the 5×5 grid with entries $1, 2, 3, ..., 25$ (in order).

B. Modelling: Wind Turbines   15 marks

A wind turbine converts wind power into electricity. The power (P kW) generated by a certain turbine at various wind speeds (v m/s) is given in the following table:

Wind speed (v m/s)	4	6	8	10	12	14	16
Power (P kW)	45	152	360	703	1215	1929	2880

1 Plot this information on the grid below.

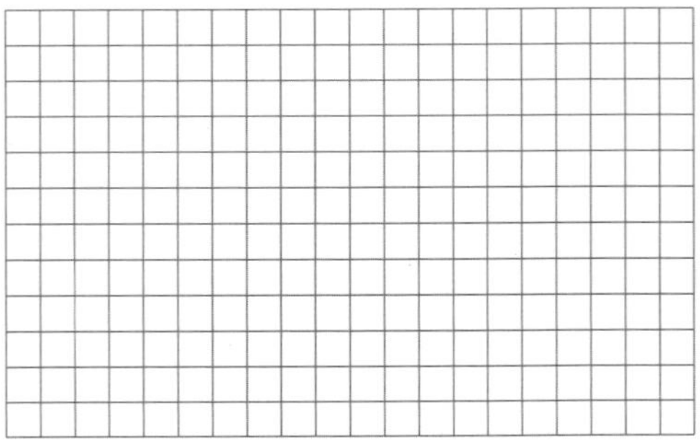

2 The power fits a model of the form $P = kv^n$. Find k and n.

3 **a** Find the power generated when the wind speed is:
  i 7.5 m/s **ii** 15 m/s
  b Find the percentage increase in power over the wind speeds in **a**.
  c Does doubling the windspeed always increase the power by the same percentage? Explain your answer.

4 In practice, we can write $k = cr^2$, where c is a constant and r is the length of the rotor blades. Given that this turbine has blades 31 m long, find the value of c.

5 A turbine of the same design has rotor blades 40 m long.
  a How much power does it generate when the wind blows at 10 m/s?
  b What windspeed is needed for the smaller turbine to generate this power?

6 Explain why, over a year, half of the total energy generated comes from much less than half of the operating time.

PRACTICE EXAM 3

Paper 2: 45 minutes / 40 marks

1 **a** Write as a single fraction: $\dfrac{2x+3}{x^2 - 3x - 4} - \dfrac{2}{x+1}$ [3]

  b Write with an integer denominator: $\dfrac{1 + \sqrt{2}}{3 - \sqrt{2}}$ [3]

2 **a** Find the gradient of the line connecting $A(-3, -2)$ and $B(7, 3)$. [1]
  b Find the equation of the perpendicular bisector of the line AB. [3]
  c Does $C(4, -3)$ lie on the perpendicular bisector? [1]
  d Find $\overrightarrow{AC}$ in component form. [1]
  e Calculate $|\overrightarrow{AC}|$, expressing your answer in simplest surd form. [2]

3 **a** Write down the next two terms of the sequence $-8, -3, 2, 7, \ldots$ [1]
  b Find the nth term of the sequence. [2]

4 List the elements of the following sets:
  a $A = \{x \mid 8 < x \leqslant 20, \ x \in \mathbb{Z}\}$ [1]
  b $B = \{\text{composite numbers between 9 and 19, inclusive}\}$ [1]
  c $A \cap B$ [1]

5 Consider the L in the diagram.

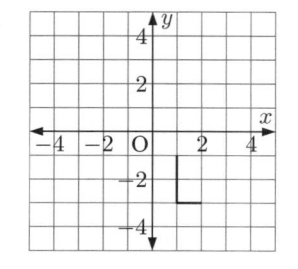

  a Enlarge the L with centre $(0, -3)$ and scale factor 2. [2]
  b Reflect the L in the line $y = x$. [2]

6 2 small buckets and 3 large buckets have total capacity 18 l. 7 small buckets and 2 large buckets have total capacity 29 l. Find the capacity of a large bucket. [5]

7 Solve for x: $(x-4)^2 = 5(14 - x)$ [3]

8 Find a and b if $y = a\sin(bx)$ has graph:

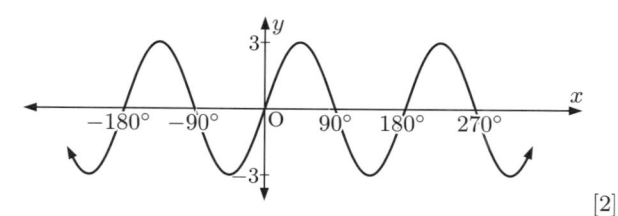

[2]

9 Find the value of x, giving reasons for your answer. [3]

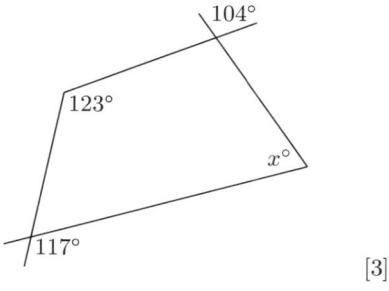

10 A marine biologist counted the number of whales and dolphins she saw each day for a month. Her sightings are summarised on this back-to-back bar chart.

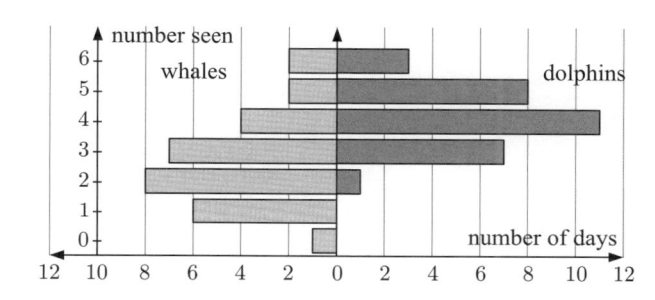

  a What was the modal number of dolphins seen? [1]
  b Describe the distribution of the whale data. [1]
  c On any one day, which animal is the biologist more likely to see? [1]

Paper 4: 2 hours 15 minutes / 120 marks

1

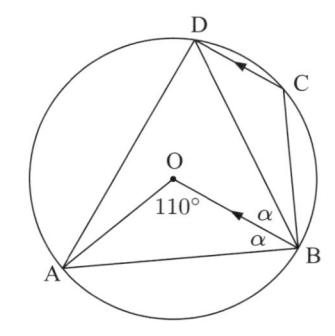

  a Find α. [2]
  b Find the size of angle ADB. [1]
  c Show that angle ADC is a right angle. [2]
  d If OB is 5 cm, find AC. [1]
  e Find $D\widehat{A}O$. [2]
  f Find the length of DC. [2]

2

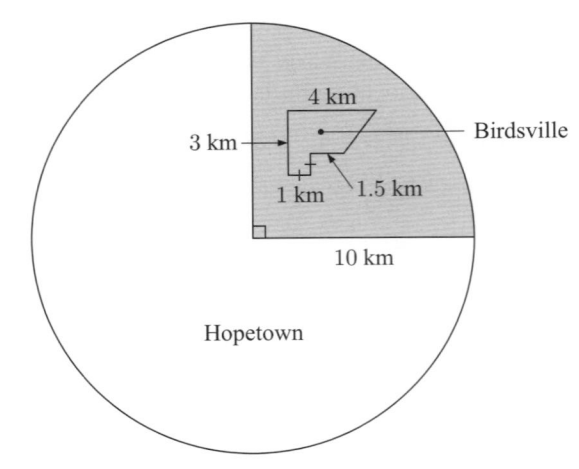

Examine the map of Hopetown. The shaded region shows the northeast suburbs, one of which is Birdsville.

a Find the area of Birdsville. [3]

b A company conducts a survey of Hopetown households. They send two questionnaires to the northeast suburbs.

  i Estimate the probability that one questionnaire goes to a Birdsville home, and the other does not.

  ii What assumption have you made in your answer? [6]

3 Tim invites his friends Wei and Hao to visit him. Wei gets lost on the way and has to stop and call Tim for directions. His progress is shown on the travel graph below.

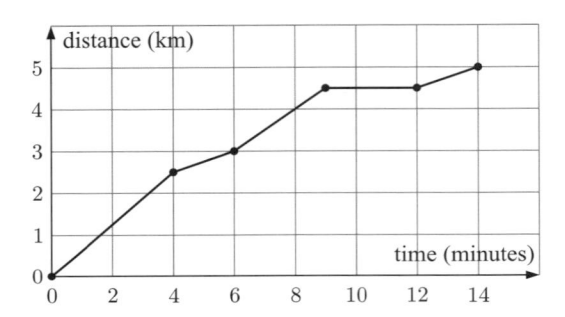

a **i** When does Wei drive fastest?
   ii What is his speed at this time? [3]

b When does he stop to get directions, and how long does this take? [2]

c What is Wei's average speed in:
   i km per minute   **ii** metres per second? [2]

d Hao leaves 5 minutes after Wei. He travels the 5.6 km from his house to Tim's at an average speed of 42 km per hour. Who arrives first? [2]

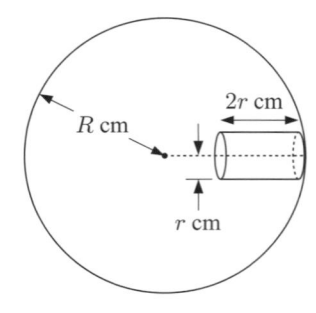

4 Lawn bowls were once given their "bias" by adding a weight into the side of the bowl. The bowl in the diagram is made of wood with a cylinder of copper inserted into it.

The spherical bowl has radius R cm, and the cylinder has radius r cm and height $2r$ cm. Each cubic centimetre of wood weighs 1.2 g and each cubic centimetre of copper weighs 9 g.

a Write down a formula for the volume of the cylinder in terms of r. [2]

b Show that the mass m of the bowl is given by $m = 1.6\pi R^3 + 15.6\pi r^3$. [4]

c Rearrange this expression to make r the subject. [2]

d Find the dimensions of the cylinder required to make a bowl with radius 6 cm and mass 1120 g. [2]

5 a Three years ago, Melissa had $15 000 invested in shares. In the first year, they went up in value by 8%. In the second year they lost 2%, and in the third year they gained 13%. What are they worth now? [2]

b Calculate the average compound change in value. [2]

c A bank offers Melissa 6.2% p.a. simple interest over 3 years if she reinvests the money. Assuming the stock market performs at the same average rate, should she take the offer? [2]

6 a Graph the region $\mathcal{R}$ defined by $2y - x \geqslant -2$, $x + 3y \leqslant 12$, $y + 2x \geqslant 9$. [5]

b Rotate the region $\mathcal{R}$ 90° clockwise about O. [3]

7 A survey of 30 students with part time jobs found that 15 earned over €10 per hour and 23 earned less than €12 per hour.

a Place this information on a Venn diagram. [3]

b How many students earn between €10 and €12 per hour? [1]

c If the minimum wage is €8 per hour and the highest paid student earns €14 per hour, estimate the mean wage for this group of students. [3]

8 a Consider $f(x) = -x^2 - 10x - 21$. Find its:
   i x-intercepts   **ii** y-intercept   **iii** vertex V [5]

b Sketch the graph of $y = f(x)$, showing all information found in **a**. [2]

c Let $g(x) = |x - 6|$. On the same set of axes, graph $y = g(x)$, and label the x-intercept P. [2]

d Which is closer to the origin, V or P? [2]

e **i** Find the equation of the line through V and P. [2]
   ii Find the midpoint of this line. [1]

9 The data table shows the number of participants in a series of aquatics camps, and the total food costs for the camp.

Participants	37	42	43	41	37	42	45	38	40
Food (£)	2986	2929	3327	2834	2888	3130	3010	2498	2782

Participants	32	34	36	41	36	26	52	30
Food (£)	2991	2119	3192	2615	2969	1837	4072	2408

a For the number of participants, find the:
   i median   **ii** range   **iii** interquartile range. [4]

b
[scatter diagram: cost of food (£) vs number of participants]

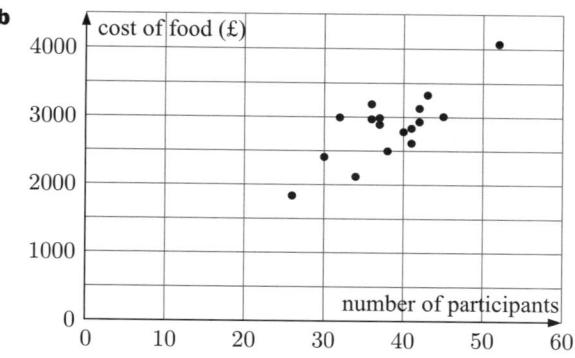

A scatter diagram for the variables is shown above. Find the mean point $(\overline{x}, \overline{y})$ and plot it on the diagram. [2]

c Comment on the correlation between the variables. [2]

d Determine the equation of the line of regression. Plot this on the graph. [2]

e Interpret the gradient of this line. [1]

f Estimate the cost of food if the number of participants is:
  i 35   **ii** 80.
  Comment on whether these predictions are reasonable. [4]

10 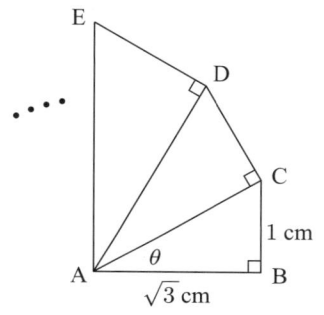 Consider the figure shown. Triangle ACD is constructed so that triangle ACD is similar to triangle ABC. Triangle ADE is constructed similar to triangle ABC and triangle ACD, and so on.

[figure: right triangle with E, D, C, A, B; AB = $\sqrt{3}$ cm, CB = 1 cm, angle θ at A]

a Find the lengths AC, AD, and AE. [3]

b **i** What type of sequence is formed by AB, AC, AD, AE, ...? [1]
  ii Find the general term of this sequence. [2]

c Find the size of θ. [1]

d Triangles are added until the hypotenuse of the last triangle lies on line segment AB. How much longer is this hypotenuse than AB? [3]

11

x	2	5	10	15
y	15	6	3	2

a Plot these points on the Cartesian plane. [2]

b Find the variation model $y = f(x)$. Comment on the proportionality of x and y. [2]

c Suppose $g(x) = 5 \times 3^x$.
  i Evaluate $g(4)$.
  ii Show that $g(x+2) = 9g(x)$.
  iii State the domain and range of $g(x)$.
  iv Write down the equations of any asymptotes of $g(x)$. [5]

d **i** Find $f(g(x))$ in simplest form.
  ii Write $g^{-1}(x)$ using a single logarithm. [5]

12 [figure: coordinate grid with triangle ABC and image AB'C']

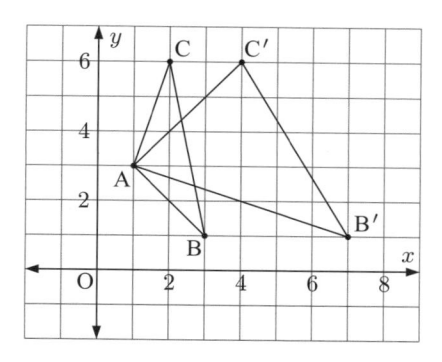

a The object triangle ABC is mapped onto the image triangle AB'C'. Describe fully the single transformation which has occurred. [2]

b Find $\overrightarrow{AB}$ and $\overrightarrow{AC}$ in component form. [2]

c Use vector methods to find D such that ABDC is a parallelogram. [2]

d Find the coordinates of D', the image of D under the transformation in **a**. [1]

e Is the image AB'D'C' a parallelogram? [1]

Paper 6:   1 hour 30 minutes / 40 marks

A. Investigation: Rolling Dice   25 marks

Alec and Bethany are playing a game. Alec has a die with 6 sides labelled 1 to 6, and Bethany has a 4-sided die labelled 1 to 4. Both players roll their die, and the higher number wins. If the numbers are equal, the game is a draw.

1 Fill in the following grid showing all possible rolls. Use A to show when Alec wins, B when Bethany wins, and D if the game is a draw.

		\multicolumn{6}{c}{Alec}
		1	2	3	4	5	6
Bethany	1	D	A				
	2	B					
	3						
	4						

2 Find the probability of:
  a Alec winning   **b** Bethany winning   **c** a draw.

3 Alec changes his die for an 8-sided die.
  a Construct and complete a table like the one in **1** for this new situation.
  b Write down the probability of:
    i Alec winning   **ii** Bethany winning   **iii** a draw.

4 Bethany switches to a 5-sided die, while Alec keeps his 8-sided die. Find the probabilities of each player winning, and of a draw.

5 If Alec has an x-sided die and Bethany has a y-sided die, with $x > y$, find the probabilities of the three different outcomes.

6 Bethany suggests that a draw should count as a win for her. For which pairs of dice does this make a 'fair' game in which the players have equal chance of winning?

7 In one final game, Alec rolls two ordinary 6-sided dice, and adds up the scores. Bethany rolls a single 12-sided die. The highest score wins, and equal scores count for a draw. Find the probability of each player winning, and of a draw.

B. Modelling: Weight by Age   15 marks

The World Health Organisation 2007 weight-for-age chart for girls provides the following mean weights:

Age (years)	5	6	7	8	9	10
Weight (kg)	18.0	20.2	22.4	25.0	28.2	31.9

Three curves which could model the data are shown in graph A:

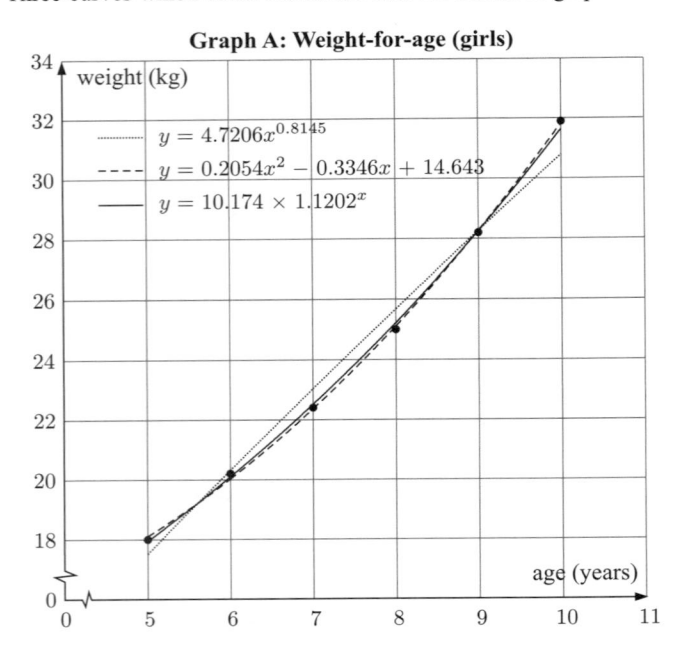

Graph A: Weight-for-age (girls)

$y = 4.7206x^{0.8145}$
$y = 0.2054x^2 - 0.3346x + 14.643$
$y = 10.174 \times 1.1202^x$

1 Which curve best models the data?

2 Use your chosen curve to predict the weight of a girl 7 years and 3 months old.

3 Predict the birth weight for a girl using each of the three models.

4 The mean birth weight of a girl is 3.2 kg.
   a Which model best predicted this?
   b Graph B includes the new birth weight data point. For these data points, find the mean age $\bar{x}$ and the mean weight $\bar{y}$.

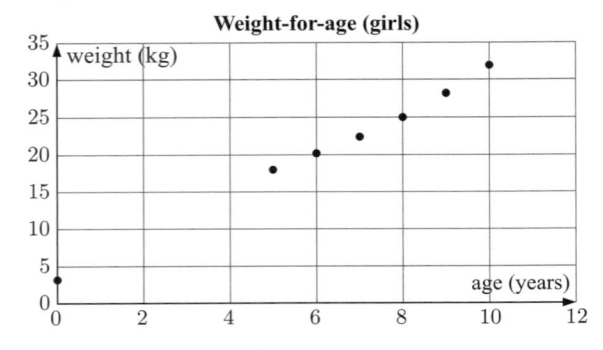

Weight-for-age (girls)

   c Find the equation of the straight line through the y-intercept and the mean point $(\bar{x}, \bar{y})$. Plot this line on graph B.
   d Comment on the appropriateness of using a linear model for this data.

5 A mathematician finds that the best curve for the data is actually a cubic, shown below.

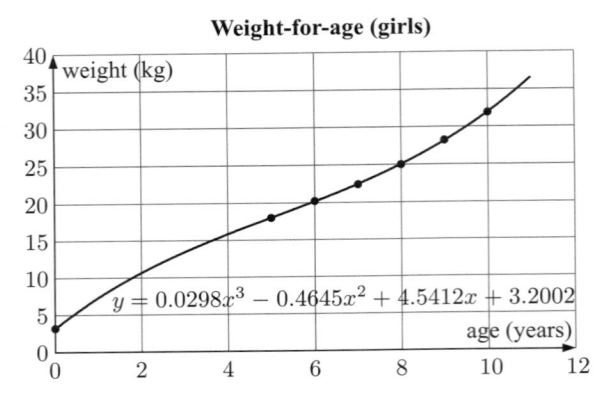

Weight-for-age (girls)

$y = 0.0298x^3 - 0.4645x^2 + 4.5412x + 3.2002$

Use this equation to predict the average weight of a 16-year-old girl. Is this a reliable estimate?

6 Think about the average weight of adults. Sketch a curve showing weight-for-age for females aged 0 to 50. Are any of the models appropriate for this data? What can you conclude?

SOLUTIONS TO TOPIC 1: NUMBER

1

Number	$\mathbb{N}$	$\mathbb{Z}$	$\mathbb{Q}$	$\mathbb{R}$
$\frac{1}{2}$	x	x	✓	✓
-5	x	✓	✓	✓
$\sqrt{-3}$	x	x	x	x
π	x	x	x	✓
4.8712	x	x	✓	✓

2 a The multiples of 6 are: 6, 12, 18, **24**, 30, ...
The multiples of 8 are: 8, 16, **24**, 32, ...
So, the LCM of 6 and 8 is 24.

b The multiples of 3 are: 3, 6, 9, 12, 15, 18, 21, 24, 27, 30, 33, 36, 39, 42, 45, 48, 51, 54, 57, **60**, 63, ...
The multiples of 4 are: 4, 8, 12, 16, 20, 24, 28, 32, 36, 40, 44, 48, 52, 56, **60**, 64, ...
The multiples of 5 are: 5, 10, 15, 20, 25, 30, 35, 40, 45, 50, 55, **60**, 65, ...
So, the LCM of 3, 4 and 5 is 60.

3 a Fifi's tax $= 25\%$ of $\$14.20$
$= \frac{25}{100} \times \$14.20$
$= \$3.55$ per hour
$\therefore$ Fifi receives $\$14.20 - \$3.55 = \$10.65$ for each hour of work.

b Fifi receives $\$10.65$ per hour.
So, Fifi must work for $\frac{4000}{10.65} \approx 376$ hours to receive $\$4000$.

4 $3^1 + 3^2 + 3^3 + 3^4 = 3 + 9 + 27 + 81$
$= 120$

5 Suppose the whole team scored x goals.
$\therefore \frac{3}{8} \times x = 42$
$\therefore x = \frac{8}{3} \times 42$
$\therefore x = 112$
So, the team scored 112 goals.

6 a €15 : €4.50
$= 15 : 4.5$
$= 30 : 9 \quad \{\times 2\}$
$= 10 : 3 \quad \{\div 3\}$

b 375 ml : 2 litres
$= 375 \text{ ml} : 2000 \text{ ml}$
$= 375 : 2000$
$= 3 : 16 \quad \{\div 125\}$

c 2 kg : 800 g
$= 2000 \text{ g} : 800 \text{ g}$
$= 2000 : 800$
$= 5 : 2 \quad \{\div 400\}$

d 3 h : 100 min
$= 180 \text{ min} : 100 \text{ min}$
$= 180 : 100$
$= 9 : 5 \quad \{\div 20\}$

7 Let the shorter side of the larger frame be x cm.
Since the dimensions are in the same ratio,
$12 : 18 = x : 24$
But $12 : 18 = 2 : 3 \quad \{\div 6\}$
$\therefore 2 : 3 = x : 24 \quad \{\times 8\}$
$\therefore x = 2 \times 8$
$\therefore x = 16$
The shorter side of the larger frame is 16 cm long.

8 1310 is 1.10 pm, 0706 is 7.06 am.
1.10 pm to 2.00 pm $= 50$ min
2.00 pm to 7.00 am the next day $= 17$ h
7.00 am to 7.06 am $= \underline{6 \text{ min}}$
17 h 56 min
The time difference is 17 hours 56 minutes.

9 a The multiplier is $105.4\% = 1.054$
$\therefore$ value after 4 years $= €4500 \times (1.054)^4$
$= €5553.60$

b Interest earned $= €5553.60 - €4500$
$= €1053.60$

10 $4\sqrt{2} - 5\sqrt{3} - 2 + 3\sqrt{3} - 5\sqrt{2}$
$= -\sqrt{2} - 2\sqrt{3} - 2$

11 Let the cost of building the drawers be £x.
$\therefore$ the profit $= 15\%$ of £x
$= £0.15x$
$\therefore x + 0.15x = 280$
$\therefore 1.15x = 280$
$\therefore x = 243.48$
The drawers cost £243.48 to build.

12 a Discount $= 10\%$ of £12 000
$= \frac{10}{100} \times £12\,000$
$= £1200$
$\therefore$ selling price $= £12\,000 - £1200$
$= £10\,800$

b Markup $= 28\%$ of $\$56$
$= \frac{28}{100} \times \$56$
$= \$15.68$
$\therefore$ selling price $= \$56 + \$15.68$
$= \$71.68$

c Loss $= 35\%$ of €950
$= \frac{35}{100} \times €950$
$= €332.50$
$\therefore$ selling price $= €950 - €332.50$
$= €617.50$

d Profit $= 20\%$ of $\$32$
$= \frac{20}{100} \times \$32$
$= \$6.40$
$\therefore$ selling price $= \$32 + \$6.40$
$= \$38.40$

13 a $(3.2 \times 10^4) \times (6.8 \times 10^3)$
$= 3.2 \times 6.8 \times 10^4 \times 10^3$
$= 21.76 \times 10^{4+3}$
$= 2.176 \times 10^1 \times 10^7$
$= 2.176 \times 10^8$

b $(4.5 \times 10^3) \div (6 \times 10^5)$
$= \frac{(4.5 \times 10^3)}{(6 \times 10^5)}$
$= \frac{4.5}{6} \times 10^{3-5}$
$= 0.75 \times 10^{-2}$
$= 7.5 \times 10^{-1} \times 10^{-2}$
$= 7.5 \times 10^{-3}$

14 a $\left|-\frac{1}{3}\right| = \frac{1}{3}$

b $|3.6801 - 2.1508| = |1.5293|$
$= 1.5293$

c $\left|\frac{-3.15}{5}\right| = |-0.63|$
$= 0.63$

15 a 236.8 million pounds
$= 236.8 \times 10^6$ pounds
$= 2.368 \times 10^2 \times 10^6$ pounds
$= 2.368 \times 10^8$ pounds

b 6.8×10^{-7} m
$= 0\overset{\frown}{0\,000\,006}.8 \div 10^7$ m
$= 0.000\,000\,68$ m

c Distance travelled in 1 second $= 260$ km
$\therefore$ distance travelled in 365 days
$= 260 \times 60 \times 60 \times 24 \times 365$ km
$= 8\overset{\frown}{199\,360\,000}$ km
$= 8.199\,36 \times 1\,000\,000\,000$ km
$= 8.199\,36 \times 10^9$ km

16 Suppose Lance's daily training ride is x km long.
$\therefore \ 60\%$ of $x = 75$
$\therefore \ 0.6x = 75$
$\therefore \ x = 125$
So, Lance rides 125 km each day. He still needs to ride another $125 - 75 = 50$ km.

17 12 years
$= 12 \times 365 \times 24 \times 60 \times 60$ seconds
$= 378\,432\,000$ seconds
$\approx 378\,000\,000$ seconds   {3 significant figures}

18 $I = Prn$, where $P = 4500$, $r = 0.12$, $n = 5$
$\therefore \ I = 4500 \times 0.12 \times 5$
$= 2700$
So, the simple interest is $2700.

19 a $(3\sqrt{7})^3 = 3\sqrt{7} \times 3\sqrt{7} \times 3\sqrt{7}$
$= 3 \times 3 \times 3 \times \sqrt{7} \times \sqrt{7} \times \sqrt{7}$
$= 27 \times 7 \times \sqrt{7}$
$= 189\sqrt{7}$

b $-4\sqrt{3} \times 2\sqrt{3} = -4 \times 2 \times \sqrt{3} \times \sqrt{3}$
$= -8 \times 3$
$= -24$

c $3\sqrt{7} \times 2\sqrt{2} = 3 \times 2 \times \sqrt{7} \times \sqrt{2}$
$= 6 \times \sqrt{7 \times 2}$
$= 6\sqrt{14}$

d $-4\sqrt{3} + 2\sqrt{3} = -2\sqrt{3}$

20 $P = 2900$, $r = 0.065$, $n = 3\frac{4}{12} = \frac{10}{3}$
$I = Prn$
$\therefore \ I = 2900 \times 0.065 \times \frac{10}{3}$
$\therefore \ I = \$628.33$
The total amount to be repaid is $\$2900 + \$628.33 = \$3528.33$

21 a
2	168
2	84
2	42
3	21
7	7
	1

$\therefore \ 168 = 2^3 \times 3 \times 7$

b
3	975
5	325
5	65
13	13
	1

$\therefore \ 975 = 3 \times 5^2 \times 13$

22 Let the number of total marks be x.
$\therefore \ 87\%$ of $x = 130\frac{1}{2}$
$\therefore \ 0.87x = 130.5$
$\therefore \ x = 150$
So, there were a total of 150 marks in the exam.

23 a $\sqrt{72} = \sqrt{36 \times 2}$
$= \sqrt{36} \times \sqrt{2}$
$= 6\sqrt{2}$

b $\sqrt{\frac{30}{27}} = \sqrt{\frac{10}{9}}$
$= \frac{\sqrt{10}}{\sqrt{9}}$
$= \frac{\sqrt{10}}{3}$

c $\sqrt{\frac{6}{5}} \times \sqrt{\frac{15}{8}}$
$= \sqrt{\frac{\overset{3}{\cancel{6}}}{\underset{1}{\cancel{5}}} \times \frac{\overset{3}{\cancel{15}}}{\underset{4}{\cancel{8}}}}$
$= \sqrt{\frac{9}{4}}$
$= \frac{\sqrt{9}}{\sqrt{4}}$
$= \frac{3}{2}$

24 Suppose the amount to be invested now is $\$x$.
$\therefore \ x \times (1.05)^6 = 10\,000$
$\therefore \ x = \frac{10\,000}{(1.05)^6} \approx 7462.15$
So, I need to invest $7462.15 now.

25 a $s = \frac{d}{t}$
$= \frac{320\,\text{m}}{7\,\text{min}}$
$= \frac{320\,\text{m}}{7 \times 60\,\text{s}}$
≈ 0.762 m/s

b $s = \frac{d}{t}$
$= \frac{320\,\text{m}}{7\,\text{min}}$
$= \frac{0.32\,\text{km}}{\frac{7}{60}\,\text{h}}$
≈ 2.743 km/h

26 a $I = 3375$, $P = 45\,000$, $n = \frac{18}{12} = 1.5$
Now $I = Prn$
$\therefore \ 3375 = 45\,000 \times r \times 1.5$
$\therefore \ 67\,500r = 3375$
$\therefore \ r = 0.05$
$\therefore$ the rate is 5% p.a. simple interest.

b $I = 1800$, $P = 2400$, $n = 10$
Now $I = Prn$
$\therefore \ 1800 = 2400 \times r \times 10$
$\therefore \ 24\,000r = 1800$
$\therefore \ r = 0.075$
$\therefore$ the rate is 7.5% p.a. simple interest.

27 a $(-3)^4 \times (-3)^3$
$= (-3)^{4+3}$
$= (-3)^7 = -2187$

b $(2^4)^3$
$= 2^{4 \times 3}$
$= 2^{12} = 4096$

c $\left(\frac{2}{7}\right)^2$
$= \frac{2^2}{7^2}$
$= \frac{4}{49}$

d $\left(\frac{4}{3}\right)^{-1}$
$= \left(\frac{3}{4}\right)^1$
$= \frac{3}{4}$

e $2^{-2} + 2^{-1} + 2^0$
$= \frac{1}{2^2} + \frac{1}{2^1} + 1$
$= \frac{1}{4} + \frac{1}{2} + 1$
$= \frac{7}{4}$

f $\left(3\frac{1}{4}\right)^{-2} = \left(\frac{13}{4}\right)^{-2}$
$= \left(\frac{4}{13}\right)^2$
$= \frac{4^2}{13^2}$
$= \frac{16}{169}$

28 a percentage change
$= \frac{\text{change}}{\text{original}} \times 100\%$
$= \frac{15 - 7}{15} \times 100\%$
$= \frac{8}{15} \times 100\%$
$\approx 53.3\%$
$\therefore$ a 53.3% decrease occurred.

b percentage change
$= \frac{\text{change}}{\text{original}} \times 100\%$
$= \frac{70\,000 - 67\,000}{67\,000} \times 100\%$
$= \frac{3000}{67\,000} \times 100\%$
$\approx 4.48\%$
$\therefore$ a 4.48% increase occurred.

c percentage change
$= \frac{\text{change}}{\text{original}} \times 100\%$
$= \frac{95 - 80}{80} \times 100\%$
$= \frac{15}{80} \times 100\%$
$\approx 18.8\%$
$\therefore$ an 18.8% increase occurred.

d percentage change
$= \frac{\text{change}}{\text{original}} \times 100\%$
$= \frac{27 - 20}{27} \times 100\%$
$= \frac{7}{27} \times 100\%$
$\approx 25.9\%$
$\therefore$ a 25.9% decrease occurred.

29 a $3\sqrt{2}(5 - \sqrt{2}) = 3\sqrt{2}(5) - 3\sqrt{2}(\sqrt{2})$
$= 15\sqrt{2} - 6$

b $-\sqrt{3}(2\sqrt{2} + \sqrt{3}) = (-\sqrt{3})(2\sqrt{2}) + (-\sqrt{3})(\sqrt{3})$
$= -2\sqrt{6} - 3$

c $(\sqrt{3} - \sqrt{2})^2 = (\sqrt{3})^2 - 2 \times \sqrt{3} \times \sqrt{2} + (\sqrt{2})^2$
$= 3 - 2\sqrt{6} + 2$
$= 5 - 2\sqrt{6}$

d $\left(\frac{1}{\sqrt{2}} + \sqrt{2}\right)^2 = \left(\frac{1}{\sqrt{2}}\right)^2 + 2 \times \frac{1}{\sqrt{2}} \times \sqrt{2} + (\sqrt{2})^2$
$= \frac{1}{2} + 2 + 2$
$= \frac{9}{2}$

e $(2\sqrt{3} - 1)(1 + 2\sqrt{3}) = (2\sqrt{3} - 1)(2\sqrt{3} + 1)$
$= (2\sqrt{3})^2 - 1^2$
$= 4 \times 3 - 1$
$= 11$

30 a $81^{\frac{1}{4}}$
$= (3^4)^{\frac{1}{4}}$
$= 3^1$
$= 3$

b $125^{-\frac{2}{3}}$
$= (5^3)^{-\frac{2}{3}}$
$= 5^{-2}$
$= \frac{1}{5^2}$
$= \frac{1}{25}$

c $64^{\frac{5}{6}}$
$= (2^6)^{\frac{5}{6}}$
$= 2^5$
$= 32$

31 speed $= \frac{\text{distance}}{\text{time}}$
$\therefore$ distance $=$ speed $\times$ time
$= 27$ km/h $\times \frac{1}{3}$ h $\{20$ min $= \frac{1}{3}$ h$\}$
$= 9$ km

32 a Value of chair $= 2100 \times (1.07)^{10}$
$= 4131.02$
So, the value of the chair now is £4131.02.

b percentage increase
$= \frac{\text{increase}}{\text{original}} \times 100\%$
$= \frac{4131.02 - 2100}{2100} \times 100\%$
$= \frac{2031.02}{2100} \times 100\%$
$\approx 96.7\%$
$\therefore$ the chair has increased in value by 96.7%.

33 Profit $=$ £900 $-$ £650 $=$ £250
$\therefore$ profit as a percentage of the cost price
$= \frac{250}{650} \times 100\%$
$\approx 38.5\%$

34 a $\frac{5 - \sqrt{3}}{2 - \sqrt{3}} = \left(\frac{5 - \sqrt{3}}{2 - \sqrt{3}}\right)\left(\frac{2 + \sqrt{3}}{2 + \sqrt{3}}\right)$
$= \frac{10 + 5\sqrt{3} - 2\sqrt{3} - 3}{4 - 3}$
$= 7 + 3\sqrt{3}$

b $\frac{\sqrt{2} - \sqrt{3}}{1 + 3\sqrt{2}} = \left(\frac{\sqrt{2} - \sqrt{3}}{1 + 3\sqrt{2}}\right)\left(\frac{1 - 3\sqrt{2}}{1 - 3\sqrt{2}}\right)$
$= \frac{\sqrt{2} - 6 - \sqrt{3} + 3\sqrt{6}}{1^2 - (3\sqrt{2})^2}$
$= \frac{\sqrt{2} - 6 - \sqrt{3} + 3\sqrt{6}}{1 - 18}$
$= \frac{6 + \sqrt{3} - \sqrt{2} - 3\sqrt{6}}{17}$

35 $I = 13\,500, \ P = 675\,000, \ r = 0.08$
Now $I = Prn$
$\therefore \ 13\,500 = 675\,000 \times 0.08 \times n$
$\therefore \ 13\,500 = 54\,000n$
$\therefore \ n = 0.25$
So, it would take 3 months to earn the interest.

36 a Let x be the original value.
A 4% increase has multiplier 1.04
A $4\frac{1}{2}$% increase has multiplier 1.045
A 5% increase has multiplier 1.05
$\therefore$ the value after 3 years
$= x \times 1.04 \times 1.045 \times 1.05$
$\approx x \times 1.141$
So, the overall effect is a 14.1% increase.

b Let x be the original value.
A $3\frac{1}{2}\%$ increase has multiplier 1.035
$\therefore$ the value after 5 years
$= x \times (1.035)^5$
$\approx x \times 1.188$
So, the overall effect is an 18.8% increase.

c Let x be the original value.
A 10% increase has multiplier 1.1
A 40% decrease has multiplier 0.6
A 15% increase has multiplier 1.15
$\therefore$ the value after 3 years
$= x \times 1.1 \times 0.6 \times 1.15$
$\approx x \times 0.759$
So, the overall effect is a 24.1% decrease.

37 a $9^{\frac{3}{2}}$
$= (3^2)^{\frac{3}{2}}$
$= 3^3$
$= 27$

b $(-1)^{\frac{3}{2}}$
$= \sqrt{(-1)^3}$
$= \sqrt{-1}$, which is not possible

c $(-1)^{\frac{2}{3}} = \sqrt[3]{(-1)^2}$
$= \sqrt[3]{1}$
$= 1$

38 a $8 \, \text{km}$

b 30 minutes {from $t = 1$ to $t = 1.5$}

c The mountain was steepest from the $4\,\text{km}$ mark to the $6\,\text{km}$ mark, as this is where both Frank and Betty are travelling the slowest.

d i average speed
$= \dfrac{\text{distance travelled}}{\text{time taken}}$
$= \dfrac{3 \, \text{km}}{1 \, \text{h}}$
$= 3 \, \text{km/h}$

ii average speed
$= \dfrac{\text{distance travelled}}{\text{time taken}}$
$= \dfrac{8 \, \text{km}}{4.5 \, \text{h}}$
$\approx 1.78 \, \text{km/h}$

39 $13^1 = 13$, $13^2 = 169$, $13^3 = 2197$, $13^4 = 28\,561$, $13^5 = 371\,293$, $13^6 = 4\,826\,809$, ...
The last digit cycles through the pattern $3, 9, 7, 1, ...$.
Now, $1313 \div 4 = 328$ remainder 1, so when calculating 13^{1313}, the cycle is repeated 328 times, and the 1 remainder takes the next cycle through to 3.
So, the last digit of 13^{1313} is 3.

40 $P = 15\,000$, $r = 0.049$, $n = 6$
Simple interest $I = Prn$
$= 15\,000 \times 0.049 \times 6$
$= 4410$ yuan
To earn the same interest, the value of the compound interest account must reach $15\,000 + 4410 = 19\,410$ yuan after 6 years.
$\therefore \ 15\,000(1+r)^6 = 19\,410$
$\therefore \ (1+r)^6 = 1.294$
$\therefore \ 1+r \approx 1.0439$
$\therefore \ r \approx 0.0439$
So, a compound interest rate of 4.39% p.a. is needed.

41 $\text{speed} = \dfrac{\text{distance}}{\text{time}}$
$\therefore \ \text{time} = \dfrac{\text{distance}}{\text{speed}}$
$\therefore \ \text{time taken} = \dfrac{420 \, \text{km}}{80 \, \text{km/h}}$
$= 5\frac{1}{4} \, \text{h}$
$= 5 \, \text{h} \, 15 \, \text{min}$

42 Let $x = 0.\overline{37} = 0.373737....$
$\therefore \ 100x = 37.373737....$
$\therefore \ 100x = 37 + x$
$\therefore \ 99x = 37$
$\therefore \ x = \frac{37}{99}$
So, $0.\overline{37} = \frac{37}{99}$, which is rational.

43 a $\sqrt[3]{13} = 13^{\frac{1}{3}}$
b $\frac{1}{\sqrt[5]{5}} = 5^{-\frac{1}{5}}$
c $\sqrt{24} = 24^{\frac{1}{2}}$
d $\frac{1}{\sqrt{29}} = 29^{-\frac{1}{2}}$

44 Sally can initially run $1.5 \, \text{km}$.
After 1 month, she can run $1.5 \times 1.09 = 1.635 \, \text{km}$
After 2 months, she can run 1.635×1.09
$= 1.5 \times 1.09^2$
$\approx 1.78 \, \text{km}$
So, after 12 months, she will be able to run
$1.5 \times 1.09^{12} \approx 4.22 \, \text{km}$

45 a $\dfrac{3 - 2\sqrt{2}}{\sqrt{2} - 1} = \left(\dfrac{3 - 2\sqrt{2}}{\sqrt{2} - 1}\right)\left(\dfrac{\sqrt{2} + 1}{\sqrt{2} + 1}\right)$
$= \dfrac{3\sqrt{2} + 3 - 4 - 2\sqrt{2}}{2 - 1}$
$= -1 + \sqrt{2}$

b $\dfrac{\frac{1}{\sqrt{2}} - 3\sqrt{2}}{\sqrt{2} - \frac{1}{\sqrt{2}}} = \left(\dfrac{\frac{1}{\sqrt{2}} - 3\sqrt{2}}{\sqrt{2} - \frac{1}{\sqrt{2}}}\right)\left(\dfrac{\sqrt{2} + \frac{1}{\sqrt{2}}}{\sqrt{2} + \frac{1}{\sqrt{2}}}\right)$
$= \dfrac{1 + \frac{1}{2} - 6 - 3}{2 - \frac{1}{2}}$
$= \dfrac{-\frac{15}{2}}{\frac{3}{2}}$
$= -5$

46 a $2 \, \text{km}$ **b** 65 minutes
c $50 - 30 = 20$ minutes
d They were walking fastest on the way home from the park. This is when the graph is steepest.

SOLUTIONS TO TOPIC 2: ALGEBRA

1 a $-x(1 + 2x)$
$= -x - 2x^2$

b $4x(1 - x^2)$
$= 4x - 4x^3$

c $-x^2(3 - \frac{1}{x})$
$= -3x^2 + x$

2 a $5x(1 - x) + 3x$
$= 5x - 5x^2 + 3x$
$= 8x - 5x^2$

b $6(x^2 - 1) - x(1 + x)$
$= 6x^2 - 6 - x - x^2$
$= 5x^2 - x - 6$

c $3(2-x) - x(2x+1)$
     $= 6 - 3x - 2x^2 - x$
     $= 6 - 4x - 2x^2$

d $5x(1-x) + x(\frac{1}{x} - 2)$
     $= 5x - 5x^2 + 1 - 2x$
     $= 3x - 5x^2 + 1$

3 a   **i** area of rectangle $A = a(b-c)$
     ii area of rectangle $B = ac$
     iii area of overall rectangle $= ab$

   b rectangle A + rectangle B = overall rectangle
     $\therefore \quad a(b-c) + ac = ab$
     $\therefore \quad a(b-c) = ab - ac$

4 a $(2+3x)(2-3x) = 2^2 - (3x)^2$
     $= 4 - 9x^2$

   b $(2x-5)^2 = (2x)^2 - 2 \times 2x \times 5 + 5^2$
     $= 4x^2 - 20x + 25$

   c $(3-2x)(1+3x) = 3 + 9x - 2x - 6x^2$
     $= 3 + 7x - 6x^2$

   d $(x+2y)^2 = x^2 + 2 \times x \times 2y + (2y)^2$
     $= x^2 + 4xy + 4y^2$

5 a $\dfrac{3a^2b}{6ab^2}$
     $= \dfrac{\cancel{3} \times \cancel{a} \times a \times \cancel{b}}{\cancel{6}_2 \times \cancel{a} \times \cancel{b} \times b}$
     $= \dfrac{a}{2b}$

   b $\dfrac{4x^2y^2}{10xy}$
     $= \dfrac{\cancel{4}^2 \times \cancel{x} \times x \times \cancel{y} \times y}{\cancel{10}_5 \times \cancel{x} \times \cancel{y}}$
     $= \dfrac{2xy}{5}$

   c $\dfrac{4-x}{4}$ cannot be simplified.

6 a $x^2y - 2xy = x \times x \times y - 2 \times x \times y$
     $= xy(x-2)$ {HCF is xy}

   b $3xy + 6x^3 = 3 \times x \times y + 3 \times 2 \times x \times x \times x$
     $= 3x(y + 2x^2)$ {HCF is $3x$}

   c $8x^2(5-2x) + 12x(2x-5)$
     $= 8x^2(5-2x) - 12x(5-2x)$
     $= 2x \times 4x(5-2x) - 3 \times 4x(5-2x)$
     $= 4x(5-2x)(2x-3)$ {HCF is $4x(5-2x)$}

7 a number line from -4 (open) to 2 (closed), x

   b number line with closed -1 going left and open 3 going right, x

8 a $\dfrac{\cancel{6}^3(\cancel{x+1})^1}{\cancel{2}_1(\cancel{x+1})_1} = 3$

   b $\dfrac{(x-1)(x+2)}{1-x}$
     $= \dfrac{-\cancel{(1-x)}^1(x+2)}{\cancel{(1-x)}_1}$
     $= -(x+2)$

   c $\dfrac{8(x+1)^2}{2(x-3)(x+1)}$
     $= \dfrac{\cancel{8}^4(x+1)\cancel{(x+1)}^1}{\cancel{2}_1(x-3)\cancel{(x+1)}_1}$
     $= \dfrac{4(x+1)}{x-3}$

9 a $(2-x)^2 - (x+1)^2$
     $= 4 - 4x + x^2 - (x^2 + 2x + 1)$
     $= 4 - 4x + x^2 - x^2 - 2x - 1$
     $= 3 - 6x$

   b $(2x+3)^2 - (x-1)(x+5)$
     $= 4x^2 + 12x + 9 - (x^2 + 5x - x - 5)$
     $= 4x^2 + 12x + 9 - x^2 - 4x + 5$
     $= 3x^2 + 8x + 14$

10 $(x-1)(2x+1)(3-x)$
   $= (2x^2 + x - 2x - 1)(3-x)$
   $= (2x^2 - x - 1)(3-x)$
   $= 6x^2 - 2x^3 - 3x + x^2 - 3 + x$
   $= -2x^3 + 7x^2 - 2x - 3$

11 a Start with 2, and each term thereafter is 3 less than the previous term.
   $u_5 = -10$, $u_6 = -13$.

   b Start with 6, and each term thereafter is half the previous term.
   $u_5 = \frac{3}{8}$, $u_6 = \frac{3}{16}$.

   c Start with 4, then 5, and the difference between successive terms increases by 1 each time.
   $u_6 = 14 + 5 = 19$, $u_7 = 19 + 6 = 25$.

   d Start with 5, and each term thereafter is 3 times the previous term.
   $u_5 = 135 \times 3 = 405$, $u_6 = 405 \times 3 = 1215$.

12 a $3(1+x) - x(1+x)$ {HCF is $(1+x)$}
     $= (1+x)(3-x)$

   b $(x-2)^2 + (x-1)(x-2)$ {HCF is $(x-2)$}
     $= (x-2)[(x-2) + (x-1)]$
     $= (x-2)(2x-3)$

   c $4(1-x)^2 - x + 1$
     $= 4(1-x)^2 + (1-x)$ {HCF is $(1-x)$}
     $= (1-x)[4(1-x) + 1]$
     $= (1-x)(5-4x)$

   d $2(a-3)(a+4) + a(a+4)$ {HCF is $(a+4)$}
     $= (a+4)[2(a-3) + a]$
     $= (a+4)(3a-6)$
     $= 3(a+4)(a-2)$

13 a $3x^2 - 27$
     $= 3(x^2 - 9)$
     $= 3(x+3)(x-3)$

   b $-8x^2 + 32$
     $= -8(x^2 - 4)$
     $= -8(x+2)(x-2)$

   c $(a-5)^2 - 4a^2$
     $= (a-5)^2 - (2a)^2$
     $= [(a-5) + 2a][(a-5) - 2a]$
     $= (3a-5)(-a-5)$
     $= -(3a-5)(a+5)$

14 $27 \times 15 = (21+6)(21-6)$
     $= 21^2 - 6^2$ {difference of squares}

15 a $\dfrac{4x-1}{5} = 2$
     $\therefore \quad 4x - 1 = 10$ {multiplying both sides by 5}
     $\therefore \quad 4x = 11$ {adding 1 to both sides}
     $\therefore \quad x = 2\frac{3}{4}$ {dividing both sides by 4}

   b $\frac{1}{3}(3-x) = -4$
     $\therefore \quad 3 - x = -12$ {multiplying both sides by 3}
     $\therefore \quad -x = -15$ {subtracting 3 from both sides}
     $\therefore \quad x = 15$ {dividing both sides by -1}

16 **a** $x^2 + 10x + 25$
$= x^2 + 2 \times x \times 5 + 5^2$
$= (x+5)^2$ {perfect square}

b $-18x^2 - 12x - 2$
$= -2(9x^2 + 6x + 1)$
$= -2\left[(3x)^2 + 2 \times 3x \times 1 + 1^2\right]$
$= -2(3x+1)^2$ {perfect square}

17 **a** $3x^2 = 48$
$\therefore x^2 = 16$ {dividing both sides by 3}
$\therefore x = \pm 4$

b $3 - 4x^2 = -13$
$\therefore -4x^2 = -16$ {subtracting 3 from both sides}
$\therefore x^2 = 4$ {dividing both sides by -4}
$\therefore x = \pm 2$

c $5x^2 - 7 = 58$
$\therefore 5x^2 = 65$ {adding 7 to both sides}
$\therefore x^2 = 13$ {dividing both sides by 5}
$\therefore x = \pm\sqrt{13}$

18 We know $(x-4)^2 \geqslant 0$ for all real x, since the square of a real number cannot be negative.
$\therefore x^2 - 8x + 16 \geqslant 0$
$\therefore x^2 + 16 \geqslant 8x$ for all real x.

19 **a** $x^2 + x - 20$ {sum $= 1$, product $= -20$}
$= (x+5)(x-4)$ $\therefore$ the numbers are 5 and -4}

b $2x^2 + 2x - 4$
$= 2(x^2 + x - 2)$ {sum $= 1$, product $= -2$}
$= 2(x+2)(x-1)$ $\therefore$ the numbers are 2 and -1}

c $6x^2 - 3x - 3 = 3(2x^2 - x - 1)$
For $2x^2 - x - 1$, $2 \times -1 = -2$
We need two factors of -2 which have a sum of -1.
These are -2 and 1.
$\therefore 3(2x^2 - x - 1) = 3(2x^2 - 2x + x - 1)$
$= 3[2x(x-1) + 1(x-1)]$
$= 3(x-1)(2x+1)$

d $-9t^2 - 30t + 24 = -3(3t^2 + 10t - 8)$
For $3t^2 + 10t - 8$, $3 \times -8 = -24$
We need two factors of -24 which have a sum of 10.
These are 12 and -2.
$\therefore -3(3t^2 + 10t - 8) = -3(3t^2 + 12t - 2t - 8)$
$= -3[3t(t+4) - 2(t+4)]$
$= -3(t+4)(3t-2)$

20 **a** $\dfrac{3x - x^2}{x^2 + x}$
$= \dfrac{{}^1\cancel{x}(3-x)}{{}_1\cancel{x}(x+1)}$
$= \dfrac{3-x}{x+1}$

b $\dfrac{2x^2 - 8}{2x^2 + 4}$
$= \dfrac{{}^1\cancel{2}(x^2 - 4)}{{}_1\cancel{2}(x^2 + 2)}$
$= \dfrac{(x+2)(x-2)}{x^2 + 2}$

c $\dfrac{3x^2 - 3y^2}{6y^2 - 6xy} = \dfrac{3(x^2 - y^2)}{6y(y-x)}$
$= \dfrac{{}^1\cancel{3}(x+y)\cancel{(x-y)}{}^1}{{}_2\cancel{6y}\cancel{(x-y)}{}_1}$
$= -\dfrac{x+y}{2y}$

21 **a** $3(2x - 1) = 5 - x$
$\therefore 6x - 3 = 5 - x$
$\therefore 7x - 3 = 5$
$\therefore 7x = 8$
$\therefore x = 1\tfrac{1}{7}$

b $2(1 - 3x) + x = 4(2x + 1)$
$\therefore 2 - 6x + x = 8x + 4$
$\therefore 2 - 5x = 8x + 4$
$\therefore 2 = 13x + 4$
$\therefore -2 = 13x$
$\therefore x = -\tfrac{2}{13}$

22 **a** $(2x^2)^3$
$= 2^3 \times (x^2)^3$
$= 8x^6$

b $\left(\dfrac{3}{4a}\right)^2$
$= \dfrac{3^2}{4^2 \times a^2}$
$= \dfrac{9}{16a^2}$

c $\left(\dfrac{x}{y^2}\right)^{-2}$
$= \left(\dfrac{y^2}{x}\right)^2$
$= \dfrac{(y^2)^2}{x^2}$
$= \dfrac{y^4}{x^2}$

23 **a** $\dfrac{2x - 5}{3} = \dfrac{x}{4}$ {LCD $= 12$}
$\therefore \dfrac{4 \times (2x - 5)}{4 \times 3} = \dfrac{3 \times x}{3 \times 4}$
$\therefore 4(2x - 5) = 3x$ {equating numerators}
$\therefore 8x - 20 = 3x$
$\therefore 5x - 20 = 0$
$\therefore 5x = 20$
$\therefore x = 4$

b $\dfrac{2 + x}{5} = \dfrac{x - 1}{2}$ {LCD $= 10$}
$\dfrac{2 \times (2 + x)}{2 \times 5} = \dfrac{5 \times (x - 1)}{5 \times 2}$
$\therefore 2(2 + x) = 5(x - 1)$ {equating numerators}
$\therefore 4 + 2x = 5x - 5$
$\therefore 9 = 3x$
$\therefore x = 3$

24 **a** $-1 \leqslant x < 4$ **b** $x < -6$ or $x \geqslant 0$

25 **a** $\dfrac{x}{3} \times \dfrac{y}{2}$
$= \dfrac{x \times y}{3 \times 2}$
$= \dfrac{xy}{6}$

b $2 \div \dfrac{3}{x}$
$= \dfrac{2}{1} \times \dfrac{x}{3}$
$= \dfrac{2 \times x}{1 \times 3}$
$= \dfrac{2x}{3}$

c $\dfrac{3n}{2} - \dfrac{4n}{5}$
$= \dfrac{3n \times 5}{2 \times 5} - \dfrac{4n \times 2}{5 \times 2}$ {LCD $= 10$}
$= \dfrac{15n}{10} - \dfrac{8n}{10}$
$= \dfrac{7n}{10}$

26 a $(x+5)^2 = 9$
$\therefore\ x+5 = \pm 3$
$\therefore\ x = -5 \pm 3$
$\therefore\ x = -8 \text{ or } -2$

b $(2x-1)^2 = 18$
$\therefore\ 2x-1 = \pm\sqrt{18}$
$\therefore\ 2x = 1 \pm \sqrt{18}$
$\therefore\ x = \dfrac{1 \pm \sqrt{18}}{2}$

c $3(2-x)^2 = 192$
$\therefore\ (2-x)^2 = 64$
$\therefore\ 2-x = \pm 8$
$\therefore\ -x = -2 \pm 8$
$\therefore\ x = 2 \mp 8$
$\therefore\ x = -6 \text{ or } 10$

27 a $(1-2x)(3+x) = 0$
$\therefore\ 1-2x = 0 \text{ or } 3+x = 0$
$\therefore\ 1 = 2x \text{ or } x = -3$
$\therefore\ x = \tfrac{1}{2} \text{ or } -3$

b $(4x-1)(2-x) = 0$
$\therefore\ 4x-1 = 0 \text{ or } 2-x = 0$
$\therefore\ 4x = 1 \text{ or } -x = -2$
$\therefore\ x = \tfrac{1}{4} \text{ or } 2$

c $2(2x-3)^2 = 0$
$\therefore\ 2x-3 = 0$
$\therefore\ 2x = 3$
$\therefore\ x = \tfrac{3}{2}$

28 a $\dfrac{x}{2} \div \dfrac{x}{3}$
$= \dfrac{x}{2} \times \dfrac{3}{x}$
$= \dfrac{{}^1\cancel{x} \times 3}{2 \times \cancel{x}_1}$
$= \dfrac{3}{2}$

b $\dfrac{2}{x} \times \dfrac{x}{y} \times \dfrac{y}{4}$
$= \dfrac{{}^1\cancel{2} \times {}^1\cancel{x} \times {}^1\cancel{y}}{{}_1\cancel{x} \times \cancel{y}_1 \times \cancel{4}_2}$
$= \tfrac{1}{2}$

c $3 - \dfrac{a}{4} + \dfrac{2a}{3}$
$= \dfrac{3 \times 12}{1 \times 12} - \dfrac{a \times 3}{4 \times 3} + \dfrac{2a \times 4}{3 \times 4}$ {LCD = 12}
$= \dfrac{36}{12} - \dfrac{3a}{12} + \dfrac{8a}{12}$
$= \dfrac{36 + 5a}{12}$

29 a $\dfrac{1}{3x} = \dfrac{7}{5}$ {LCD = $15x$}
$\therefore\ \dfrac{5 \times 1}{5 \times 3x} = \dfrac{7 \times 3x}{5 \times 3x}$
$\therefore\ 5 = 21x$ {equating numerators}
$\therefore\ x = \tfrac{5}{21}$

b $\dfrac{2}{x+2} - \dfrac{3x}{x-1} = -\dfrac{10}{3}$ {LCD = $3(x+2)(x-1)$}
$\therefore\ \dfrac{2 \times 3(x-1)}{(x+2) \times 3(x-1)}$
$\quad - \dfrac{3x \times 3(x+2)}{(x-1) \times 3(x+2)} = -\dfrac{10(x+2)(x-1)}{3(x+2)(x-1)}$
$\therefore\ 6(x-1) - 9x(x+2) = -10(x+2)(x-1)$
{equating numerators}
$\therefore\ 6x - 6 - 9x^2 - 18x = -10(x^2 - x + 2x - 2)$
$\therefore\ -9x^2 - 12x - 6 = -10x^2 - 10x + 20$
$\therefore\ x^2 - 2x - 26 = 0$

$\therefore\ x = \dfrac{2 \pm \sqrt{(-2)^2 - 4(1)(-26)}}{2}$
$\therefore\ x = \dfrac{2 \pm \sqrt{4 + 104}}{2}$
$\therefore\ x = \dfrac{2 \pm 2\sqrt{27}}{2}$
$\therefore\ x = 1 \pm \sqrt{27} \text{ or } (1 \pm 3\sqrt{3})$

30 a $u_n = 4n - 1$
$\therefore\ u_1 = 4(1) - 1 = 3$
$u_2 = 4(2) - 1 = 7$
$u_3 = 4(3) - 1 = 11$
$u_4 = 4(4) - 1 = 15$

b $u_n = 3n^2 - 2$
$\therefore\ u_1 = 3(1)^2 - 2 = 1$
$u_2 = 3(2)^2 - 2 = 10$
$u_3 = 3(3)^2 - 2 = 25$
$u_4 = 3(4)^2 - 2 = 46$

c $u_n = n(n^2 - 3)$
$\therefore\ u_1 = 1(1^2 - 3) = -2$
$u_2 = 2(2^2 - 3) = 2$
$u_3 = 3(3^2 - 3) = 18$
$u_4 = 4(4^2 - 3) = 52$

31 a $\dfrac{x^2 - 2x - 3}{x^2 - 5x + 6}$
$= \dfrac{(x+1)\cancel{(x-3)}^1}{(x-2)\cancel{(x-3)}_1}$
$= \dfrac{x+1}{x-2}$

b $\dfrac{x^2 - x - 2}{2x^2 - 4x}$
$= \dfrac{(x+1)\cancel{(x-2)}^1}{2x\cancel{(x-2)}_1}$
$= \dfrac{x+1}{2x}$

c $\dfrac{2x^2 + 5x - 3}{-2x^2 + 7x - 3}$
$= \dfrac{{}^1\cancel{(2x-1)}(x+3)}{{}_1\cancel{-(2x-1)}(x-3)}$
$= \dfrac{x+3}{3-x}$

32 a $\dfrac{x}{3} + x - 2$ {LCD = 3}
$= \dfrac{x}{3} + \dfrac{3x}{3} - \dfrac{6}{3}$
$= \dfrac{4x - 6}{3}$

b $\dfrac{2}{x} - 3 + \dfrac{x}{3}$ {LCD = $3x$}
$= \dfrac{3 \times 2}{3 \times x} - \dfrac{3 \times 3x}{1 \times 3x} + \dfrac{x \times x}{3 \times x}$
$= \dfrac{6}{3x} - \dfrac{9x}{3x} + \dfrac{x^2}{3x}$
$= \dfrac{x^2 - 9x + 6}{3x}$

33 Let x be the smallest odd integer.
$\therefore\ $ the next is $x+2$ and the largest is $x+4$.
So, $x + (x+2) + (x+4) = 51$
$\therefore\ 3x + 6 = 51$
$\therefore\ 3x = 45$
$\therefore\ x = 15$
$\therefore\ $ the smallest of the integers is 15.

34 a $3x - 2x^2 = 0$
$\therefore\ x(3 - 2x) = 0$
$\therefore\ x = 0\ \text{or}\ 3 - 2x = 0$
$\therefore\ x = 0\ \text{or}\ 3 = 2x$
$\therefore\ x = 0\ \text{or}\ \frac{3}{2}$

b $x^2 - 25 = 0$
$\therefore\ x^2 = 25$
$\therefore\ x = \pm 5$

c $x^2 + 8x + 16 = 0$
$\therefore\ (x + 4)^2 = 0$ {perfect square}
$\therefore\ x + 4 = 0$
$\therefore\ x = -4$

35 a $4x - 2y = 3$
$\therefore\ -2y = -4x + 3$
$\therefore\ y = \frac{-4x + 3}{-2}$
$\therefore\ y = 2x - \frac{3}{2}$

b $ax + by + c = 0$
$\therefore\ by + c = -ax$
$\therefore\ by = -ax - c$
$\therefore\ y = \frac{-ax - c}{b}$
or $y = -\frac{a}{b}x - \frac{c}{b}$

36 a $x^2 + 7x + 12 = 0$ {sum $= 7$, product $= 12$.
$\therefore$ numbers are 3 and 4}
$\therefore\ (x + 3)(x + 4) = 0$
$\therefore\ x + 3 = 0\ \text{or}\ x + 4 = 0$
$\therefore\ x = -3\ \text{or}\ -4$

b $x^2 = 10x - 25$
$\therefore\ x^2 - 10x + 25 = 0$
$\therefore\ (x - 5)^2 = 0$ {perfect square}
$\therefore\ x - 5 = 0$
$\therefore\ x = 5$

c $x^2 - 5x = 24$
$\therefore\ x^2 - 5x - 24 = 0$ {sum $= -5$, product $= -24$
$\therefore$ numbers are -8 and 3}
$\therefore\ (x - 8)(x + 3) = 0$
$\therefore\ x - 8 = 0\ \text{or}\ x + 3 = 0$
$\therefore\ x = 8\ \text{or}\ -3$

37 a Start with 5, and each term thereafter is 6 more than the previous term.
$u_5 = 29,\ u_6 = 35$

b Start with 2, then 5, and the difference between successive terms increases by 2 each time.
$u_6 = 26 + 11 = 37,\ u_7 = 37 + 13 = 50$

c Start with 4, and each term thereafter is 3 times more than the previous term.
$u_5 = 108 \times 3 = 324,\ u_6 = 324 \times 3 = 972$

d Start with 3, and each term thereafter is 9 more than the previous term.
$u_5 = 39,\ u_6 = 48$

38 a $6 - 5x \leqslant 1$
$\therefore\ -5x \leqslant -5$
$\therefore\ x \geqslant 1$ {reverse the sign}

b $4(3 - x) > 2$
$\therefore\ 3 - x > \frac{1}{2}$
$\therefore\ -x > -\frac{5}{2}$
$\therefore\ x < \frac{5}{2}$ {reverse the sign}

39 $\frac{3x - 1}{4} + \frac{x + 3}{2} = \frac{4x + 3}{3}$ {LCD $= 12$}
$\therefore\ \frac{3 \times (3x - 1)}{3 \times 4} + \frac{6 \times (x + 3)}{6 \times 2} = \frac{4 \times (4x + 3)}{4 \times 3}$
$\therefore\ 3(3x - 1) + 6(x + 3) = 4(4x + 3)$ {equating numerators}
$\therefore\ 9x - 3 + 6x + 18 = 16x + 12$
$\therefore\ 15x + 15 = 16x + 12$
$\therefore\ 15 = x + 12$
$\therefore\ x = 3$

40 a C   **b** A   **c** B

41 a $3x^2 - 5x - 2 = 0$
$\therefore\ 3x^2 - 6x + x - 2 = 0$ {splitting middle term}
$\therefore\ 3x(x - 2) + 1(x - 2) = 0$
$\therefore\ (x - 2)(3x + 1) = 0$
$\therefore\ x - 2 = 0\ \text{or}\ 3x + 1 = 0$
$\therefore\ x = 2\ \text{or}\ -\frac{1}{3}$

b $11x = 2x^2 + 15$
$\therefore\ 2x^2 - 11x + 15 = 0$ {splitting middle term}
$\therefore\ 2x^2 - 5x - 6x + 15 = 0$
$\therefore\ x(2x - 5) - 3(2x - 5) = 0$
$\therefore\ (2x - 5)(x - 3) = 0$
$\therefore\ 2x - 5 = 0\ \text{or}\ x - 3 = 0$
$\therefore\ x = \frac{5}{2}\ \text{or}\ 3$

c $16x^2 = 24x - 9$
$\therefore\ 16x^2 - 24x + 9 = 0$
$\therefore\ (4x)^2 - 2 \times 4x \times 3 + 3^2 = 0$
$\therefore\ (4x - 3)^2 = 0$ {perfect square}
$\therefore\ 4x - 3 = 0$
$\therefore\ x = \frac{3}{4}$

42 a $\frac{a}{x} \times \frac{x}{b}$
$= \frac{a \times x^1}{{}_1 x \times b}$
$= \frac{a}{b}$

b $\frac{b}{m^2} \div \frac{b^2}{m}$
$= \frac{b}{m^2} \times \frac{m}{b^2}$
$= \frac{{}^1 b \times m^1}{{}_1 m \times m \times b \times b_1}$
$= \frac{1}{bm}$

c $\frac{x}{3} + \frac{x}{7} + \frac{x}{5}$ {LCD $= 105$}
$= \frac{35 \times x}{35 \times 3} + \frac{15 \times x}{15 \times 7} + \frac{21 \times x}{21 \times 5}$
$= \frac{35x}{105} + \frac{15x}{105} + \frac{21x}{105}$
$= \frac{71x}{105}$

43 $T = \dfrac{k\sqrt{l}}{\pi}$

$\therefore\ T\pi = k\sqrt{l}$

$\therefore\ \dfrac{T\pi}{k} = \sqrt{l}$

$\therefore\ l = \left(\dfrac{T\pi}{k}\right)^2$ {squaring both sides}

When $k = 2\pi$ and $T = 6$, $l = \left(\dfrac{6\pi}{2\pi}\right)^2$
$= 3^2$
$= 9$

44 a

3, 9, 18, 30, 45, ...

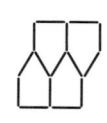

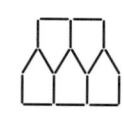

b

5, 9, 12, 15, 18, ...

45 Let x be the number.

When 11 is added to the number and the result is divided by 3, we get $\dfrac{x+11}{3}$. One less than the number is $x - 1$.

$\therefore\ \dfrac{x+11}{3} = x - 1$

$\therefore\ x + 11 = 3x - 3$

$\therefore\ 11 = 2x - 3$

$\therefore\ 14 = 2x$

$\therefore\ x = 7$ So, the number is 7.

46 a $x(x-1) - 3(x+7) = 0$

$\therefore\ x^2 - x - 3x - 21 = 0$

$\therefore\ x^2 - 4x - 21 = 0$

$\therefore\ (x+3)(x-7) = 0$

$\therefore\ x = -3$ or 7

b $2x(x+1) - 3(x+2) = 0$

$\therefore\ 2x^2 + 2x - 3x - 6 = 0$

$\therefore\ 2x^2 - x - 6 = 0$

$\therefore\ 2x^2 - 4x + 3x - 6 = 0$ {splitting}

$\therefore\ 2x(x-2) + 3(x-2) = 0$

$\therefore\ (x-2)(2x+3) = 0$

$\therefore\ x = 2$ or $-\dfrac{3}{2}$

47 a $x^3 - 26 = 0$

$\therefore\ x^3 = 26$

$\therefore\ x = \sqrt[3]{26}$

$\therefore\ x \approx 2.96$

b $\dfrac{x}{3} = \dfrac{4}{x}$ {LCD = $3x$}

$\therefore\ \dfrac{x \times x}{3 \times x} = \dfrac{3 \times 4}{3 \times x}$

$\therefore\ x^2 = 12$ {equating numerators}

$\therefore\ x = \pm\sqrt{12}$

$\therefore\ x \approx \pm 3.46$

c $x^4 = 40$

$\therefore\ x = \pm\sqrt[4]{40}$

$\therefore\ x \approx \pm 2.51$

d $\dfrac{x}{5} = -\dfrac{3}{x}$ {LCD = $5x$}

$\therefore\ \dfrac{x \times x}{5 \times x} = -\dfrac{5 \times 3}{5 \times x}$

$\therefore\ x^2 = -15$ {equating numerators}

which has no real solutions as x^2 cannot be < 0.

48 a $\dfrac{3x}{(x+1)(x-2)} - \dfrac{2}{x+1}$

$= \dfrac{3x}{(x+1)(x-2)} - \left(\dfrac{2}{x+1}\right)\left(\dfrac{x-2}{x-2}\right)$

   {LCD = $(x+1)(x-2)$}

$= \dfrac{3x - 2(x-2)}{(x+1)(x-2)}$

$= \dfrac{3x - 2x + 4}{(x+1)(x-2)}$

$= \dfrac{x+4}{(x+1)(x-2)}$

b i The expression is undefined when
$(x+1)(x-2) = 0$ {denominator = 0}
$\therefore\ x = -1$ or 2

ii The expression is zero when
$x + 4 = 0$ {numerator = 0}
$\therefore\ x = -4$

49 Using technology:

a $x \approx -1.638$ or 3.052

b $x \approx -1.232$ or 1.550

50

Flower	Cost	Number	Value
Lily	€5	x	€$5x$
Orchid spike	€2	$x + 4$	€$2(x+4)$
			€43

So, $5x + 2(x+4) = 43$

$\therefore\ 5x + 2x + 8 = 43$

$\therefore\ 7x + 8 = 43$

$\therefore\ 7x = 35$

$\therefore\ x = 5$

$\therefore$ Elliot bought 5 lilies and 9 orchid spikes.

51 a $(3p^2q)^{-1} = \dfrac{1}{3p^2q}$

b $(3x^2y^3)^3$
$= 3^3 \times (x^2)^3 \times (y^3)^3$
$= 27x^6y^9$

c $\left(\dfrac{4x^{-1}}{3y^2}\right)^3 = \left(\dfrac{4}{3xy^2}\right)^3$

$= \dfrac{4^3}{3^3 \times x^3 \times (y^2)^3}$

$= \dfrac{64}{27x^3y^6}$

52 a $u_1 = 3 \times 1$, $u_2 = 3 \times 2$, $u_3 = 3 \times 3$, ...

$\therefore\ u_n = 3n$

b **i** $u_1 = 3+4$, $u_2 = 6+4$, $u_3 = 9+4$, ...
Each term is 4 more than in the sequence in **a**.
$\therefore$ $u_n = 3n+4$
ii $u_1 = 3-4$, $u_2 = 6-4$, $u_3 = 9-4$, ...
Each term is 4 less than in the sequence in **a**.
$\therefore$ $u_n = 3n-4$

53 **a** $E = \frac{1}{2}mv^2$ where $m = 2.4$, $v = 9.3$
$\therefore$ $E = \frac{1}{2} \times 2.4 \times 9.3^2$
≈ 104
So, the basketball has kinetic energy 104 J.

b $E = \frac{1}{2}mv^2$ where $E = 270$, $m = 0.15$
$\therefore$ $\frac{1}{2} \times 0.15 \times v^2 = 270$
$\therefore$ $0.075v^2 = 270$
$\therefore$ $v^2 = 3600$
$\therefore$ $v = 60$ {as $v > 0$}
So, the bullet has speed 60 m/s.

54 **a** $5^x = \frac{1}{125}$
$\therefore$ $5^x = 5^{-3}$
$\therefore$ $x = -3$

b $8 \times 3^x = 72$
$\therefore$ $3^x = 9$
$\therefore$ $3^x = 3^2$
$\therefore$ $x = 2$

c $32^x = \left(\frac{1}{4}\right)^{x-1}$
$\therefore$ $(2^5)^x = (2^{-2})^{x-1}$
$\therefore$ $2^{5x} = 2^{-2(x-1)}$
$\therefore$ $2^{5x} = 2^{-2x+2}$
$\therefore$ $5x = -2x+2$
$\therefore$ $7x = 2$
$\therefore$ $x = \frac{2}{7}$

55 Let the number be x.
$\therefore$ $\frac{x}{2} + \frac{2}{x} = 2$ {LCD $= 2x$}
$\therefore$ $\frac{x^2}{2x} + \frac{4}{2x} = \frac{4x}{2x}$
$\therefore$ $x^2 + 4 = 4x$ {equating numerators}
$\therefore$ $x^2 - 4x + 4 = 0$
$\therefore$ $(x-2)^2 = 0$ {perfect square}
$\therefore$ $x = 2$
$\therefore$ the number is 2.

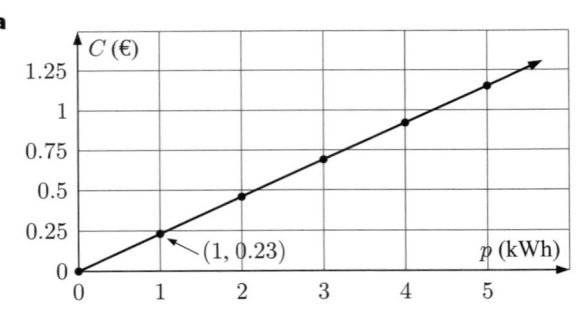

56 **a**

b The graph of C against p is a straight line through the origin, so C and p are directly proportional.

c **i** Gradient $= \frac{0.23 - 0}{1 - 0} = 0.23$
So, the proportionality constant is 0.23.
ii The law connecting C and p is $C = 0.23p$.

57 **a** €$W = 12 \times 38 \times 21$
$= €9576$

b €$W = 12 \times h \times 21$
$= €252h$

c €$W = 12 \times h \times p$
$= €12hp$

d €$W = n \times h \times p$
$= €nhp$

58 **a** If $y = 2x - 3$ and $y = 3x + 1$, then
$2x - 3 = 3x + 1$ {equating ys}
$\therefore$ $-3 = x + 1$
$\therefore$ $x = -4$
and so $y = 2(-4) - 3$ {using $y = 2x - 3$}
$= -11$
So, the simultaneous solution is $x = -4$, $y = -11$.
Check: In $y = 3x + 1$, $y = 3(-4) + 1 = -11$ ✓

b If $y = x - 5$ and $y = -2x + 1$, then
$x - 5 = -2x + 1$ {equating ys}
$\therefore$ $3x - 5 = 1$
$\therefore$ $3x = 6$
$\therefore$ $x = 2$
and so $y = 2 - 5$ {using $y = x - 5$}
$= -3$
So, the simultaneous solution is $x = 2$, $y = -3$.
Check: In $y = -2x + 1$, $y = -2(2) + 1 = -3$ ✓

59 **a** $2x - 1 \geqslant 0$
$\therefore$ $2x \geqslant 1$
$\therefore$ $x \geqslant \frac{1}{2}$

b $\frac{x-2}{5} \leqslant 3$
$\therefore$ $x - 2 \leqslant 15$
$\therefore$ $x \leqslant 17$

c $2x - 3 < 3x + 5$
$\therefore$ $-x - 3 < 5$
$\therefore$ $-x < 8$
$\therefore$ $x > -8$ {reverse the sign}

60 **a** To change x from 9 to 17.2, we multiply by $\frac{17.2}{9}$.
Since $y \propto x$, we also multiply y by $\frac{17.2}{9}$.
$\therefore$ $y = 211.5 \times \frac{17.2}{9} = 404.2$

x	9	17.2
y	211.5	?

b To change y from 211.5 to 329, we multiply by $\frac{329}{211.5}$.
Since $y \propto x$, we also multiply x by $\frac{329}{211.5}$.
$\therefore$ $x = 9 \times \frac{329}{211.5} = 14$

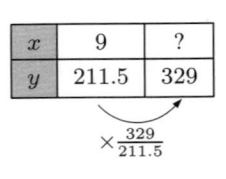

x	9	?
y	211.5	329

61 **a** $2x^2 - 11x + 13 = 0$
$\therefore$ $x = \frac{11 \pm \sqrt{(-11)^2 - 4(2)(13)}}{2(2)}$
$\therefore$ $x = \frac{11 \pm \sqrt{121 - 104}}{4}$
$\therefore$ $x = \frac{11 \pm \sqrt{17}}{4}$

b $3x + \frac{1}{x} = -6$
$\therefore$ $3x^2 + 1 = -6x$ {multiply both sides by x}
$\therefore$ $3x^2 + 6x + 1 = 0$
$\therefore$ $x = \frac{-6 \pm \sqrt{6^2 - 4(3)(1)}}{2(3)}$

$$\therefore\ x = \frac{-6 \pm \sqrt{36-12}}{6}$$
$$= \frac{-6 \pm \sqrt{24}}{6}$$
$$= \frac{-3 \pm \sqrt{6}}{3}$$

62 **a** $u_1 = 1^3$, $u_2 = 2^3$, $u_3 = 3^3$, $u_4 = 4^3$, ...
$\therefore\ u_n = n^3$

b **i** $u_1 = 1-2$, $u_2 = 8-2$, $u_3 = 27-2$, $u_4 = 64-2$, ...
Each term is 2 less than in the sequence in **a**.
$\therefore\ u_n = n^3 - 2$

ii $u_1 = \frac{1}{1^3}$, $u_2 = \frac{1}{2^3}$, $u_3 = \frac{1}{3^3}$, $u_4 = \frac{1}{4^3}$, ...
Each term is the reciprocal of the term in **a**.
$\therefore\ u_n = \frac{1}{n^3}$

iii $u_1 = \frac{1}{2^3}$, $u_2 = \frac{2}{3^3}$, $u_3 = \frac{3}{4^3}$, $u_4 = \frac{4}{5^3}$, ...
The numerator of the nth term is n, and the denominator of the nth term is the $(n+1)$th term of the sequence in **a**.
$\therefore\ u_n = \frac{n}{(n+1)^3}$

63 If the volume is V and the pipe's radius is r, then $V \propto r^2$.
When the radius is increased by 25%,
r is multiplied by 1.25
$\therefore\ r^2$ is multiplied by $(1.25)^2$
$\therefore\ V$ is multiplied by $(1.25)^2 = 1.5625$
So, the volume of stormwater flow V is increased by 56.25%.

64 **a** $\dfrac{\frac{3}{1-x} - \frac{x}{2}}{x-5} = \dfrac{\frac{6}{2(1-x)} - \frac{x(1-x)}{2(1-x)}}{x-5}$
$= \dfrac{6 - x(1-x)}{2(1-x)(x-5)}$
$= \dfrac{x^2 - x + 6}{2(1-x)(x-5)}$

b **i** The expression is undefined when
$2(1-x)(x-5) = 0$ {denominator $= 0$}
$\therefore\ x = 1$ or 5

ii The expression is zero when
$x^2 - x + 6 = 0$ {numerator $= 0$}
$\therefore\ x = \dfrac{1 \pm \sqrt{(-1)^2 - 4(1)(6)}}{2}$
$= \dfrac{1 \pm \sqrt{-23}}{2}$

So, the expression is never zero.

65 **a** $F = ma$ $\therefore\ a = \dfrac{F}{m}$

b $a = \dfrac{F}{m}$ where $F = 14.3$, $m = 2.4$
$\therefore\ a = \frac{14.3}{2.4} \approx 5.96$
So, the acceleration of the object is 5.96 m/s^2.

66 **a** $2x + 5 \geqslant 3 - 4x$
$\therefore\ 6x + 5 \geqslant 3$
$\therefore\ 6x \geqslant -2$
$\therefore\ x \geqslant -\frac{1}{3}$

b $2 - (3-x) < 4(x-1) - 3$
$\therefore\ 2 - 3 + x < 4x - 4 - 3$
$\therefore\ x - 1 < 4x - 7$
$\therefore\ -3x - 1 < -7$
$\therefore\ -3x < -6$
$\therefore\ x > 2$

67 **a** **i** When $x = 1$, $y = 3$, so $\dfrac{y}{\sqrt{x}} = \dfrac{3}{\sqrt{1}} = 3$
When $x = 16$, $y = 12$, so $\dfrac{y}{\sqrt{x}} = \dfrac{12}{\sqrt{16}} = 3$
When $x = 81$, $y = 27$, so $\dfrac{y}{\sqrt{x}} = \dfrac{27}{\sqrt{81}} = 3$

ii $\dfrac{y}{\sqrt{x}} = 3$ in each case, so $y = 3\sqrt{x}$.
Hence, $y \propto \sqrt{x}$.

iii $y = 3\sqrt{x}$

iv When $x = 196$, $y = a$
$\therefore\ a = 3 \times \sqrt{196} = 42$

b **i** When $x = 25$, $y = 2.5$, so $\dfrac{y}{\sqrt{x}} = \dfrac{2.5}{\sqrt{25}} = 0.5$
When $x = 36$, $y = 3$, so $\dfrac{y}{\sqrt{x}} = \dfrac{3}{\sqrt{36}} = 0.5$
When $x = 49$, $y = 3.5$, so $\dfrac{y}{\sqrt{x}} = \dfrac{3.5}{\sqrt{49}} = 0.5$

ii $\dfrac{y}{\sqrt{x}} = 0.5$ in each case, so $y = 0.5\sqrt{x}$.
Hence, $y \propto \sqrt{x}$.

iii $y = 0.5\sqrt{x}$

iv When $y = 6.5$, $x = a$
$\therefore\ 6.5 = 0.5\sqrt{a}$
$\therefore\ 13 = \sqrt{a}$
$\therefore\ a = 169$

68 **a** $y = 2x - 1$ .... (1)
$2x - 4y = -4$ .... (2)
Substituting (1) into (2),
$2x - 4(2x - 1) = -4$
$\therefore\ 2x - 8x + 4 = -4$
$\therefore\ -6x + 4 = -4$
$\therefore\ -6x = -8$
$\therefore\ x = \frac{4}{3}$
Substituting $x = \frac{4}{3}$ into (1) gives
$y = 2(\frac{4}{3}) - 1 = \frac{5}{3}$.
So, the solution is $x = \frac{4}{3}$, $y = \frac{5}{3}$.
Check: In (2), $2(\frac{4}{3}) - 4(\frac{5}{3}) = \frac{8}{3} - \frac{20}{3} = -4$ ✓

b $x = -2y - 3$ .... (1)
$3y - 4 = 2x$ .... (2)
Substituting (1) into (2),
$3y - 4 = 2(-2y - 3)$
$\therefore\ 3y - 4 = -4y - 6$
$\therefore\ 7y - 4 = -6$
$\therefore\ 7y = -2$
$\therefore\ y = -\frac{2}{7}$
Substituting $y = -\frac{2}{7}$ into (1) gives
$x = -2(-\frac{2}{7}) - 3 = -\frac{17}{7}$.
So, the solution is $x = -\frac{17}{7}$, $y = -\frac{2}{7}$.
Check: In (2), LHS $= 3\left(-\frac{2}{7}\right) - 4 = -\frac{34}{7}$,
RHS $= 2\left(-\frac{17}{7}\right) = -\frac{34}{7}$ ✓

69 **a** $6x^2 - 17x - 57 = 0$

$$\therefore x = \frac{17 \pm \sqrt{(-17)^2 - 4(6)(-57)}}{2(6)}$$

$$\therefore x = \frac{17 \pm \sqrt{289 + 1368}}{12}$$

$$\therefore x = \frac{17 \pm \sqrt{1657}}{12}$$

b $2x^2 + 5x + 4 = 0$

$$\therefore x = \frac{-5 \pm \sqrt{5^2 - 4(2)(4)}}{2(2)}$$

$$\therefore x = \frac{-5 \pm \sqrt{25 - 32}}{4}$$

$$\therefore x = \frac{-5 \pm \sqrt{-7}}{4}$$

Since $\sqrt{-7}$ does not exist, $2x^2 + 5x + 4 = 0$ has no real solutions.

c $\frac{13}{x} - 7 = 2x$

$\therefore 13 - 7x = 2x^2$ {multiply both sides by x}

$\therefore 2x^2 + 7x - 13 = 0$

$$\therefore x = \frac{-7 \pm \sqrt{7^2 - 4(2)(-13)}}{2(2)}$$

$$\therefore x = \frac{-7 \pm \sqrt{49 + 104}}{4}$$

$$\therefore x = \frac{-7 \pm \sqrt{153}}{4}$$

70 **a** $H = \frac{d}{2x^2 y}$

$\therefore Hy = \frac{d}{2x^2}$

$\therefore y = \frac{d}{2Hx^2}$

b **i** When $H = 25$, $d = 100$, and $x = 2$,

$y = \frac{100}{2 \times 25 \times 2^2} = \frac{100}{200} = \frac{1}{2}$

ii When $H = 18$, $d = 144$, and $x = 2$,

$y = \frac{144}{2 \times 18 \times 2^2} = \frac{144}{144} = 1$

71 **a** $3^{2x+1} = 81$

$\therefore 3^{2x+1} = 3^4$

$\therefore 2x + 1 = 4$

$\therefore 2x = 3$

$\therefore x = \frac{3}{2}$

b $7^x = 1$

$\therefore 7^x = 7^0$

$\therefore x = 0$

c $3^{x^2-3} = 9^x$

$\therefore 3^{x^2-3} = (3^2)^x$

$\therefore 3^{x^2-3} = 3^{2x}$

$\therefore x^2 - 3 = 2x$

$\therefore x^2 - 2x - 3 = 0$

$\therefore (x-3)(x+1) = 0$

$\therefore x = 3$ or -1

72 **a**

x	15	3	12	30
y	4	20	5	2
xy	60	60	60	60

$\therefore xy = 60$ for all points

$\therefore y = \frac{60}{x}$

$\therefore y$ is inversely proportional to x.

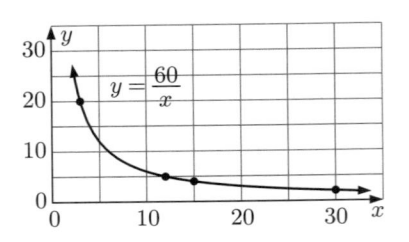

b

x	2	4	6	12
y	7.5	7	6.5	6
xy	15	28	39	72

The value of xy is not constant, so y is not inversely proportional to x.

c

x	4	20	80	10
y	25	5	1.25	10
xy	100	100	100	100

$\therefore xy = 100$ for all points

$\therefore y = \frac{100}{x}$

$\therefore y$ is inversely proportional to x.

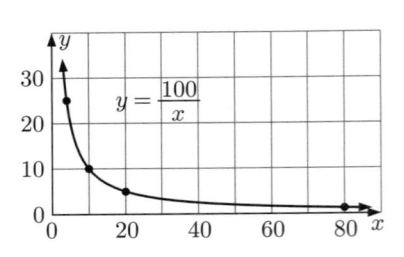

73 **a** $\frac{x}{x-1} = \frac{8}{x+2}$

$\therefore x(x+2) = 8(x-1)$

$\therefore x^2 + 2x = 8x - 8$

$\therefore x^2 - 6x + 8 = 0$

$\therefore (x-4)(x-2) = 0$

$\therefore x = 4$ or 2

b $\frac{5}{x+3} - \frac{x}{x+1} = 3$ {LCD $= (x+3)(x+1)$}

$\therefore \frac{5(x+1)}{(x+3)(x+1)} - \frac{x(x+3)}{(x+1)(x+3)} = \frac{3(x+3)(x+1)}{(x+3)(x+1)}$

$\therefore 5(x+1) - x(x+3) = 3(x+3)(x+1)$

$\therefore 5x + 5 - x^2 - 3x = 3(x^2 + 4x + 3)$

$\therefore -x^2 + 2x + 5 = 3x^2 + 12x + 9$

$\therefore -4x^2 - 10x - 4 = 0$

$\therefore -2(2x^2 + 5x + 2) = 0$

$\therefore -2(2x+1)(x+2) = 0$

$\therefore x = -\frac{1}{2}$ or -2

74 **a** $5x - 3y = 12$ .... (1)

$\underline{-2x + 3y = 3}$ .... (2)

Adding, $3x = 15$

$\therefore x = 5$

Substituting $x = 5$ into (2) gives

$-2(5) + 3y = 3$

$\therefore 3y = 13$

$\therefore y = \frac{13}{3}$

So, the solution is $x = 5$, $y = \frac{13}{3}$.

Check: In (1), $5x - 3y = 5(5) - 3\left(\frac{13}{3}\right) = 12$ ✓

b Consider $\quad 3x - 4y = 6 \quad \ldots (1)$
$\qquad\qquad\quad 4x + y = 27 \quad \ldots (2)$

$\therefore \quad 3x - 4y = 6$
$\quad \underline{16x + 4y = 108} \quad \{4 \times (2)\}$
Adding, $\quad 19x = 114$
$\therefore \quad x = 6$

Substituting $x = 6$ into (2) gives
$\quad 4(6) + y = 27$
$\therefore \quad y = 3$

So, the solution is $x = 6$, $y = 3$.
Check: In (1), $3x - 4y = 3(6) - 4(3) = 6$ ✓

75 a The difference table is:

n	1	2	3	4	5	6
u_n	1	7	13	19	25	31
$\Delta 1$	6	6	6	6	6	

The $\Delta 1$ values are constant, so the sequence is linear.
$\therefore \; u_n = an + b$ with $a = 6$ and $a + b = 1$
$\therefore \quad 6 + b = 1$
$\therefore \quad b = -5$
$\therefore$ the general term is $u_n = 6n - 5$

b The difference table is:

n	1	2	3	4	5	6
u_n	-1	0	3	8	15	24
$\Delta 1$	1	3	5	7	9	
$\Delta 2$	2	2	2	2		

The $\Delta 2$ values are constant, so the sequence is quadratic with general term $u_n = an^2 + bn + c$.
$\quad 2a = 2, \quad$ so $\quad a = 1$
$\quad 3a + b = 1, \quad$ so $\quad 3 + b = 1$
$\qquad\qquad\qquad\qquad \therefore \; b = -2$
$a + b + c = -1, \quad$ so $\quad 1 - 2 + c = -1$
$\qquad\qquad\qquad\qquad \therefore \; c = 0$
$\therefore$ the general term is $u_n = n^2 - 2n$

c The difference table is:

n	1	2	3	4	5	6
u_n	-2	-1	-14	-53	-130	-257
$\Delta 1$	1	-13	-39	-77	-127	
$\Delta 2$	-14	-26	-38	-50		
$\Delta 3$	-12	-12	-12			

The $\Delta 3$ values are constant, so the sequence is cubic with general term $u_n = an^3 + bn^2 + cn + d$.
$\quad 6a = -12, \quad$ so $\quad a = -2$
$\quad 12a + 2b = -14, \quad$ so $\quad -24 + 2b = -14$
$\qquad\qquad\qquad\qquad\qquad \therefore \; b = 5$
$\quad 7a + 3b + c = 1, \quad$ so $\quad -14 + 15 + c = 1$
$\qquad\qquad\qquad\qquad\qquad \therefore \; c = 0$
$a + b + c + d = -2, \quad$ so $\quad -2 + 5 + d = -2$
$\qquad\qquad\qquad\qquad\qquad \therefore \; d = -5$
$\therefore$ the general term is $\; u_n = -2n^3 + 5n^2 - 5$

76 a

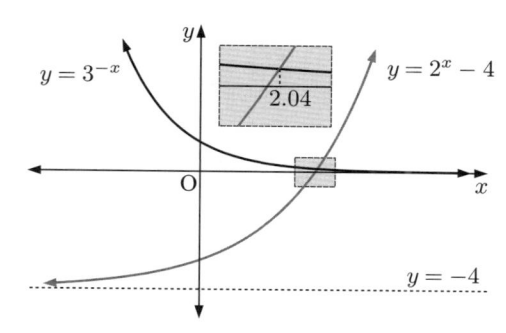

The x-coordinate of the intersection point is $x \approx 2.04$.
$3^{-x} \geqslant 2^x - 4$ when the graph of $y = 3^{-x}$ is on or above the graph of $y = 2^x - 4$.
This occurs when $x \leqslant 2.04$.

b

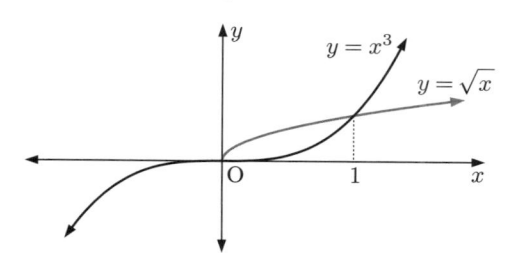

The x-coordinates of the intersection points are $x = 0$ and $x = 1$.
$x^3 \leqslant \sqrt{x}$ when the graph of $y = x^3$ is on or below the graph of $y = \sqrt{x}$.
This occurs when $0 \leqslant x \leqslant 1$.

c

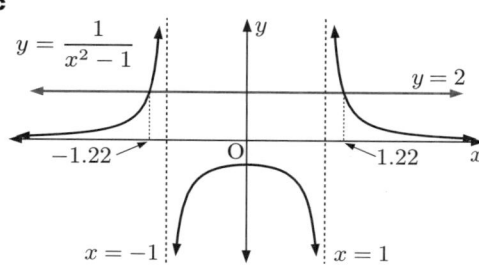

The x-coordinates of the intersection points are $x \approx -1.22$ and $x \approx 1.22$.
$\dfrac{1}{x^2 - 1} > 2$ when the graph of $y = \dfrac{1}{x^2 - 1}$ is above the line $y = 2$.
This occurs when $-1.22 < x < -1$ and $1 < x < 1.22$.

77 Let x cm be the length of the rectangle. The width of the rectangle is $(x - 3)$ cm. 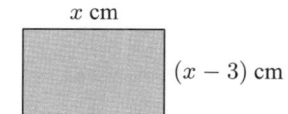
area $= 270 \text{ cm}^2$
$\therefore \; x(x - 3) = 270$
$\therefore \; x^2 - 3x = 270$
$\therefore \; x^2 - 3x - 270 = 0$
$\therefore \; (x - 18)(x + 15) = 0$
$\therefore \; x = 18 \quad \{$as $x > 0\}$
So, the length of the rectangle is 18 cm.

78 a $\quad u_n = 4 \times 3^n$
$\therefore \; u_1 = 4 \times 3^1 = 12$
$\quad u_2 = 4 \times 3^2 = 36$
$\quad u_3 = 4 \times 3^3 = 108$
$\quad u_4 = 4 \times 3^4 = 324$
$\quad u_5 = 4 \times 3^5 = 972$

b $u_n = 5 \times \left(-\frac{1}{2}\right)^{n-1}$
$\therefore u_1 = 5 \times \left(-\frac{1}{2}\right)^0 = 5$
$u_2 = 5 \times \left(-\frac{1}{2}\right)^1 = -\frac{5}{2}$
$u_3 = 5 \times \left(-\frac{1}{2}\right)^2 = \frac{5}{4}$
$u_4 = 5 \times \left(-\frac{1}{2}\right)^3 = -\frac{5}{8}$
$u_5 = 5 \times \left(-\frac{1}{2}\right)^4 = \frac{5}{16}$

79 $a \times 3^n = 108$ .... (1)
$a \times 6^n = 864$ .... (2)
Dividing (2) by (1) gives
$\frac{a \times 6^n}{a \times 3^n} = \frac{864}{108}$
$\therefore \left(\frac{6}{3}\right)^n = 8$
$\therefore 2^n = 2^3$
$\therefore n = 3$
Using (1), $a \times 3^3 = 108$
$\therefore a = \frac{108}{27}$
$\therefore a = 4$
So, $a = 4, \ n = 3$.

80 a The variation model is $y = kx^n$, where k and n are constants.
Now $k \times 2^n = 62.5$ .... (1)
and $k \times 10^n = 0.5$ .... (2)
Dividing (2) by (1) gives $\frac{k \times 10^n}{k \times 2^n} = \frac{0.5}{62.5}$
$\therefore \left(\frac{10}{2}\right)^n = \frac{1}{125}$
$\therefore 5^n = 5^{-3}$
$\therefore n = -3$
Using (1), $k \times 2^{-3} = 62.5$
$\therefore \frac{k}{8} = 62.5$
$\therefore k = 500$
So, the variation model is $y = 500x^{-3}$ or $y = \frac{500}{x^3}$.
Check: When $x = 5$, $y = \frac{500}{5^3} = \frac{500}{125} = 4$ ✓

b The variation model is $p = kt^n$, where k and n are constants.
Now $k \times 2^n = 12$ .... (1)
and $k \times 4^n = 48$ .... (2)
Dividing (2) by (1) gives $\frac{k \times 4^n}{k \times 2^n} = \frac{48}{12}$
$\therefore \left(\frac{4}{2}\right)^n = 4$
$\therefore 2^n = 2^2$
$\therefore n = 2$
Using (1), $k \times 2^2 = 12$
$\therefore k = \frac{12}{4}$
$\therefore k = 3$
So, the variation model is $p = 3t^2$.
Check: When $t = 6$, $p = 3(6)^2 = 108$ ✓

81 a $A = 2\pi r^2 + 2\pi rh$
$\therefore 2\pi rh = A - 2\pi r^2$
$\therefore h = \frac{A - 2\pi r^2}{2\pi r}$
$= \frac{A}{2\pi r} - r$

b i When $A = 240\pi$ and $r = 6$,
$h = \frac{240\pi}{2\pi \times 6} - 6$
$= \frac{240}{12} - 6$
$= 14$

ii When $A \approx 1385.2$ and $r = 8.9$,
$h \approx \frac{1385.2}{2\pi \times 8.9} - 8.9 \approx 15.9$

82 a Each term is the reciprocal of the square of the term number.
$u_6 = \frac{1}{6^2} = \frac{1}{36}, \ u_7 = \frac{1}{7^2} = \frac{1}{49}$

b Start with 100, and each term thereafter is half the previous term.
$u_5 = 6\frac{1}{4}, \ u_6 = 3\frac{1}{8}$

c Start with 10, and each term thereafter is $3\frac{1}{2}$ more than the previous term.
$u_5 = 24, \ u_6 = 27\frac{1}{2}$

d Start with 1, then 4, and the difference between successive terms increases by 3 each time.
$u_6 = 31 + 15 = 46, \ u_7 = 46 + 18 = 64$

83 $Q \propto \frac{1}{t^3}$

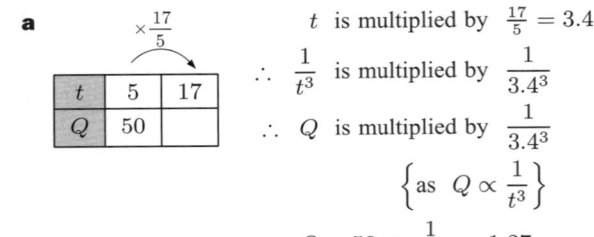

a

t	5	17
Q	50	

$\times \frac{17}{5}$

t is multiplied by $\frac{17}{5} = 3.4$
$\therefore \frac{1}{t^3}$ is multiplied by $\frac{1}{3.4^3}$
$\therefore Q$ is multiplied by $\frac{1}{3.4^3}$
$\left\{\text{as } Q \propto \frac{1}{t^3}\right\}$
$\therefore Q = 50 \times \frac{1}{3.4^3} \approx 1.27$

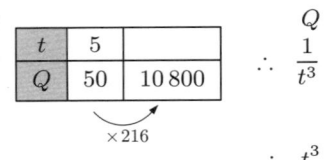

b

t	5	
Q	50	10 800

$\times 216$

Q is multiplied by 216
$\therefore \frac{1}{t^3}$ is multiplied by 216
$\left\{\text{as } Q \propto \frac{1}{t^3}\right\}$
$\therefore t^3$ is multiplied by $\frac{1}{216}$
$\therefore t$ is multiplied by $\sqrt[3]{\frac{1}{216}} = \frac{1}{6}$
$\therefore t = 5 \times \frac{1}{6} = \frac{5}{6}$

84 Let the kite have width x cm and length $(x + 25)$ cm.
area of kite $= \frac{1}{2}$(width $\times$ length)
$\therefore \frac{1}{2}x(x + 25) = 3838$
$\therefore x(x + 25) = 7676$
$\therefore x^2 + 25x - 7676 = 0$
$\therefore (x + 101)(x - 76) = 0$
$\therefore$ since $x > 0$, $x = 76$
$\therefore$ the kite has width 76 cm and length $76 + 25 = 101$ cm.

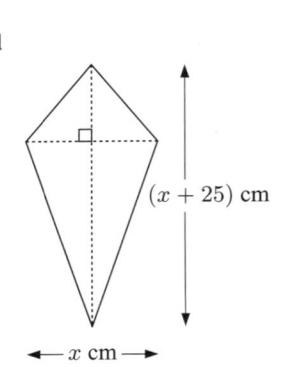

85 a

4, 8, 12, 16, 20, ...

b

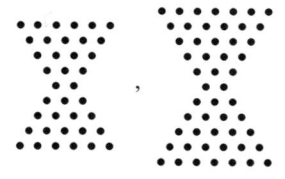

2, 8, 16, 26, 38, 52, ...

86 $I \propto \dfrac{1}{d}$

a d is multiplied by 4

$\therefore \ \dfrac{1}{d}$ is multiplied by $\dfrac{1}{4}$

$\therefore \ I$ is multiplied by $\dfrac{1}{4}$

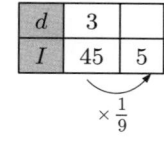

$\therefore \ I = 45 \times \dfrac{1}{4} = 11.25$

	×4	
d	3	12
I	45	

When the depth is 12 m, the light intensity is 11.25 units.

b I is multiplied by $\dfrac{1}{9}$

$\therefore \ \dfrac{1}{d}$ is multiplied by $\dfrac{1}{9}$

$\therefore \ d$ is multiplied by 9

$\therefore \ d = 3 \times 9 = 27$

d	3	
I	45	5

×$\dfrac{1}{9}$

The light intensity is 5 units when the depth is 27 m.

87 Using technology:

a $x \approx 4.192$ **b** $x \approx 0.862$ **c** $x \approx 1.774$

88 Using equal lengths:

$2x + 2 = 3y$ .... (1)

$5x - y = x + 4y \quad \therefore \ 4x = 5y$

$\therefore \ x = \dfrac{5}{4}y$ .... (2)

Substituting (2) into (1) gives

$2 \times \dfrac{5}{4}y + 2 = 3y$

$\therefore \ \dfrac{5}{2}y + 2 = 3y$

$\therefore \ 2 = \dfrac{1}{2}y$

$\therefore \ y = 4$

Substituting $y = 4$ into (2) gives

$x = \dfrac{5}{4}(4) = 5$

So, $x = 5$ and $y = 4$

Check: In (1), LHS $= 2(5) + 2 = 12$
RHS $= 3(4) = 12 \ \checkmark$

89 a $a = 0.45t^4$ **b** $Q \approx 19.9x^{0.402}$

90 a To get each term we multiply the previous one by 2.

$u_1 = 9 \times 2^0$
$u_2 = 9 \times 2^1$ So, $u_5 = 9 \times 2^4 = 144$
$u_3 = 9 \times 2^2$ $u_6 = 9 \times 2^5 = 288$
$u_4 = 9 \times 2^3$ and $u_n = 9 \times 2^{n-1}$

b To get each term we multiply the previous one by $-\dfrac{1}{4}$.

$u_1 = 96 \times \left(-\dfrac{1}{4}\right)^0$
$u_2 = 96 \times \left(-\dfrac{1}{4}\right)^1$ So, $u_5 = 96 \times \left(-\dfrac{1}{4}\right)^4 = \dfrac{3}{8}$
$u_3 = 96 \times \left(-\dfrac{1}{4}\right)^2$ $u_6 = 96 \times \left(-\dfrac{1}{4}\right)^5 = -\dfrac{3}{32}$
$u_4 = 96 \times \left(-\dfrac{1}{4}\right)^3$ and $u_n = 96 \times \left(-\dfrac{1}{4}\right)^{n-1}$

c To get each term we multiply the previous one by $\dfrac{1}{2}$.

$u_1 = 1 \times \left(\dfrac{1}{2}\right)^0$
$u_2 = 1 \times \left(\dfrac{1}{2}\right)^1$ So, $u_5 = 1 \times \left(\dfrac{1}{2}\right)^4 = \dfrac{1}{16}$
$u_3 = 1 \times \left(\dfrac{1}{2}\right)^2$ $u_6 = 1 \times \left(\dfrac{1}{2}\right)^5 = \dfrac{1}{32}$
$u_4 = 1 \times \left(\dfrac{1}{2}\right)^3$ and $u_n = 1 \times \left(\dfrac{1}{2}\right)^{n-1}$

91 a $V = \pi r^2 h$

$\therefore \ V = kr^2$ where $k = \pi h$ is constant

$\therefore \ V$ is directly proportional to the square of r.

b If the radius is increased by 40%,

r is multiplied by 1.4

$\therefore \ r^2$ is multiplied by $(1.4)^2$

$\therefore \ V$ is multiplied by $(1.4)^2 = 1.96$

So, the volume of the cylinder would increase by 96%.

92

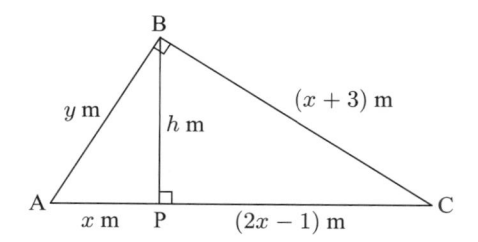

Using Pythagoras for each triangle:

In $\triangle$BPC, $h^2 = (x+3)^2 - (2x-1)^2$

In $\triangle$APB, $y^2 = x^2 + h^2$
$= x^2 + (x+3)^2 - (2x-1)^2$

In $\triangle$ABC, $(x+3)^2 + y^2 = (3x-1)^2$

$\therefore \ (x+3)^2 + x^2 + (x+3)^2 - (2x-1)^2 = (3x-1)^2$

$\therefore \ 2(x^2 + 6x + 9) + x^2 - (4x^2 - 4x + 1) = 9x^2 - 6x + 1$

$\therefore \ 2x^2 + 12x + 18 + x^2 - 4x^2 + 4x - 1 = 9x^2 - 6x + 1$

$\therefore \ -x^2 + 16x + 17 = 9x^2 - 6x + 1$

$\therefore \ -10x^2 + 22x + 16 = 0$

$\therefore \ -2(5x^2 - 11x - 8) = 0$

$\therefore \ x = \dfrac{11 \pm \sqrt{(-11)^2 - 4(5)(-8)}}{2(5)}$

$\therefore \ x = \dfrac{11 \pm \sqrt{121 + 160}}{10}$

$\therefore \ x = \dfrac{11 \pm \sqrt{281}}{10}$

$\therefore \ x \approx -0.576$ or 2.78

But $x > 0$, so $x \approx 2.78$

93

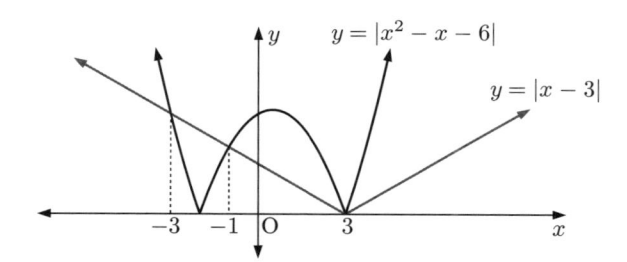

The x-coordinates of the intersection points are $x = -3, -1$ and 3.

$|x^2 - x - 6| > |x - 3|$ when the graph of $y = |x^2 - x - 6|$ is above the graph of $y = |x - 3|$.

This occurs when $x < -3$, $-1 < x < 3$, or $x > 3$.

94 $F \propto \dfrac{1}{d^2}$

If the distance is decreased by 55%,
d is multiplied by 0.45

$\therefore \dfrac{1}{d^2}$ is multiplied by $\dfrac{1}{(0.45)^2}$

$\therefore F$ is multiplied by $\dfrac{1}{(0.45)^2} \approx 4.94$

So, the force between the particles is increased by 394%.

95 a The difference table is:

n	1	2	3	4	5	6
u_n	22	15	8	1	-6	-13
$\Delta 1$	-7	-7	-7	-7	-7	

The $\Delta 1$ values are constant, so the sequence is linear.

$\therefore u_n = an + b$ with $a = -7$ and $a + b = 22$
$\therefore -7 + b = 22$
$\therefore b = 29$

$\therefore$ the general term is $u_n = -7n + 29$

b The difference table is:

n	1	2	3	4	5	6
u_n	-3	3	19	45	81	127
$\Delta 1$	6	16	26	36	46	
$\Delta 2$	10	10	10	10		

The $\Delta 2$ values are constant, so the sequence is quadratic with general term $u_n = an^2 + bn + c$.

$2a = 10$, so $a = 5$
$3a + b = 6$, so $15 + b = 6$ and $\therefore b = -9$
$a + b + c = -3$, so $5 - 9 + c = -3$ and $\therefore c = 1$

$\therefore$ the general term is $u_n = 5n^2 - 9n + 1$

c The difference table is:

n	1	2	3	4	5	6
u_n	0	5	20	51	104	185
$\Delta 1$	5	15	31	53	81	
$\Delta 2$	10	16	22	28		
$\Delta 3$	6	6	6			

The $\Delta 3$ values are constant, so the sequence is cubic with general term $u_n = an^3 + bn^2 + cn + d$.

$6a = 6$, so $a = 1$
$12a + 2b = 10$, so $12 + 2b = 10$
$\therefore b = -1$
$7a + 3b + c = 5$, so $7 - 3 + c = 5$
$\therefore c = 1$
$a + b + c + d = 0$, so $1 - 1 + 1 + d = 0$
$\therefore d = -1$

$\therefore$ the general term is $u_n = n^3 - n^2 + n - 1$.

96 $y \propto \sqrt{x}$

$\therefore y = k\sqrt{x}$ where k is a constant
From the graph, when $x = 4$, $y = 60$
$\therefore 60 = k \times \sqrt{4}$ and so $k = 30$
The model is $y = 30\sqrt{x}$.
Check: When $x = 16$, $y = 30 \times \sqrt{16} = 120$ ✓

97 a $y = \dfrac{2x-3}{1-x}$

$\therefore y(1-x) = 2x - 3$
$\therefore y - xy = 2x - 3$
$\therefore y + 3 = 2x + xy$
$\therefore y + 3 = x(y+2)$
$\therefore x = \dfrac{y+3}{y+2}$

b i When $y = -\dfrac{4}{3}$,

$x = \dfrac{-\dfrac{4}{3} + 3}{-\dfrac{4}{3} + 2}$

$= \dfrac{\dfrac{5}{3}}{\dfrac{2}{3}}$

$= \dfrac{5}{2}$

ii When $y = -\dfrac{5}{3}$,

$x = \dfrac{-\dfrac{5}{3} + 3}{-\dfrac{5}{3} + 2}$

$= \dfrac{\dfrac{4}{3}}{\dfrac{1}{3}}$

$= 4$

98 Jo can wrap x glasses per hour.

$\therefore$ it takes her $\dfrac{150}{x}$ hours to complete the job.

If she wrapped $(x + 10)$ glasses per hour, she would take $\left(\dfrac{150}{x} - 0.1\right)$ hours to complete the job.

$\{6 \text{ minutes} = 0.1 \text{ hours}\}$

$\therefore (x+10)\left(\dfrac{150}{x} - 0.1\right) = 150$

$\therefore 150 - 0.1x + \dfrac{1500}{x} - 1 = 150$

$\therefore -0.1x + \dfrac{1500}{x} - 1 = 0$

$\therefore -0.1x^2 - x + 1500 = 0$

Using technology, $x \approx 117.6$ $\{x > 0\}$

So, Jo can wrap 117 glasses each hour.

99 a $u_1 = 4$, $u_2 = 13$, $u_3 = 24$

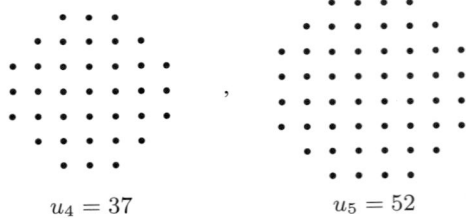

$u_4 = 37 \qquad u_5 = 52$

b The nth figure consists of a square of size $(n+1)$ in the centre, and an additional $(n-1)$ dots along each edge of the square.

$\therefore u_n = (n+1)^2 + 4(n-1)$
$= n^2 + 2n + 1 + 4n - 4$
$= n^2 + 6n - 3$

c $u_{45} = 45^2 + 6 \times 45 - 3$
$= 2292$

So, 2292 dots are needed to make up the 45th pattern.

100 a Using technology, $C \approx 15.0 t^{1.16}$

b Yes, $r^2 \approx 1$

c In 2007, $t = 15$
$\therefore C \approx 15.0 \times 15^{1.16}$
≈ 347.1

So, approximately 347 000 CPUs were sold in 2007.

101 a $u_1 = 6$
$u_2 = 6 + 24 = 30$
$u_3 = 6 + 24 + 54 = 84$
$u_4 = 6 + 24 + 54 + 96 = 180$
$u_5 = 2 \times 3 + 4 \times 6 + 6 \times 9 + 8 \times 12 + 10 \times 15$
$ = 6 + 24 + 54 + 96 + 150$
$ = 330$
$u_6 = 2 \times 3 + 4 \times 6 + 6 \times 9 + 8 \times 12 + 10 \times 15$
$ + 12 \times 18$
$ = 6 + 24 + 54 + 96 + 150 + 216$
$ = 546$

b The difference table is:

n	1	2	3	4	5	6
u_n	6	30	84	180	330	546
$\Delta 1$	24	54	96	150	216	
$\Delta 2$		30	42	54	66	
$\Delta 3$			12	12	12	

The $\Delta 3$ values are constant, so the sequence is cubic with general term $u_n = an^3 + bn^2 + cn + d$.
$6a = 12$, so $a = 2$
$12a + 2b = 30$, so $24 + 2b = 30$
$\therefore b = 3$
$7a + 3b + c = 24$, so $14 + 9 + c = 24$
$\therefore c = 1$
$a + b + c + d = 6$, so $2 + 3 + 1 + d = 6$
$\therefore d = 0$
$\therefore$ the general term is $u_n = 2n^3 + 3n^2 + n$.

c $u_{75} = 2 \times 75^3 + 3 \times 75^2 + 75$
$\phantom{u_{75}} = 860\,700$

102 Suppose the supermarket sold x 500 g packs and y 1 kg packs.
$\therefore\ 0.5x + y = 66$ {weight}
or $y = 66 - 0.5x$ .... (1)
and $3.25x + 5.85y = 404.95$ .... (2) {value}
Substituting (1) into (2) gives
$3.25x + 5.85(66 - 0.5x) = 404.95$
$\therefore\ 3.25x + 386.1 - 2.925x = 404.95$
$\therefore\ 0.325x = 18.85$
$\therefore\ x = 58$
Substituting $x = 58$ into (1) gives
$y = 66 - 0.5(58) = 37$
So, the supermarket sold 58 of the 500 g packs, and 37 of the 1 kg packs.
Check: In (2), $3.25(58) + 5.85(37) = 404.95$ ✓

103 a Using technology, $k \approx 0.5222$, $n \approx 3.0012$,
$V \approx 0.522 d^{3.00}$

b When $d = 5$, $V \approx 0.5222 \times 5^{3.0012}$
$ \approx 65.4$ ✓

c $V = \frac{4}{3}\pi r^3$
$ = \frac{4}{3}\pi \left(\frac{d}{2}\right)^3$
$ = \frac{4}{3}\pi \times \frac{d^3}{8}$
$ \approx 0.524 d^3$
which is a close approximation to $V \approx 0.522 d^{3.00}$.

d i When $d = 4$,
$V \approx 0.5222 \times 4^{3.0012}$
$ \approx 33.5$

ii When $V = 500$,
$500 \approx 0.5222 \times d^{3.0012}$
$\therefore\ d^{3.0012} \approx \frac{500}{0.5222}$
$\therefore\ d \approx \left(\frac{500}{0.5222}\right)^{\frac{1}{3.0012}}$
$ \approx 9.85$

104 a $u_1 = 1$, $u_2 = 7$, $u_3 = 16$

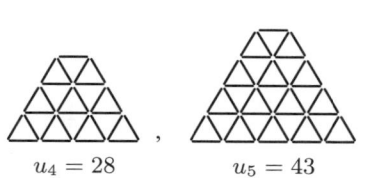

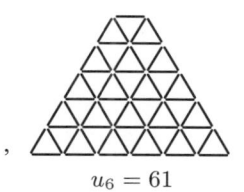

$u_4 = 28$, $u_5 = 43$, $u_6 = 61$

b The difference table is:

n	1	2	3	4	5	6
u_n	1	7	16	28	43	61
$\Delta 1$	6	9	12	15	18	
$\Delta 2$		3	3	3	3	

The $\Delta 2$ values are constant, so the sequence is quadratic with general term $u_n = an^2 + bn + c$.
$2a = 3$, so $a = \frac{3}{2}$
$3a + b = 6$, so $\frac{9}{2} + b = 6$
$\therefore b = \frac{3}{2}$
$a + b + c = 1$, so $\frac{3}{2} + \frac{3}{2} + c = 1$
$\therefore c = -2$
$\therefore$ the general term is $u_n = \frac{3}{2}n^2 + \frac{3}{2}n - 2$.

c $u_{20} = \frac{3}{2} \times 20^2 + \frac{3}{2} \times 20 - 2 = 628$
$\therefore$ there are 628 matchsticks in the 20th figure.

SOLUTIONS TO TOPIC 3: FUNCTIONS

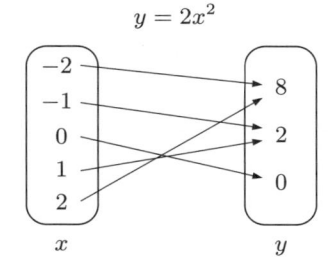

1 a $y = 2x^2$

b Range $= \{0, 2, 8\}$ **c** many-one
d The relation is a function.

2 a Domain $= \{x \mid -4 \leqslant x \leqslant 4\}$
Range $= \{y \mid -1 \leqslant y \leqslant 1\}$
The graph is a function.

b Domain $= \{x \mid -4 \leqslant x \leqslant 4\}$
Range $= \{y \mid -4 \leqslant y \leqslant 4\}$
The graph fails the vertical line test, so it is not a function.

3 $f(x) = \dfrac{3x-2}{x^2}$

a $f(1)$
$= \dfrac{3(1)-2}{1^2}$
$= \dfrac{1}{1}$
$= 1$

b $f(2)$
$= \dfrac{3(2)-2}{2^2}$
$= \dfrac{4}{4}$
$= 1$

c $f(3)$
$= \dfrac{3(3)-2}{3^2}$
$= \dfrac{7}{9}$

4 a This is a function, since no two ordered pairs have the same x-coordinate.

b This is not a function, since $(1, 1)$ and $(1, 4)$ have the same x-coordinate.

5 a $m(x) = -1$
$\therefore x^2 - 5 = -1$
$\therefore x^2 = 4$
$\therefore x = \pm 2$

b $m(x) = 11$
$\therefore x^2 - 5 = 11$
$\therefore x^2 = 16$
$\therefore x = \pm 4$

6 $f(x) = 3 - 4x$

a $f(a)$
$= 3 - 4a$

b $f(-a)$
$= 3 - 4(-a)$
$= 3 + 4a$

c $f(1-x)$
$= 3 - 4(1-x)$
$= 3 - 4 + 4x$
$= 4x - 1$

d $f(x^2)$
$= 3 - 4(x^2)$
$= 3 - 4x^2$

e $f(x^2 + \tfrac{3}{4}) = 3 - 4(x^2 + \tfrac{3}{4})$
$= 3 - 4x^2 - 3$
$= -4x^2$

7 a $x^2 + y^2 = 1$

Mapping diagram: $x \in \{-1, 0, 1\}$ to $y \in \{-1, 0, 1\}$

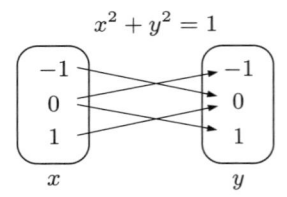

b Range $= \{-1, 0, 1\}$ **c** many-many

d The relation is not a function.

8 $f(x) = 3^{-x} + 1$

a $f(0)$
$= 3^0 + 1$
$= 1 + 1$
$= 2$

b $f(-2)$
$= 3^{-(-2)} + 1$
$= 3^2 + 1$
$= 10$

c $f(3)$
$= 3^{-3} + 1$
$= \tfrac{1}{27} + 1$
$= 1\tfrac{1}{27}$

9 $f(x) = 2x + 1$, $g(x) = x^2$

a $f(2) = 2(2) + 1$
$= 5$

b $g(2) = 2^2$
$= 4$

c $f(g(2)) = f(4)$
$= 2(4) + 1$
$= 9$

d $g(f(2)) = g(5)$
$= 5^2$
$= 25$

e $f(f(2)) = f(5)$
$= 2(5) + 1$
$= 11$

f $g(g(2)) = g(4)$
$= 4^2$
$= 16$

10 a $y = \dfrac{16}{x}$ is undefined when $x = 0$, so $x = 0$ is a vertical asymptote.

As x gets very large, $\dfrac{16}{x}$ approaches 0, so $y = 0$ is a horizontal asymptote.

b i When $x = 256$, $y = \dfrac{16}{256} = \dfrac{1}{16}$
ii When $x = -256$, $y = \dfrac{16}{-256} = -\dfrac{1}{16}$

c If $y = \dfrac{16}{x}$, then $x = \dfrac{16}{y}$

i When $y = 256$, $x = \dfrac{16}{256} = \dfrac{1}{16}$
ii When $y = -256$, $x = \dfrac{16}{-256} = -\dfrac{1}{16}$

d Graph of $y = \dfrac{16}{x}$ with asymptotes $y = 0$ and $x = 0$.

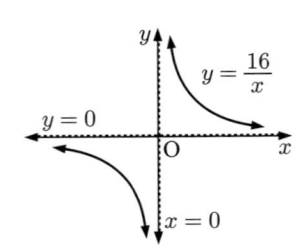

11 a $81^{\frac{1}{2}} = 9 \Leftrightarrow \log_{81} 9 = \tfrac{1}{2}$

b $2^{10} = 1024 \Leftrightarrow \log_2 1024 = 10$

c $32^{-\frac{1}{5}} = \tfrac{1}{2} \Leftrightarrow \log_{32} \tfrac{1}{2} = -\tfrac{1}{5}$

d $8\sqrt{2} = 2^{3.5}$
$\therefore 2^{3.5} = 8\sqrt{2} \Leftrightarrow \log_2 8\sqrt{2} = 3.5$

12 a If $x = -2$,
$|2x - x^2|$
$= |2(-2) - (-2)^2|$
$= |-4 - 4|$
$= |-8|$
$= 8$

b If $x = -2$,
$\dfrac{|x|}{3-2x}$
$= \dfrac{|-2|}{3-2(-2)}$
$= \dfrac{2}{3+4}$
$= \dfrac{2}{7}$

c If $x = -2$, $\dfrac{-x^2}{|2+3x|} = \dfrac{-(-2)^2}{|2+3(-2)|}$
$= \dfrac{-4}{|2-6|}$
$= \dfrac{-4}{4}$
$= -1$

13 $f(x) = -2x^2 + 5x + 1$

a $f(3) = -2(3)^2 + 5(3) + 1$
$= -2(9) + 15 + 1$
$= -2$

b If $f(a) = 4$ then $-2a^2 + 5a + 1 = 4$
$\therefore 2a^2 - 5a + 3 = 0$
$\therefore (2a-3)(a-1) = 0$
$\therefore a = \tfrac{3}{2}$ or 1

14 a Let $\log_2 \sqrt{2} = x$
$\therefore 2^x = \sqrt{2}$
$\therefore 2^x = 2^{\frac{1}{2}}$
$\therefore x = \tfrac{1}{2}$
$\therefore \log_2 \sqrt{2} = \tfrac{1}{2}$

b Let $\log_{10} 10\,000\,000 = x$
$\therefore 10^x = 10\,000\,000$
$\therefore 10^x = 10^7$
$\therefore x = 7$
$\therefore \log_{10} 10\,000\,000 = 7$

c Let $\log_5 3125 = x$
$\therefore\ 5^x = 3125$
$\therefore\ 5^x = 5^5$
$\therefore\ x = 5$
$\therefore\ \log_5 3125 = 5$

d Let $\log_7 1 = x$
$\therefore\ 7^x = 1$
$\therefore\ 7^x = 7^0$
$\therefore\ x = 0$
$\therefore\ \log_7 1 = 0$

e Let $\log_3\left(\dfrac{1}{81\sqrt{3}}\right) = x$
$\therefore\ 3^x = \dfrac{1}{81\sqrt{3}}$
$\therefore\ 3^x = \dfrac{1}{3^4 \times 3^{\frac{1}{2}}}$
$\therefore\ 3^x = 3^{-4\frac{1}{2}}$
$\therefore\ x = -4\frac{1}{2}$
$\therefore\ \log_3\left(\dfrac{1}{81\sqrt{3}}\right) = -4\frac{1}{2}$

f Let $\log_{\sqrt{2}} 64 = x$
$\therefore\ (\sqrt{2})^x = 64$
$\therefore\ \left(2^{\frac{1}{2}}\right)^x = 2^6$
$\therefore\ 2^{\frac{x}{2}} = 2^6$
$\therefore\ \frac{x}{2} = 6$
$\therefore\ x = 12$
$\therefore\ \log_{\sqrt{2}} 64 = 12$

15 a $y = 2x^2 - 3x - 1$

x	-3	-2	-1	0	1	2	3
y	26	13	4	-1	-2	1	8

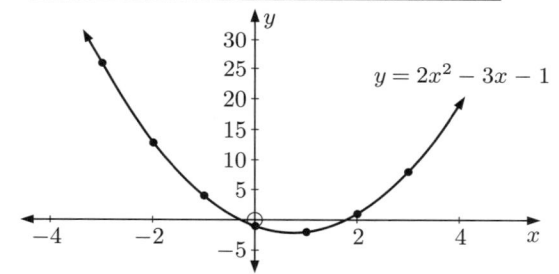

b $y = -x^2 + x + 5$

x	-3	-2	-1	0	1	2	3
y	-7	-1	3	5	5	3	-1

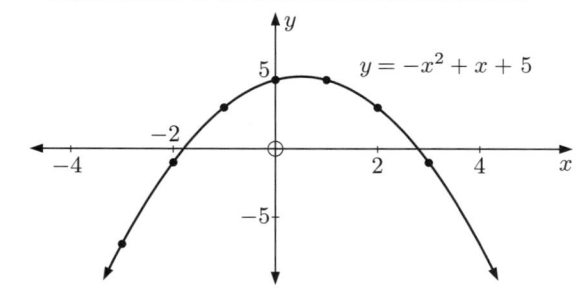

16 a $\log_3 81 = 4 \Leftrightarrow 3^4 = 81$
b $\log_7 7\sqrt{7} = 1.5 \Leftrightarrow 7^{1.5} = 7\sqrt{7}$
c $\log_{\sqrt{5}} 125 = 6 \Leftrightarrow (\sqrt{5})^6 = 125$

17 a $|3 - x| = 5$
$\therefore\ 3 - x = 5$ or $3 - x = -5$
$\therefore\ -x = 2$ or $-x = -8$
$\therefore\ x = -2$ or 8

b $3 - |x| = 2$
$\therefore\ |x| = 1$
$\therefore\ x = \pm 1$

18 $f(x) = 2 - x,\ g(x) = 3 + 2x$

a i $f(g(x))$
$= f(3 + 2x)$
$= 2 - (3 + 2x)$
$= -1 - 2x$

ii $g(f(x))$
$= g(2 - x)$
$= 3 + 2(2 - x)$
$= 7 - 2x$

iii $f(f(x))$
$= f(2 - x)$
$= 2 - (2 - x)$
$= x$

iv $g(g(x))$
$= g(3 + 2x)$
$= 3 + 2(3 + 2x)$
$= 9 + 4x$

b $f(g(x)) = g(f(x))$ when
$-1 - 2x = 7 - 2x$
$\therefore\ -1 = 7$ which is never true.
$\therefore\ f(g(x)) \neq g(f(x))$ for all x.

c $f(f(x)) = g(g(x))$ when
$x = 9 + 4x$
$\therefore\ -3x = 9$
$\therefore\ x = -3$

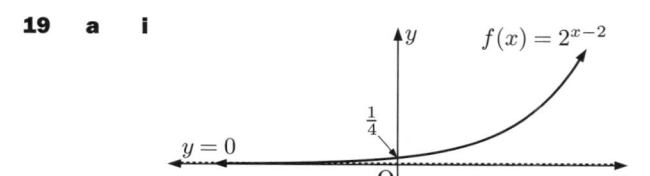

19 a i

ii $f(0) = 2^{-2} = \frac{1}{4}$
So, the y-intercept is $\frac{1}{4}$.

iii As x gets smaller, $f(x)$ approaches 0, so $y = 0$ is a horizontal asymptote.

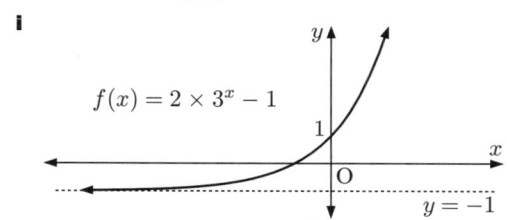

b i

ii $f(0) = 2 \times 3^0 - 1 = 1$
So, the y-intercept is 1.

iii As x gets smaller, $f(x)$ approaches -1, so $y = -1$ is a horizontal asymptote.

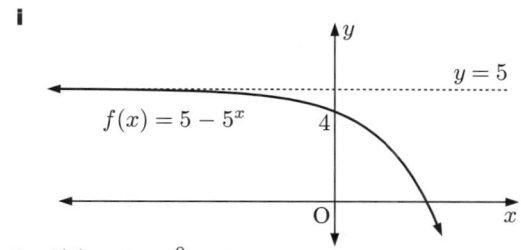

c i

ii $f(0) = 5 - 5^0 = 4$
So, the y-intercept is 4.

iii As x gets smaller, $f(x)$ approaches 5, so $y = 5$ is a horizontal asymptote.

20 a $y = 3^{\frac{1}{2}x}$
$\therefore\ \log_3 y = \frac{1}{2}x$

b $y = 5^{-a}$
$\therefore\ \log_5 y = -a$

c $y = 4 \times 2^n$
$\therefore\ y = 2^{n+2}$
$\therefore\ \log_2 y = n + 2$

21 a $P = \log_3 t$
$\therefore\ 3^P = t$

b $y = \tfrac{1}{4}\log_2 x$
$\therefore\ 4y = \log_2 x$
$\therefore\ 2^{4y} = x$

c $Q = \log_5\left(\dfrac{t}{3}\right)$
$\therefore\ 5^Q = \dfrac{t}{3}$

22 a We draw $y = x^2$ and translate it by $\begin{pmatrix} 1 \\ -2 \end{pmatrix}$.

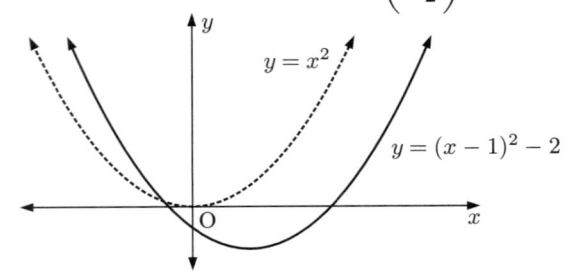

The vertex is at $(1, -2)$.

b The vertex is at $(-2, 1)$.

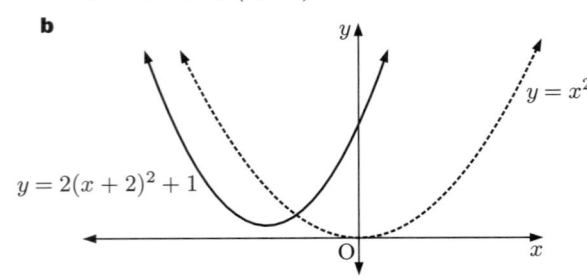

c The vertex is at $(3, -4)$.

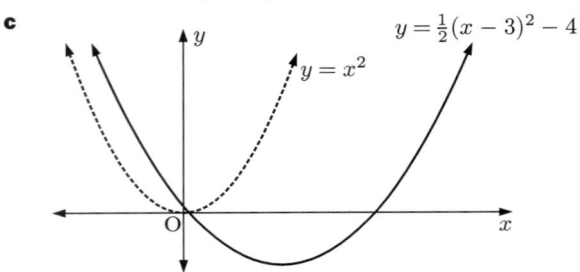

d The vertex is at $(-1, 6)$.

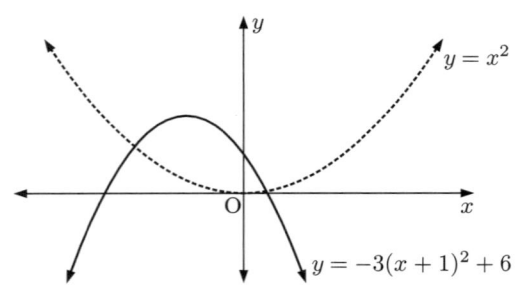

23 a $y = 7^{x+1}$
$\therefore\ \log_7 y = x + 1$
$\therefore\ x = \log_7 y - 1$

b $d = \log_5(3x)$
$\therefore\ 5^d = 3x$
$\therefore\ x = \dfrac{5^d}{3}$

c $z = \log_6\left(\dfrac{x}{2}\right)$
$\therefore\ 6^z = \dfrac{x}{2}$
$\therefore\ x = 2 \times 6^z$

d $T = \tfrac{2}{3} \times 2^{10x}$
$\therefore\ \dfrac{3T}{2} = 2^{10x}$
$\therefore\ \log_2\left(\dfrac{3T}{2}\right) = 10x$
$\therefore\ x = \dfrac{\log_2\left(\tfrac{3T}{2}\right)}{10}$

24 a $40 = 10^{\log_{10} 40}$
$\therefore\ 40 \approx 10^{1.60}$

b $0.004 = 10^{\log_{10} 0.004}$
$\therefore\ 0.004 \approx 10^{-2.40}$

c $500 = 10^{\log_{10} 500}$
$\therefore\ 500 \approx 10^{2.70}$

d $0.05 = 10^{\log_{10} 0.05}$
$\therefore\ 0.05 \approx 10^{-1.30}$

25 a $y = -2\sin\left(\tfrac{1}{2}x\right)$ has period $\dfrac{360°}{\tfrac{1}{2}} = 720°$ and amplitude $|-2| = 2$.

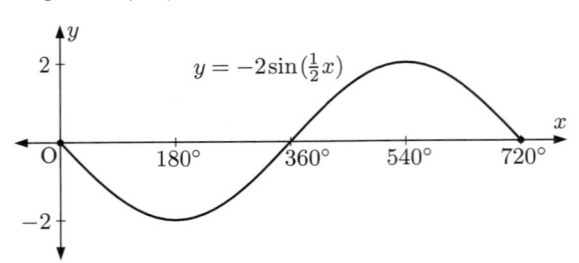

b $y = 3\cos(3x)$ has period $\dfrac{360°}{3} = 120°$ and amplitude $|3| = 3$.

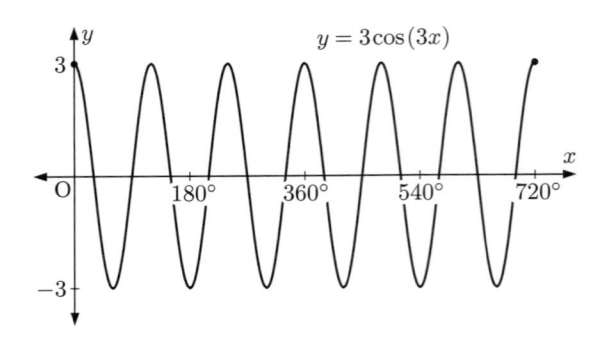

26 a The inverse of
$y = 2x - 5$
is $x = 2y - 5$
$\therefore\ 2y = x + 5$
$\therefore\ y = \dfrac{x+5}{2}$
$\therefore\ f^{-1}(x) = \dfrac{x+5}{2}$

b The inverse of
$y = -x^3 + 2$
is $x = -y^3 + 2$
$\therefore\ y^3 = 2 - x$
$\therefore\ y = \sqrt[3]{2-x}$
$\therefore\ f^{-1}(x) = \sqrt[3]{2-x}$

c The inverse of $y = \dfrac{3}{1-x}$ is $x = \dfrac{3}{1-y}$
$\therefore\ 1 - y = \dfrac{3}{x}$
$\therefore\ y = 1 - \dfrac{3}{x}$
$\therefore\ f^{-1}(x) = 1 - \dfrac{3}{x}$

27 a

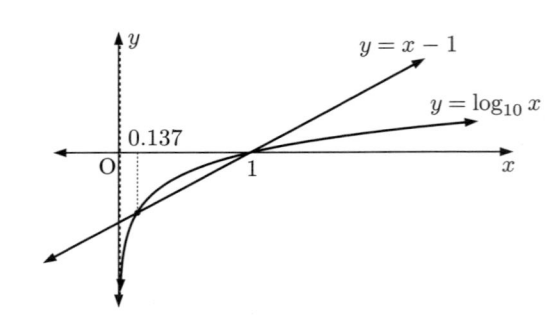

The solutions to $\log_{10} x = x - 1$
are $x \approx 0.137,\ x = 1$

b

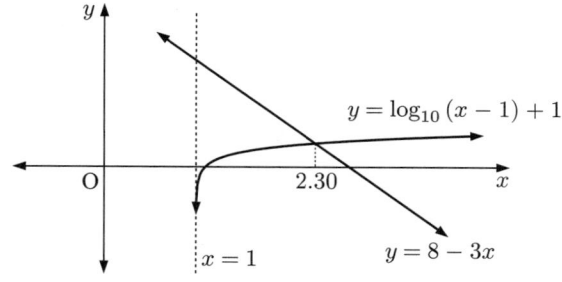

[Graph showing $y = \log_{10}(x-1) + 1$ and $y = 8 - 3x$ intersecting near $x = 2.30$, with vertical asymptote $x = 1$.]

The solution to $\log_{10}(x-1) + 1 = 8 - 3x$ is $x \approx 2.30$

c

[Graph showing $y = 3^{-x}$ and $y = \log_{10}\left(\frac{x}{2}\right) + 1$ intersecting near $x = 0.632$.]

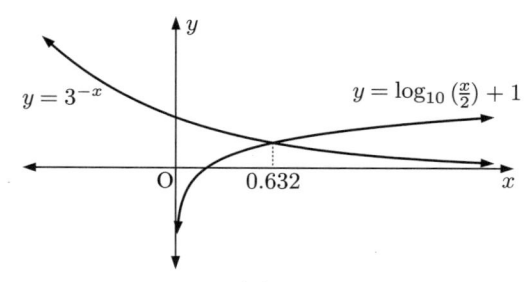

The solution to $\log_{10}\left(\frac{x}{2}\right) + 1 = 3^{-x}$ is $x \approx 0.632$

28 a When $x = 0$, $y = -(0)^2 + 6(0) - 5$
$= -5$
$\therefore$ the y-intercept is -5.
When $y = 0$, $-x^2 + 6x - 5 = 0$
$\therefore -(x^2 - 6x + 5) = 0$
$\therefore -(x - 1)(x - 5) = 0$
$\therefore x = 1$ or 5
$\therefore$ the x-intercepts are 1 and 5.

b When $x = 0$, $y = 2(0 - 1)(0 + 5)$
$= -10$
$\therefore$ the y-intercept is -10.
When $y = 0$, $2(x - 1)(x + 5) = 0$
$\therefore x = 1$ or -5
$\therefore$ the x-intercepts are 1 and -5.

29 a $y = 3 - |x|$
$\therefore y = \begin{cases} 3 - x & \text{if } x \geqslant 0 \\ 3 + x & \text{if } x < 0. \end{cases}$

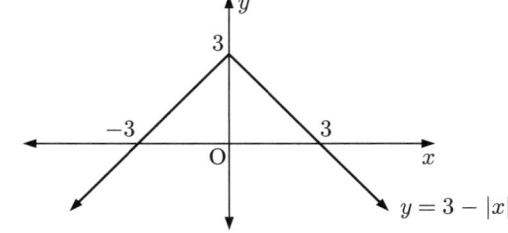

[Graph of $y = 3 - |x|$, inverted V with peak at $(0,3)$, x-intercepts at -3 and 3.]

b $y = 2|x| - 4$
$\therefore y = \begin{cases} 2x - 4 & \text{if } x \geqslant 0 \\ -2x - 4 & \text{if } x < 0. \end{cases}$

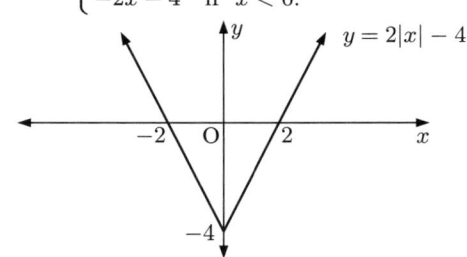

[Graph of $y = 2|x| - 4$, V-shape with vertex at $(0,-4)$, x-intercepts at -2 and 2.]

30 speed of cyclist $= \dfrac{\text{distance}}{\text{time}} = \dfrac{13.6 \text{ km}}{48 \text{ min}} = \dfrac{13.6 \text{ km}}{0.8 \text{ h}}$
$= 17 \text{ km/h}$

$\therefore 17 = 45 - g - \dfrac{g^2}{8}$

$\therefore \dfrac{g^2}{8} + g - 28 = 0$

$\therefore g^2 + 8g - 224 = 0$

$\therefore g = \dfrac{-8 \pm \sqrt{8^2 - 4(1)(-224)}}{2(1)}$

$= \dfrac{-8 \pm \sqrt{960}}{2}$

$\approx 11.5 \quad \{g > 0\}$

The angle of incline of the mountain is $11.5°$.

31 a $\log_3 7 + \log_3 4$
$= \log_3(7 \times 4)$
$= \log_3 28$

b $5 + \log_2 3$
$= \log_2 2^5 + \log_2 3$
$= \log_2(32 \times 3)$
$= \log_2 96$

c $\log_2 11 - \log_2 7$
$= \log_2\left(\frac{11}{7}\right)$

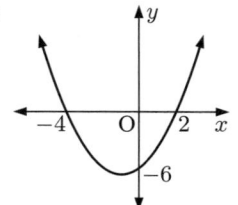

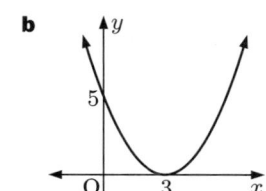

d $3\log_5 a + 4\log_5 b$
$= \log_5 a^3 + \log_5 b^4$
$= \log_5\left(a^3 b^4\right)$

32 a [Parabola opening upward with x-intercepts at -4 and 2, y-intercept at -6.]

b [Parabola opening upward with vertex at $(3, 0)$, y-intercept at 5.]

33 Using the horizontal line test, the functions in **b** and **d** have inverses.

34 a $\log_2 35$
$= \log_2(5 \times 7)$
$= \log_2 5 + \log_2 7$
$= p + q$

b $\log_2(1.4)$
$= \log_2\left(\frac{7}{5}\right)$
$= \log_2 7 - \log_2 5$
$= q - p$

c $\log_2 125 = \log_2 5^3$
$= 3 \times \log_2 5$
$= 3p$

35 a When $x = 0$, $T = 65 \times 0.95^0 = 65$
The initial temperature of the stew was $65\,°C$.

b i When $x = 3$, $T = 65 \times 0.95^3 \approx 55.7$
The temperature after 3 minutes is $55.7\,°C$.

ii When $x = 10$, $T = 65 \times 0.95^{10} \approx 38.9$
The temperature after 10 minutes is $38.9\,°C$.

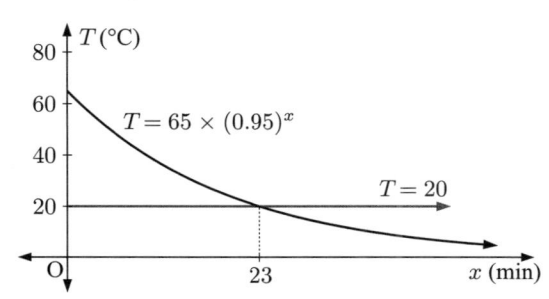

c [Graph of $T = 65 \times (0.95)^x$ decreasing from 65, and horizontal line $T = 20$, intersecting at $x = 23$.]

The graphs of $T = 65 \times (0.95)^x$ and $T = 20$ meet when $x \approx 23.0$

$\therefore$ it takes about 23.0 minutes for the stew's temperature to drop to $20\,°C$.

36 a When $x = 0$, $y = (0+4)(0-2) = -8$
∴ the y-intercept is -8.
When $y = 0$, $(x+4)(x-2) = 0$
∴ $x = -4$ or 2
∴ the x-intercepts are -4 and 2.

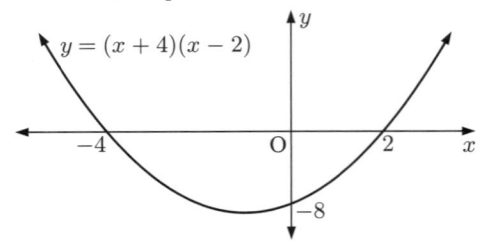

[Graph: $y = (x+4)(x-2)$, parabola through $(-4, 0)$, $(2, 0)$, $(0, -8)$]

b When $x = 0$, $y = -2(0-1)^2 = -2$
∴ the y-intercept is -2.
When $y = 0$, $-2(x-1)^2 = 0$
∴ $x = 1$
∴ the x-intercept is 1.

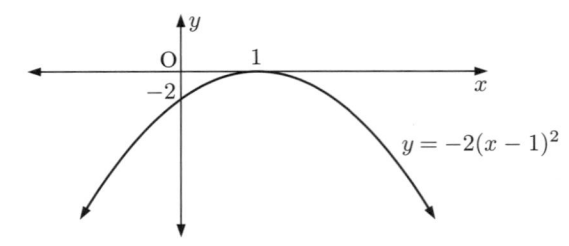

[Graph: $y = -2(x-1)^2$, downward parabola with vertex $(1, 0)$, y-intercept -2]

c When $x = 0$, $y = 2(0)^2 + 3(0) - 5 = -5$
∴ the y-intercept is -5.
When $y = 0$, $2x^2 + 3x - 5 = 0$
∴ $(2x+5)(x-1) = 0$
∴ $x = -\frac{5}{2}$ or 1
∴ the x-intercepts are $-\frac{5}{2}$ and 1.

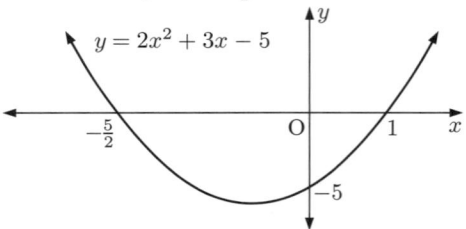

[Graph: $y = 2x^2 + 3x - 5$, upward parabola through $-\frac{5}{2}$, 1, y-intercept -5]

d When $x = 0$, $y = -6(0)^2 + 13(0) - 5 = -5$
∴ the y-intercept is -5.
When $y = 0$, $-6x^2 + 13x - 5 = 0$
∴ $6x^2 - 13x + 5 = 0$
∴ $6x^2 - 10x - 3x + 5 = 0$
∴ $2x(3x - 5) - 1(3x - 5) = 0$
∴ $(3x - 5)(2x - 1) = 0$
∴ $x = \frac{5}{3}$ or $\frac{1}{2}$
∴ the x-intercepts are $\frac{5}{3}$ and $\frac{1}{2}$.

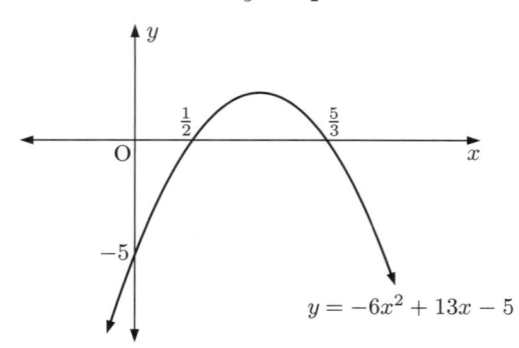

[Graph: $y = -6x^2 + 13x - 5$, downward parabola through $\frac{1}{2}$, $\frac{5}{3}$, y-intercept -5]

37 a $\log_3 y = \frac{1}{2} \log_3 a$
∴ $\log_3 y = \log_3 a^{\frac{1}{2}}$
∴ $y = a^{\frac{1}{2}}$ or $\sqrt{a}$

b $\log_{10} y = a - \log_{10} b$
∴ $\log_{10} y = \log_{10} 10^a - \log_{10} b$
∴ $\log_{10} y = \log_{10}\left(\frac{10^a}{b}\right)$
∴ $y = \frac{10^a}{b}$

c $\log_7 y = -\log_7 b$
∴ $\log_7 y = \log_7 b^{-1}$
∴ $y = b^{-1}$ or $\frac{1}{b}$

d $\log_2 y = \log_2 a + 3 \log_2 b$
∴ $\log_2 y = \log_2 a + \log_2 b^3$
∴ $\log_2 y = \log_2(ab^3)$
∴ $y = ab^3$

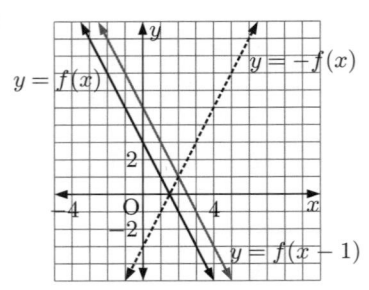

38 a [Graph showing $y = f(x)$, $y = -f(x)$, $y = f(x-1)$]

b i horizontal translation of $\begin{pmatrix} 1 \\ 0 \end{pmatrix}$

ii reflection in the x-axis.

39 a $(3, -4)$ lies on the graph of $y = \frac{k}{x}$.
∴ $-4 = \frac{k}{3}$ and so $k = -12$
∴ the graph has equation $y = -\frac{12}{x}$.

b $(-5, -2)$ lies on the graph of $y = \frac{k}{x}$.
∴ $-2 = \frac{k}{-5}$ and so $k = 10$
∴ the graph has equation $y = \frac{10}{x}$.

40 a $\dfrac{\log_3 81}{\log_3 \sqrt{3}}$
$= \dfrac{\log_3 3^4}{\log_3 3^{\frac{1}{2}}}$
$= \dfrac{4}{\frac{1}{2}}$
$= 8$

b $\dfrac{\log_2 \left(\frac{1}{4}\right)}{\log_2 16}$
$= \dfrac{\log_2 2^{-2}}{\log_2 2^4}$
$= \dfrac{-2}{4}$
$= -\frac{1}{2}$

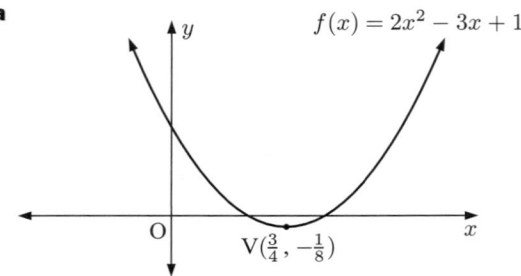

41 a [Graph: $f(x) = 2x^2 - 3x + 1$, upward parabola with vertex $V\left(\frac{3}{4}, -\frac{1}{8}\right)$]

∴ the vertex has coordinates $\left(\frac{3}{4}, -\frac{1}{8}\right)$.

b

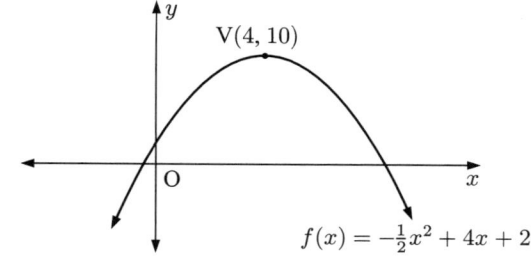

$$f(x) = -\tfrac{1}{2}x^2 + 4x + 2$$

$\therefore$ the vertex has coordinates $(4, 10)$.

42 a We first graph $f(x) = x + 3$.

The graph cuts the x-axis when $y = 0$
$$\therefore \ x + 3 = 0$$
$$\therefore \ x = -3$$

To obtain the graph of $f(x) = |x + 3|$ we reflect all points with $x < -3$ in the x-axis.

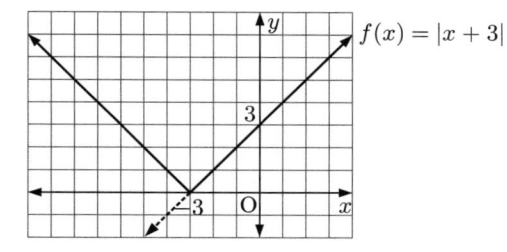

b We first graph $f(x) = 3 - 2x$.

The graph cuts the x-axis when $y = 0$
$$\therefore \ 3 - 2x = 0$$
$$\therefore \ x = \tfrac{3}{2}$$

To obtain the graph of $f(x) = |3 - 2x|$ we reflect all points with $x > \tfrac{3}{2}$ in the x-axis.

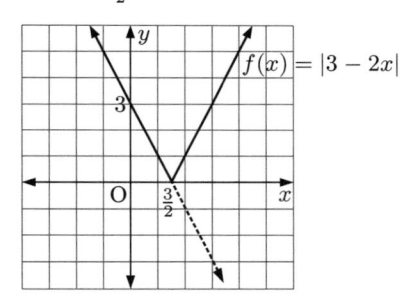

43 a $y = 2^{-x}$ has inverse function $x = 2^{-y}$
$$\therefore \ -y = \log_2 x$$
$$\therefore \ y = -\log_2 x$$
$$\therefore \ f^{-1}(x) = -\log_2 x$$

b $y = 4\log_5 x$ has inverse function $x = 4\log_5 y$
$$\therefore \ \tfrac{x}{4} = \log_5 y$$
$$\therefore \ y = 5^{\tfrac{x}{4}}$$
$$\therefore \ f^{-1}(x) = 5^{\tfrac{x}{4}}$$

c $y = \log_{\sqrt{3}} x$ has inverse function $x = \log_{\sqrt{3}} y$
$$\therefore \ y = \left(\sqrt{3}\right)^x$$
$$\therefore \ f^{-1}(x) = \left(\sqrt{3}\right)^x$$

d $y = 4 \times 10^x$ has inverse function $x = 4 \times 10^y$
$$\therefore \ \tfrac{x}{4} = 10^y$$
$$\therefore \ y = \log_{10}\left(\tfrac{x}{4}\right)$$
$$\therefore \ f^{-1}(x) = \log_{10}\left(\tfrac{x}{4}\right)$$

44 a $y = (x-2)(x+1)(x-5)$
has x-intercepts 2, -1, and 5.
When $x = 0$, $y = (-2)(1)(-5) = 10$
$\therefore$ the y-intercept is 10.

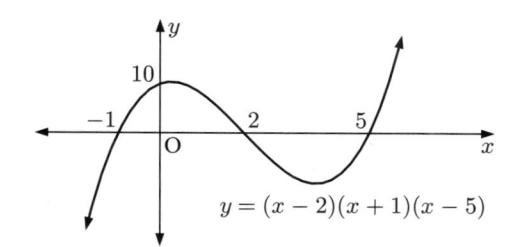

b $y = -(x+2)^2(x-3)$ cuts the x-axis at $x = 3$, and touches the x-axis at $x = -2$.
When $x = 0$, $y = -(2)^2(-3) = 12$
$\therefore$ the y-intercept is 12.

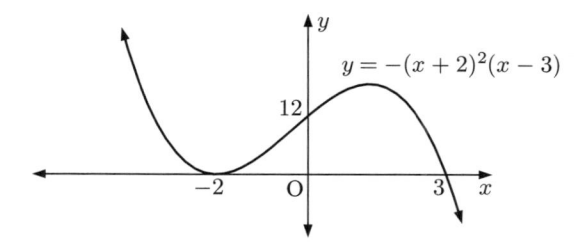

c $y = 2x(x-1)(x-2)$ has x-intercepts 0, 1, and 2.
When $x = 0$, $y = 2(0)(-1)(-2) = 0$
$\therefore$ the y-intercept is 0.

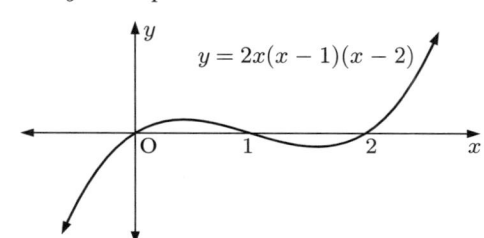

d $y = -3(2x+1)(2x-1)(2x-5)$
has x-intercepts $-\tfrac{1}{2}$, $\tfrac{1}{2}$, and $\tfrac{5}{2}$.
When $x = 0$, $y = -3(1)(-1)(-5) = -15$
$\therefore$ the y-intercept is -15.

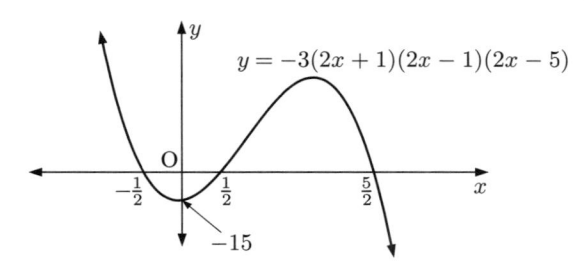

45 Using technology:

a $P \approx 1.93 \times 2.76^t$, $r^2 \approx 0.9999$

b $Q \approx 24.9 \times (0.750)^x$, $r^2 \approx 1$

46 a i When $x = 0$, $y = 10$
$\therefore$ the y-intercept is 10.
When $y = 0$, $x^2 - 7x + 10 = 0$
$\therefore \ (x-2)(x-5) = 0$
$\therefore \ x = 2$ or 5
$\therefore$ the x-intercepts are 2 and 5.

ii The line of symmetry is halfway between the x-intercepts
∴ since $\frac{7}{2}$ is the average of 2 and 5, the line of symmetry is $x = \frac{7}{2}$.

iii When $x = \frac{7}{2}$, $y = \left(\frac{7}{2}\right)^2 - 7\left(\frac{7}{2}\right) + 10$
$= \frac{49}{4} - \frac{49}{2} + 10$
$= -\frac{9}{4}$
∴ the vertex is at $\left(\frac{7}{2}, -\frac{9}{4}\right)$.

iv

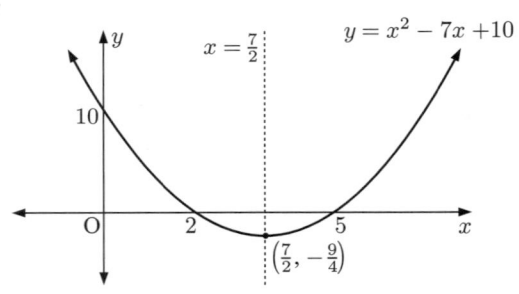

Graph of $y = x^2 - 7x + 10$ with line of symmetry $x = \frac{7}{2}$, x-intercepts at 2 and 5, y-intercept at 10, vertex at $\left(\frac{7}{2}, -\frac{9}{4}\right)$.

b i When $x = 0$, $y = -8$
∴ the y-intercept is -8.
When $y = 0$, $-2x^2 + 8x - 8 = 0$
∴ $-2(x^2 - 4x + 4) = 0$
∴ $-2(x - 2)^2 = 0$
∴ $x = 2$
∴ the x-intercept is 2.

ii Since there is only one x-intercept, and this is when $x = 2$, the line of symmetry is $x = 2$.

iii When $x = 2$, $y = -2(2)^2 + 8(2) - 8$
$= -8 + 16 - 8$
$= 0$
∴ the vertex is at $(2, 0)$.

iv

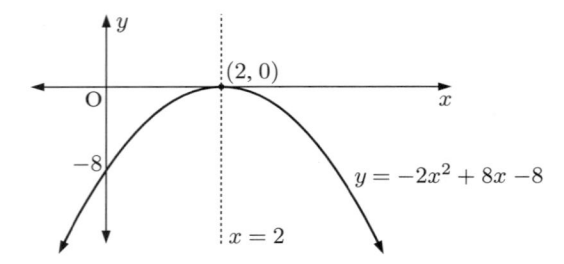

Graph of $y = -2x^2 + 8x - 8$, vertex $(2, 0)$, y-intercept -8, line of symmetry $x = 2$.

c i When $x = 0$, $y = -15$
∴ the y-intercept is -15.
When $y = 0$, $3x^2 - 15 = 0$
∴ $3x^2 = 15$
∴ $x^2 = 5$
∴ $x = \pm\sqrt{5}$
∴ the x-intercepts are $-\sqrt{5}$ and $\sqrt{5}$.

ii The line of symmetry is halfway between the x-intercepts
∴ since the average of $-\sqrt{5}$ and $\sqrt{5}$ is 0, the line of symmetry is $x = 0$.

iii When $x = 0$, $y = -15$
∴ the vertex is at $(0, -15)$.

iv

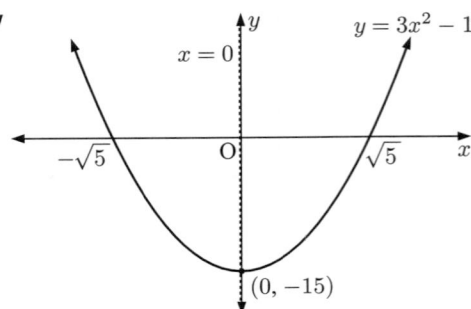

Graph of $y = 3x^2 - 15$, line of symmetry $x = 0$, x-intercepts $\pm\sqrt{5}$, $(0, -15)$.

d i When $x = 0$, $y = -9$
∴ the y-intercept is -9.
When $y = 0$, $6x^2 - 15x - 9 = 0$
∴ $3(2x^2 - 5x - 3) = 0$
∴ $3(2x + 1)(x - 3) = 0$
∴ $x = -\frac{1}{2}$ or 3
∴ the x-intercepts are $-\frac{1}{2}$ and 3.

ii The line of symmetry is halfway between the x-intercepts
∴ since the average of $-\frac{1}{2}$ and 3 is $\frac{5}{4}$, the line of symmetry is $x = \frac{5}{4}$.

iii When $x = \frac{5}{4}$, $y = 6\left(\frac{5}{4}\right)^2 - 15\left(\frac{5}{4}\right) - 9$
$= \frac{75}{8} - \frac{75}{4} - 9$
$= -\frac{147}{8}$
∴ the vertex is at $\left(\frac{5}{4}, -\frac{147}{8}\right)$.

iv

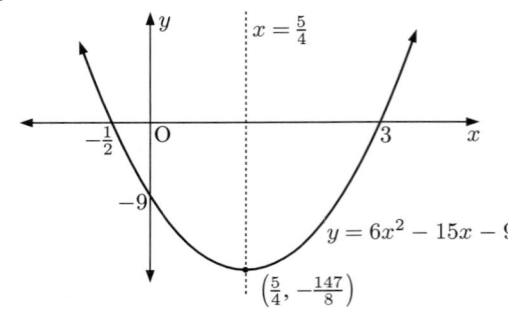

Graph of $y = 6x^2 - 15x - 9$, line of symmetry $x = \frac{5}{4}$, x-intercepts $-\frac{1}{2}$ and 3, y-intercept -9, vertex $\left(\frac{5}{4}, -\frac{147}{8}\right)$.

47 a

Graphs of $p(x) = x^2$, $y = p(x+3)$, and $y = p(x) - 2$.

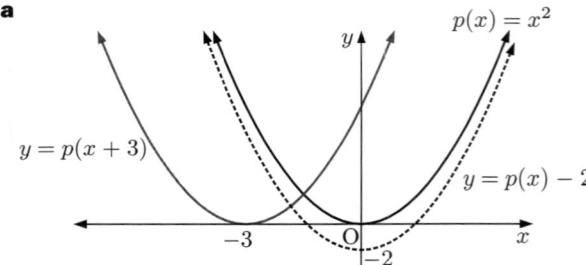

b A translation of $\begin{pmatrix} 3 \\ -2 \end{pmatrix}$.

48 a

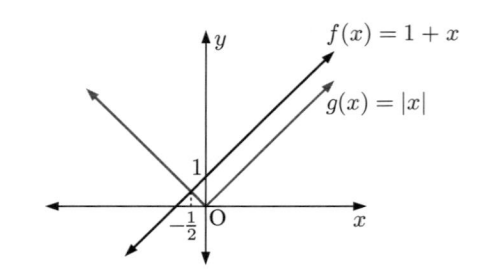

Graphs of $f(x) = 1 + x$ and $g(x) = |x|$.

b $f(x) = g(x)$ when $x = -\frac{1}{2}$

c $f(g(x))$ $\qquad$ $g(f(x))$
$= f(|x|)$ $\qquad$ $= g(1+x)$
$= 1 + |x|$ $\qquad$ $= |1+x|$

d
$f(g(x)) = 1 + |x|$
$g(f(x)) = |1+x|$

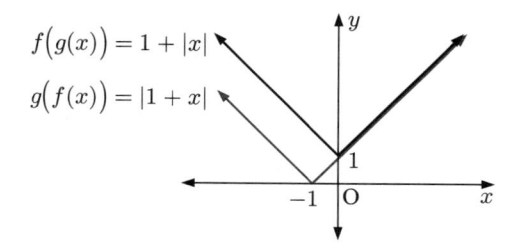

e $f(g(x)) = g(f(x))$ for all $x \geqslant 0$.

49 a $\log 4 + \log 5$
$= \log(4 \times 5)$
$= \log 20$

b $3 - \log_{10} 5$
$= \log_{10} 10^3 - \log_{10} 5$
$= \log_{10} \left(\frac{10^3}{5}\right)$
$= \log_{10} 200$

c $2\log 3 + 3\log 4$
$= \log 3^2 + \log 4^3$
$= \log(3^2 \times 4^3)$
$= \log 576$

d $\frac{1}{4}\log 81 + \log\left(\frac{1}{3}\right)$
$= \log 81^{\frac{1}{4}} + \log\left(\frac{1}{3}\right)$
$= \log 3 + \log\left(\frac{1}{3}\right)$
$= \log\left(3 \times \frac{1}{3}\right)$
$= \log 1$
$= 0$

e $5\log 2 - 3\log 5$
$= \log 2^5 - \log 5^3$
$= \log 32 - \log 125$
$= \log\left(\frac{32}{125}\right)$

f $\log_{10} 25 + \log_{10} 4$
$= \log_{10}(25 \times 4)$
$= \log_{10} 100$
$= 2$

50 a i When $x = 0$, $y = -3(-2)(3) = 18$
$\therefore$ the y-intercept is 18.
When $y = 0$, $-3(x-2)(x+3) = 0$
$\therefore$ $x = 2$ or -3
$\therefore$ the x-intercepts are 2 and -3.

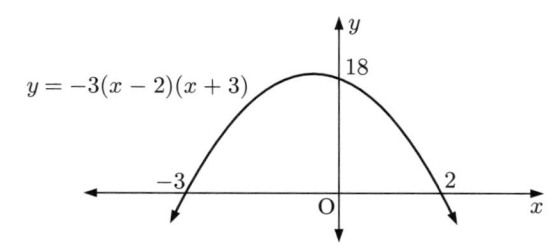

$y = -3(x-2)(x+3)$

ii The line of symmetry is halfway between the x-intercepts
$\therefore$ since $-\frac{1}{2}$ is the average of -3 and 2, the line of symmetry is $x = -\frac{1}{2}$.
When $x = -\frac{1}{2}$, $y = -3(-\frac{1}{2} - 2)(-\frac{1}{2} + 3)$
$= -3 \times -\frac{5}{2} \times \frac{5}{2}$
$= \frac{75}{4}$
$\therefore$ the vertex is at $\left(-\frac{1}{2}, \frac{75}{4}\right)$.

b i When $x = 0$, $y = 2(1)(7) = 14$
$\therefore$ the y-intercept is 14.
When $y = 0$, $2(x+1)(7-x) = 0$
$\therefore$ $x = -1$ or 7
$\therefore$ the x-intercepts are -1 and 7.

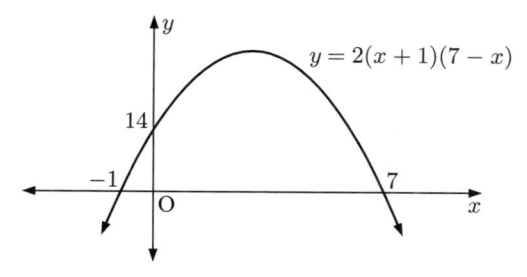

$y = 2(x+1)(7-x)$

ii The line of symmetry is halfway between the x-intercepts
$\therefore$ since 3 is the average of -1 and 7, the line of symmetry is $x = 3$.
When $x = 3$, $y = 2(3+1)(7-3)$
$= 2 \times 4 \times 4$
$= 32$
$\therefore$ the vertex is at $(3, 32)$.

c i When $x = 0$, $y = 4(-3)^2 = 36$
$\therefore$ the y-intercept is 36.
When $y = 0$, $4(x-3)^2 = 0$
$\therefore$ $x = 3$
$\therefore$ the x-intercept is 3.

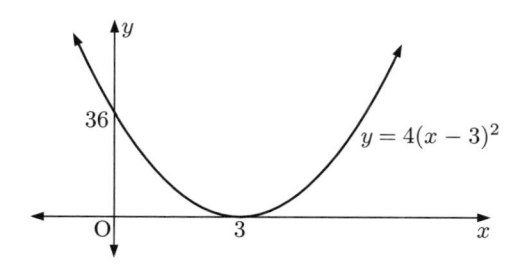

$y = 4(x-3)^2$

ii Since 3 is the only x-intercept, the line of symmetry is $x = 3$.
When $x = 3$, $y = 4(3-3)^2$
$= 0$
$\therefore$ the vertex is at $(3, 0)$.

d i When $x = 0$, $y = 2(0)^2 - 7(0) + 3 = 3$
$\therefore$ the y-intercept is 3.
When $y = 0$, $2x^2 - 7x + 3 = 0$
$\therefore$ $(2x-1)(x-3) = 0$
$\therefore$ $x = \frac{1}{2}$ or 3
$\therefore$ the x-intercepts are $\frac{1}{2}$ and 3.

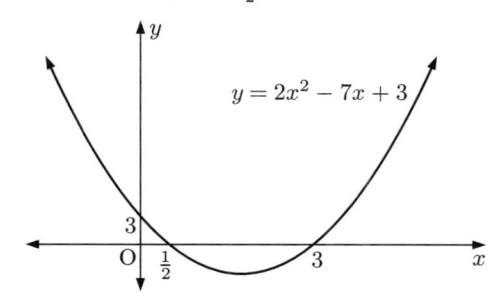

$y = 2x^2 - 7x + 3$

ii The line of symmetry is halfway between the x-intercepts
$\therefore$ since $\frac{7}{4}$ is the average of $\frac{1}{2}$ and 3, the line of symmetry is $x = \frac{7}{4}$.
When $x = \frac{7}{4}$, $y = 2\left(\frac{7}{4}\right)^2 - 7\left(\frac{7}{4}\right) + 3$
$= \frac{49}{8} - \frac{49}{4} + 3$
$= -\frac{25}{8}$
$\therefore$ the vertex is at $\left(\frac{7}{4}, -\frac{25}{8}\right)$.

51 **a** $\log\left(\sqrt[5]{2}\right)$ **b** $\log\left(\frac{1}{81}\right)$ **c** $\log 20$
  $= \log\left(2^{\frac{1}{5}}\right)$ $= \log\left(\frac{1}{3^4}\right)$ $= \log(2 \times 10)$
  $= \frac{1}{5}\log 2$ $= \log\left(3^{-4}\right)$ $= \log 2 + \log 10$
             $= -4\log 3$ $= \log 2 + 1$

52 **a** The graph has amplitude 5 $\therefore\ a = 5$
     The graph has period $540°$
     $\therefore\ \dfrac{360°}{b} = 540°$
     $\therefore\ b = \dfrac{360°}{540°} = \dfrac{2}{3}$

   b The graph has amplitude 3, and has been reflected in the x-axis $\therefore\ a = -3$
     The graph has period $720°$
     $\therefore\ \dfrac{360°}{b} = 720°$
     $\therefore\ b = \dfrac{360°}{720°} = \dfrac{1}{2}$

53 **a** As $h = -1$ and $k = 2$, $f(x) = a(x+1)^2 + 2$
     But $f(0) = 3$, so $a(1)^2 + 2 = 3$
     $\therefore\ a + 2 = 3$
     $\therefore\ a = 1$
     So, $f(x) = (x+1)^2 + 2$
     or $= x^2 + 2x + 3$

   b The x-intercept $-\frac{3}{2}$ comes from the linear factor $2x + 3$.
     The x-intercept $\frac{1}{2}$ comes from the linear factor $2x - 1$.
     $\therefore\ f(x) = a(2x+3)(2x-1)$
     But $f(0) = -3$, so $a(3)(-1) = -3$
     $\therefore\ -3a = -3$
     $\therefore\ a = 1$
     So, $f(x) = (2x+3)(2x-1)$

   c As $h = 3$ and $k = 2$, $f(x) = a(x-3)^2 + 2$
     But $f(5) = -2$, so $a(5-3)^2 + 2 = -2$
     $\therefore\ 4a + 2 = -2$
     $\therefore\ 4a = -4$
     $\therefore\ a = -1$
     So, $f(x) = -(x-3)^2 + 2$
     or $= -x^2 + 6x - 7$

   d The x-intercept 0 comes from the linear factor x.
     The x-intercept 8 comes from the linear factor $x - 8$.
     $\therefore\ f(x) = ax(x-8)$
     But $f(7) = -5$, so $a(7)(-1) = -5$
     $\therefore\ -7a = -5$
     $\therefore\ a = \frac{5}{7}$
     So, $f(x) = \frac{5}{7}x(x-8)$

54 **a** gradient $= \dfrac{3-0}{0-(-2)} = \dfrac{3}{2}$
     $\therefore\ a = \frac{3}{2}$
     $\therefore\ f(x) = \left|\frac{3}{2}x + b\right|$
     But $f(-2) = 0$ $\therefore\ \left|\frac{3}{2}(-2) + b\right| = 0$
     $\therefore\ |-3 + b| = 0$
     $\therefore\ -3 + b = 0$
     $\therefore\ b = 3$
     So, $f(x) = \left|\frac{3}{2}x + 3\right|$

   b gradient $= \dfrac{2-0}{0-4} = -\dfrac{1}{2}$
     $\therefore\ a = -\frac{1}{2}$
     $\therefore\ f(x) = \left|-\frac{1}{2}x + b\right|$
     But $f(4) = 0$ $\therefore\ \left|-\frac{1}{2}(4) + b\right| = 0$
     $\therefore\ |-2 + b| = 0$
     $\therefore\ -2 + b = 0$
     $\therefore\ b = 2$
     So, $f(x) = \left|-\frac{1}{2}x + 2\right|$ or $f(x) = \left|\frac{1}{2}x - 2\right|$

55 **a** $y = a^2 b$
     $\therefore\ \log y = \log(a^2 b)$
     $\therefore\ \log y = \log a^2 + \log b$
     $\therefore\ \log y = 2\log a + \log b$

   b $Q = \dfrac{m\sqrt{n}}{p}$
     $\therefore\ \log Q = \log\left(\dfrac{m \times n^{\frac{1}{2}}}{p}\right)$
     $\therefore\ \log Q = \log m + \log n^{\frac{1}{2}} - \log p$
     $\therefore\ \log Q = \log m + \frac{1}{2}\log n - \log p$

   c $P = \sqrt[3]{s^2 t}$
     $\therefore\ P = (s^2 t)^{\frac{1}{3}}$
     $\therefore\ P = s^{\frac{2}{3}} t^{\frac{1}{3}}$
     $\therefore\ \log P = \log\left(s^{\frac{2}{3}} \times t^{\frac{1}{3}}\right)$
     $\therefore\ \log P = \log s^{\frac{2}{3}} + \log t^{\frac{1}{3}}$
     $\therefore\ \log P = \frac{2}{3}\log s + \frac{1}{3}\log t$

56 **a** The vertex is at $(-2, -4)$ $\therefore\ y = a(x+2)^2 - 4$
     But when $x = 0$, $y = -3$
     $\therefore\ a(2)^2 - 4 = -3$
     $\therefore\ 4a - 4 = -3$
     $\therefore\ 4a = 1$
     $\therefore\ a = \frac{1}{4}$
     So, $y = \frac{1}{4}(x+2)^2 - 4$

   b The vertex is at $(-1, 0)$ $\therefore\ y = a(x+1)^2$
     But when $x = 0$, $y = \frac{3}{2}$
     $\therefore\ a(1)^2 = \frac{3}{2}$
     $\therefore\ a = \frac{3}{2}$
     So, $y = \frac{3}{2}(x+1)^2$

   c The vertex is at $(1, 8)$ $\therefore\ y = a(x-1)^2 + 8$
     But when $x = 3$, $y = 0$
     $\therefore\ a(3-1)^2 + 8 = 0$
     $\therefore\ 4a + 8 = 0$
     $\therefore\ 4a = -8$
     $\therefore\ a = -2$
     So, $y = -2(x-1)^2 + 8$

   d The vertex is at $(1, -2)$ $\therefore\ y = a(x-1)^2 - 2$
     But when $x = 0$, $y = -2\frac{1}{4}$
     $\therefore\ a(-1)^2 - 2 = -2\frac{1}{4}$
     $\therefore\ a - 2 = -2\frac{1}{4}$
     $\therefore\ a = -\frac{1}{4}$
     So, $y = -\frac{1}{4}(x-1)^2 - 2$

57 **a** $\log_{10} T = x - 2$
$\therefore T = 10^{x-2}$ or $T = \dfrac{10^x}{100}$

b $\log_{10} K = \tfrac{1}{3}x - 1$
$\therefore K = 10^{\frac{1}{3}x - 1}$ or $K = \dfrac{10^{\frac{1}{3}x}}{10}$

c $\log_{10} Q \approx 0.903 + x$
$\therefore Q \approx 10^{0.903 + x}$
$\therefore Q \approx 10^{0.903} \times 10^x$
$\therefore Q \approx 8 \times 10^x$

d $\log p = -\tfrac{1}{2} \log q$
$= \log q^{-\frac{1}{2}}$
$= \log \left(\dfrac{1}{\sqrt{q}} \right)$
$\therefore p = \dfrac{1}{\sqrt{q}}$

e $\quad 2 + \log_{10} y = \log_{10} x$
$\therefore \log_{10} 10^2 + \log_{10} y = \log_{10} x$
$\therefore \log_{10}(100 \times y) = \log_{10} x$
$\therefore x = 100y$

f $\log A = \log c + \tfrac{1}{3} \log d$
$= \log c + \log d^{\frac{1}{3}}$
$= \log \left(c \times d^{\frac{1}{3}} \right)$
$\therefore A = c \times \sqrt[3]{d}$

58 a i

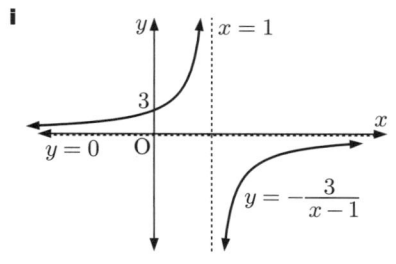

[graph showing $y = -\dfrac{3}{x-1}$ with asymptotes $x = 1$ and $y = 0$, passing through $(0, 3)$]

ii horizontal asymptote $y = 0$, vertical asymptote $x = 1$

b i

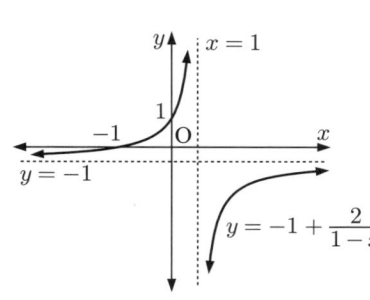

[graph showing $y = -1 + \dfrac{2}{1-x}$ with asymptotes $x = 1$ and $y = -1$, passing through $(-1, 0)$ and $(0, 1)$]

ii horizontal asymptote $y = -1$, vertical asymptote $x = 1$

59 a When $n = 0$, $F = 7500 \times (1.076)^0$
$= 7500$
$\therefore$ the original investment was €7500.

b i When $n = 4$, $F = 7500 \times (1.076)^4$
$\approx 10\,053.34$
$\therefore$ the value of the investment after 4 years was €10 053.34.

ii When $n = 10$, $F = 7500 \times (1.076)^{10}$
$\approx 15\,602.13$
$\therefore$ the value of the investment after 10 years was €15 602.13.

c

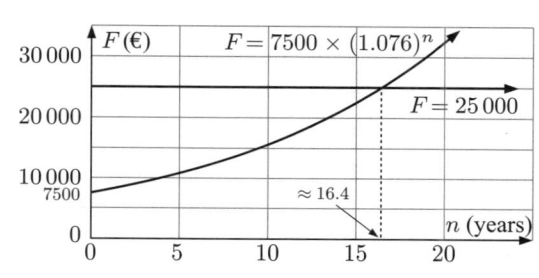

[graph of $F = 7500 \times (1.076)^n$ and $F = 25\,000$ meeting at $n \approx 16.4$]

The graphs of $F = 7500 \times (1.076)^n$ and $F = 25\,000$ meet when $n \approx 16.4$
$\therefore$ it will take about 16.4 years for the value of the investment to reach €25 000.

60 a $10^x = 30$
$\therefore x = \dfrac{\log_{10} 30}{\log_{10} 10}$
$\therefore x \approx 1.48$

b $10^x = 0.000\,031\,416$
$\therefore x = \dfrac{\log_{10} 0.000\,031\,416}{\log_{10} 10}$
$\therefore x \approx -4.50$

c $9^x = 45$
$\therefore x = \dfrac{\log_{10} 45}{\log_{10} 9}$
$\therefore x \approx 1.73$

d $7^x = 0.004\,17$
$\therefore x = \dfrac{\log_{10} 0.004\,17}{\log_{10} 7}$
$\therefore x \approx -2.82$

e $(0.375)^x = 8$
$\therefore x = \dfrac{\log_{10} 8}{\log_{10} 0.375}$
$\therefore x \approx -2.12$

f $(0.9)^x = 7.29$
$\therefore x = \dfrac{\log_{10} 7.29}{\log_{10} 0.9}$
$\therefore x \approx -18.9$

61 a i Stretch with invariant x-axis, scale factor 3.
 ii $h(x) = 3g(x)$

b i Translation of $\begin{pmatrix} -2 \\ -3 \end{pmatrix}$.
 ii $h(x) = g(x + 2) - 3$

62 a As $h = 4$ and $k = 3$, $f(x) = a(x-4)^2 + 3$
But $f(0) = 0$, so $a(-4)^2 + 3 = 0$
$\therefore 16a + 3 = 0$
$\therefore 16a = -3$
$\therefore a = -\tfrac{3}{16}$
$\therefore f(x) = -\tfrac{3}{16}(x-4)^2 + 3$
or $= -\tfrac{3}{16}(x^2 - 8x + 16) + 3$
$= -\tfrac{3}{16}x^2 + \tfrac{3}{2}x$

b As $h = 6$ and $k = 16$, $f(x) = a(x-6)^2 + 16$
But $f(8) = 0$, so $a(8-6)^2 + 16 = 0$
$\therefore 4a + 16 = 0$
$\therefore 4a = -16$
$\therefore a = -4$
$\therefore f(x) = -4(x-6)^2 + 16$
or $= -4(x^2 - 12x + 36) + 16$
$= -4x^2 + 48x - 128$

c The x-intercept 3 comes from the linear factor $x - 3$.
The x-intercept 5 comes from the linear factor $x - 5$.
$\therefore f(x) = a(x-3)(x-5)$
But $f\left(\tfrac{9}{2}\right) = 2$, so $a\left(\tfrac{9}{2} - 3\right)\left(\tfrac{9}{2} - 5\right) = 2$
$\therefore a\left(\tfrac{3}{2}\right)\left(-\tfrac{1}{2}\right) = 2$
$\therefore -\tfrac{3}{4}a = 2$
$\therefore a = -\tfrac{8}{3}$
So, $f(x) = -\tfrac{8}{3}(x-3)(x-5)$

63 **a** $(0, 2)$ is a point on the cubic
$$\therefore \quad d = 2$$
$(4, 0)$ is a point on the cubic
$$\therefore \quad \tfrac{1}{10} \times 4^3 + 4c + 2 = 0$$
$$\therefore \quad 4c = -2 - 6.4$$
$$\therefore \quad c = \frac{-8.4}{4} = -2.1$$

b $(2, 4)$ is a point on the cubic
$$\therefore \quad 4 = 8a + 4b + 4 + 3$$
$$\therefore \quad 8a + 4b + 3 = 0$$
$$\therefore \quad 8a + 4b = -3 \quad \text{.... (1)}$$
$(-2, 0)$ is a point on the cubic
$$\therefore \quad 0 = -8a + 4b - 4 + 3$$
$$\therefore \quad 8a - 4b = -1 \quad \text{.... (2)}$$
Adding (1) and (2) gives
$$8a + 4b = -3$$
$$8a - 4b = -1$$
$$\overline{16a \quad\quad = -4}$$
$$\therefore \quad a = -\tfrac{1}{4}$$
Substituting in (1) gives
$$8(-\tfrac{1}{4}) + 4b = -3$$
$$\therefore \quad -2 + 4b = -3$$
$$\therefore \quad 4b = -1$$
$$\therefore \quad b = -\tfrac{1}{4}$$

64 **a**

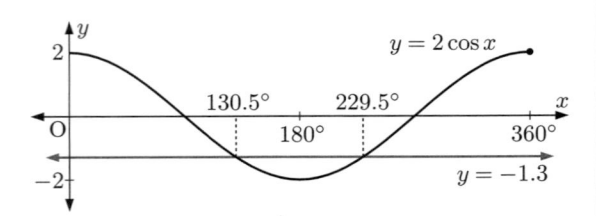

[Graph showing $y = 2\cos x$ with intersection line $y = -1.3$ at $130.5°$, $180°$, $229.5°$, $360°$]

$\therefore$ the solutions to $2\cos x = -1.3$
are $x \approx 130.5°, 229.5°$

b

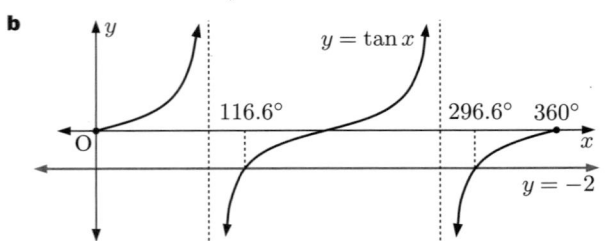

[Graph showing $y = \tan x$ with line $y = -2$ intersecting at $116.6°$, $296.6°$, $360°$]

$\therefore$ the solutions to $\tan x = -2$
are $x \approx 116.6°, 296.6°$

c

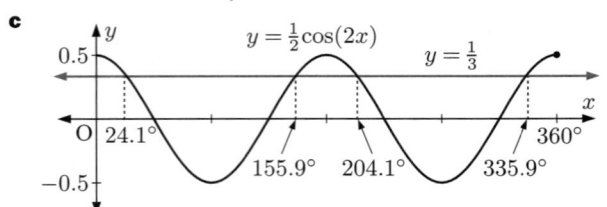

[Graph showing $y = \tfrac{1}{2}\cos(2x)$ with line $y = \tfrac{1}{3}$ intersecting at $24.1°$, $155.9°$, $204.1°$, $335.9°$, $360°$]

$\therefore$ the solutions to $\tfrac{1}{2}\cos(2x) = \tfrac{1}{3}$
are $x \approx 24.1°, 155.9°, 204.1°, 335.9°$

d

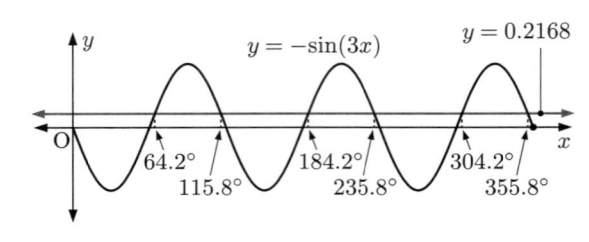

[Graph showing $y = -\sin(3x)$ with line $y = 0.2168$ intersecting at $64.2°$, $115.8°$, $184.2°$, $235.8°$, $304.2°$, $355.8°$]

$\therefore$ the solutions to $-\sin(3x) = 0.2168$
are $x \approx 64.2°, 115.8°, 184.2°, 235.8°, 304.2°, 355.8°$

65 **a** The inverse of $y = \dfrac{x-2}{x+1}$ is $x = \dfrac{y-2}{y+1}$
$$\therefore \quad x(y+1) = y - 2$$
$$\therefore \quad xy + x = y - 2$$
$$\therefore \quad xy - y = -x - 2$$
$$\therefore \quad y(x-1) = -x - 2$$
$$\therefore \quad y = -\frac{x+2}{x-1}$$
$$\therefore \quad f^{-1}(x) = -\frac{x+2}{x-1}$$

b

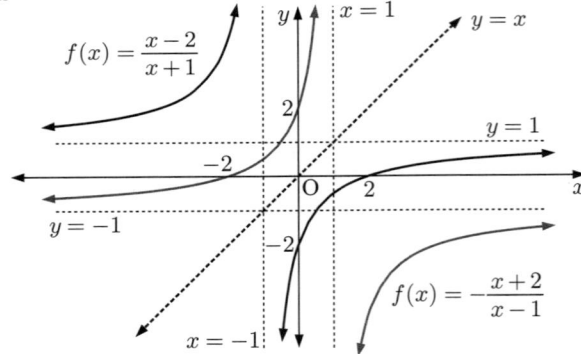

[Graph showing $f(x) = \dfrac{x-2}{x+1}$ with asymptotes $x = -1$, $y = 1$, and its inverse $f(x) = -\dfrac{x+2}{x-1}$ with asymptotes $x = 1$, $y = -1$, reflected in line $y = x$]

c $y = f^{-1}(x)$ is a reflection of $y = f(x)$ in the line $y = x$.

66 **a** $\log_7 x = 50$
$$\therefore \quad \frac{\log_{10} x}{\log_{10} 7} = 50$$
$$\therefore \quad \log_{10} x = 50 \log_{10} 7$$
$$\therefore \quad \log_{10} x = \log_{10} 7^{50}$$
$$\therefore \quad x = 7^{50}$$

b $\log_3 x = 2$
$$\therefore \quad \frac{\log_{10} x}{\log_{10} 3} = 2$$
$$\therefore \quad \log_{10} x = 2\log_{10} 3$$
$$\therefore \quad \log_{10} x = \log_{10} 3^2$$
$$\therefore \quad \log_{10} x = \log_{10} 9$$
$$\therefore \quad x = 9$$

c $\log_2 \left(\tfrac{x}{2}\right) = -2$
$$\therefore \quad \frac{\log_{10}\left(\tfrac{x}{2}\right)}{\log_{10} 2} = -2$$
$$\therefore \quad \log_{10}\left(\tfrac{x}{2}\right) = -2\log_{10} 2$$
$$\therefore \quad \log_{10}\left(\tfrac{x}{2}\right) = \log_{10}(2^{-2})$$
$$\therefore \quad \tfrac{x}{2} = \tfrac{1}{4}$$
$$\therefore \quad x = \tfrac{1}{2}$$

67 **a** **i** $h(2) = 20.2(2) - 4.9(2)^2$
$$= 20.8$$
The projectile is 20.8 m high after 2 seconds.

  ii $h(4) = 20.2(4) - 4.9(4)^2$
$$= 2.4$$
The projectile is 2.4 m high after 4 seconds.

b $h(t) = 0$ when $20.2t - 4.9t^2 = 0$
$$\therefore \quad t(20.2 - 4.9t) = 0$$
$$\therefore \quad t = 0 \text{ or } \tfrac{20.2}{4.9}$$
$\therefore$ it takes $\tfrac{20.2}{4.9} \approx 4.12$ seconds for the projectile to return to the ground.

68 **a**, **b**

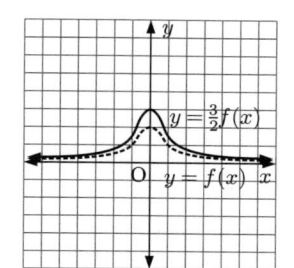

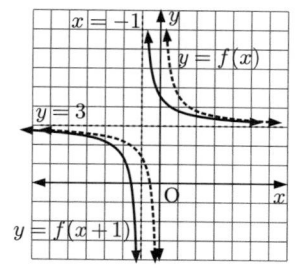

[graphs]

69 **a**

[graph of $f(x) = \dfrac{x^2 - 4}{x^2 + 2x - 3}$ with asymptotes $x = -3$, $x = 1$, $y = 1$, and y-intercept $\tfrac{4}{3}$]

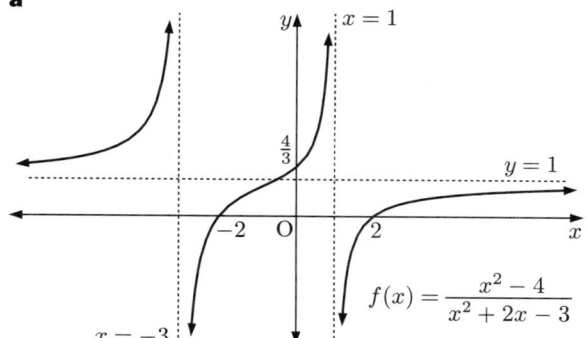

b $x = -3$ and $x = 1$ are vertical asymptotes.
$y = 1$ is a horizontal asymptote.

c When $x = 0$, $y = \dfrac{0^2 - 4}{0^2 + 2(0) - 3} = \dfrac{4}{3}$

$\therefore$ the y-intercept is $\tfrac{4}{3}$.

When $y = 0$, $x^2 - 4 = 0$ {numerator $= 0$}
$\therefore\ x^2 = 4$
$\therefore\ x = \pm 2$

$\therefore$ the x-intercepts are -2 and 2.

d There are no turning points.

70 **a** When $x = 1$, $y = 1^3 + 1^2 + 1 - 3 = 0$
$\therefore$ the function has an x-intercept of 1.

b $y = x^3 + x^2 + x - 3 = (x - 1)(x^2 + bx + c)$
Comparing constant terms, $-3 = -c$
$\therefore\ c = 3$
Comparing coefficients of x^2, $1 = b - 1$
$\therefore\ b = 2$
$\therefore\ y = (x - 1)(x^2 + 2x + 3)$

c $y = 0$ when $(x - 1)(x^2 + 2x + 3) = 0$
$\therefore\ x = 1$ or $x = \dfrac{-2 \pm \sqrt{2^2 - 4(1)(3)}}{2(1)} = \dfrac{-2 \pm \sqrt{-8}}{2}$
$\therefore$ the function has no other x-intercepts.

d

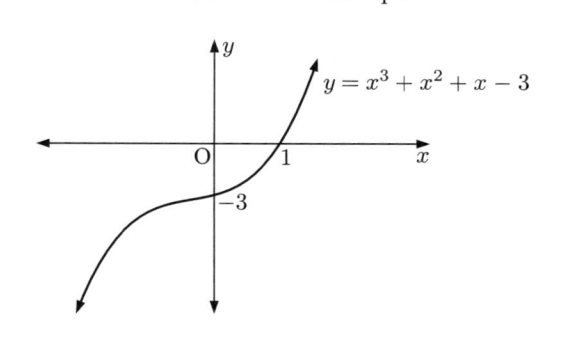

[graph of $y = x^3 + x^2 + x - 3$]

71

[graph of $W = 5 \times 2^{-0.067t}$]

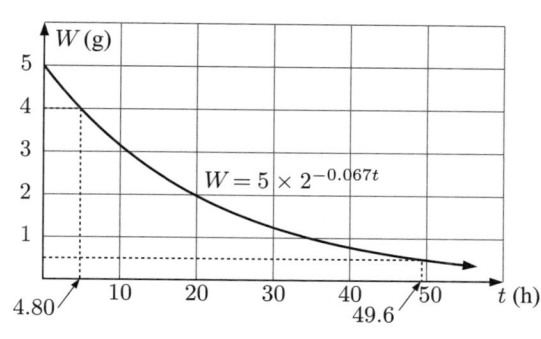

a The weight will reach 4 grams after 4.80 hours.

b The weight will reach 0.5 grams after 49.6 hours.

72 **a**

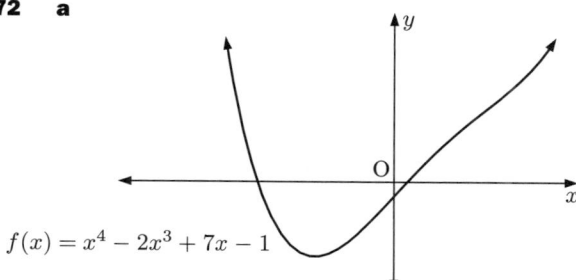

[graph of $f(x) = x^4 - 2x^3 + 7x - 1$]

b The zeros are $x \approx -1.484,\ 0.144$.

c There is a local minimum at $(-0.861,\ -5.201)$.

73 **a** $V \approx 15\,022.78 \times (0.799\,14)^t$, $r^2 \approx 1$

b **i** When $t = 5$, $V \approx 15\,022.78 \times (0.799\,14)^5$
≈ 4896
In 2005, the value of the car was approximately £4896.

ii When $t = 15$, $V \approx 15\,022.78 \times (0.799\,14)^{15}$
≈ 520
In 2015, the value of the car will be approximately £520.

c The estimate in **b i** is more reliable, as it is an interpolation.

74 **a** The image of $y = 2^x$ under the translation $\begin{pmatrix} -3 \\ 2 \end{pmatrix}$ is $y = 2^{x+3} + 2$.

b The image of $y = 3^x$ under a reflection in the line $y = x$, is $x = 3^y$ {inverse}.

75 **a** $V(10) = 10^2 - 40(10) + 500 = 200$
The value of the cabinet 10 years after its purchase is €200.

b

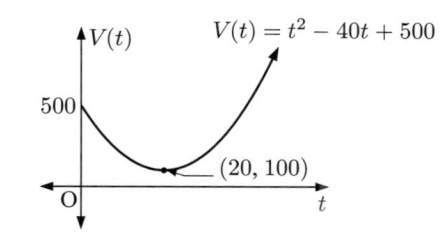

[graph of $V(t) = t^2 - 40t + 500$ with point $(20, 100)$ and y-intercept 500]

c The cabinet's value is at a minimum after 20 years. The minimum value is €100.

d No, the value of the cabinet is unlikely to continue increasing as rapidly as t gets large.

76 a

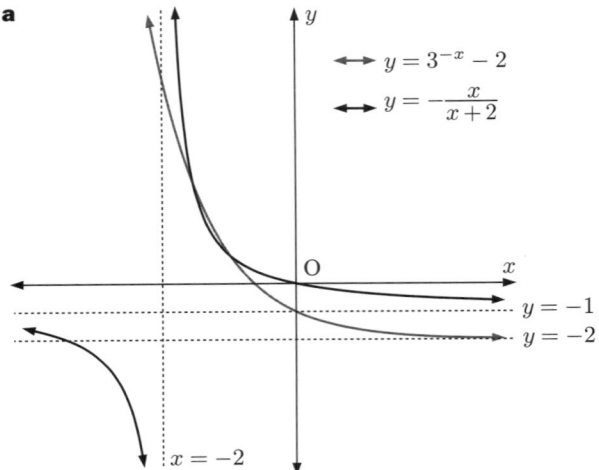

[Graph showing $y = 3^{-x} - 2$ and $y = -\dfrac{x}{x+2}$ with asymptotes $y = -1$, $y = -2$, and $x = -2$]

b Using technology, the graphs meet at $(-1.56, 3.56)$ and $(-1, 1)$.

c
$$3^{-x} \geqslant 2 - \dfrac{x}{x+2}$$
$$\therefore \ 3^{-x} - 2 \geqslant -\dfrac{x}{x+2}$$
From the graph, this occurs when $-1.56 \leqslant x \leqslant -1$.

77 a The graph has amplitude $\frac{1}{2}$ $\therefore \ a = \frac{1}{2}$.
The graph has period $120°$
$\therefore \ \dfrac{360°}{b} = 120°$ and so $b = \dfrac{360°}{120°} = 3$

b The graph has amplitude 4, and it has been reflected in the x-axis $\therefore \ a = -4$.
The graph has period $1080°$
$\therefore \ \dfrac{360°}{b} = 1080°$ and so $b = \dfrac{360°}{1080°} = \dfrac{1}{3}$

78 The graph touches the x-axis at 3,
$\therefore \ y = (px + q)(x - 3)^2$ for some p and q
$= (px + q)(x^2 - 6x + 9)$
$= px^3 - 6px^2 + 9px + qx^2 - 6qx + 9q$
$= px^3 + (q - 6p)x^2 + (9p - 6q)x + 9q$

Comparing the coefficients with
$y = 2x^3 + bx^2 + cx + 18$ gives:

- For x^3, $p = 2$
- For the constant term, $9q = 18$
$\therefore \ q = 2$
- For x^2, $b = q - 6p$
$= 2 - 6(2)$
$= -10$
- For x, $c = 9p - 6q$
$= 9(2) - 6(2)$
$= 6$

So, $b = -10$ and $c = 6$.

SOLUTIONS TO TOPIC 4: GEOMETRY

1 a obtuse angle **b** reflex angle
 c right angle **d** acute angle

2 a isosceles triangle **b** trapezium
 c kite **d** parallelogram

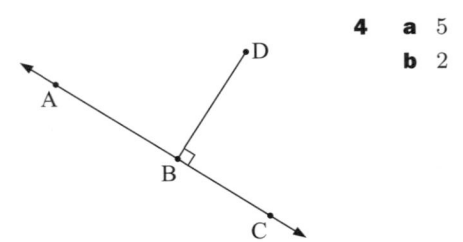

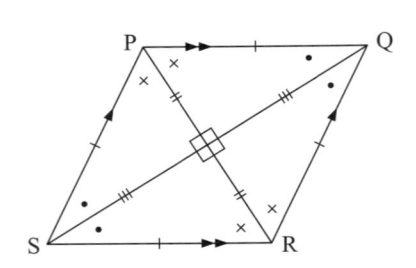

3 [Diagram showing rays from point B to A, C, D with right angle at B]

4 a 5
 b 2

5 [Diagram of rhombus PQRS with diagonals]

6 a $x^2 = 4^2 + 4^2$ {Pythagoras}
$\therefore \ x^2 = 16 + 16$
$\therefore \ x^2 = 32$
$\therefore \ x = \sqrt{32}$ {as $x > 0$}
$\therefore \ x \approx 5.66$

b $x^2 + 2.6^2 = 4.1^2$ {Pythagoras}
$\therefore \ x^2 + 6.76 = 16.81$
$\therefore \ x^2 = 10.05$
$\therefore \ x = \sqrt{10.05}$ {as $x > 0$}
$\therefore \ x \approx 3.17$

c $x^2 = 5.8^2 + 3.2^2$ {Pythagoras}
$\therefore \ x^2 = 33.64 + 10.24$
$\therefore \ x^2 = 43.88$
$\therefore \ x = \sqrt{43.88}$ {as $x > 0$}
$\therefore \ x \approx 6.62$

7 a $8^2 = 64$
and $4^2 + 7^2 = 16 + 49 = 65$
$\therefore \ 4^2 + 7^2 \neq 8^2$
So, $\{4, 7, 8\}$ is not a Pythagorean triple.

b $13^2 = 169$
and $5^2 + 12^2 = 25 + 144 = 169$
$\therefore \ 5^2 + 12^2 = 13^2$
So, $\{5, 12, 13\}$ is a Pythagorean triple.

8 The triangle has one line of symmetry.

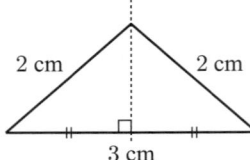

[Isosceles triangle with two sides of 2 cm and base 3 cm]

9 a The triangles have two pairs of equal sides and a pair of equal angles, but the equal angle is not the included angle. So, we cannot say that the triangles are congruent.

b The triangles are right angled, and the hypotenuses and one pair of sides are equal.
So, the triangles are congruent. {RHS}

10 a $x = 115$ {corresponding angles}
$y + 115 = 180$ {co-interior angles}
$\therefore \ y = 65$

b $\quad y = 100 \quad$ {vertically opposite angles}
$\quad x + 100 = 180 \quad$ {angles on a line}
$\quad \therefore \ x = 80$

c $\quad x + 20 = 90 \quad$ {right angle}
$\quad \therefore \ x = 70$
$\quad y + 40 = 180 \quad$ {angles on a line}
$\quad \therefore \ y = 140$

d $\quad 2x + 3x + 4x = 360 \quad$ {angles at a point}
$\quad \therefore \ 9x = 360$
$\quad \therefore \ x = 40$

11 a The 3 cm side in the small figure has been enlarged to 4.5 cm in the large figure.
$\quad \therefore \ $ scale factor $k = \dfrac{4.5}{3} = 1.5$

b $\quad \dfrac{x}{6} = k$
$\quad \therefore \ \dfrac{x}{6} = 1.5$
$\quad \therefore \ x = 9$

12 a Alternate angles are equal.
$\quad \therefore \ $ AB is parallel to CD.

b $\quad 110° + 110° \neq 180°$
$\quad \therefore \ $ co-interior angles are not supplementary.
$\quad \therefore \ $ AB is not parallel to CD.

13 a $\quad (2x)^2 + x^2 = 85^2 \quad$ {Pythagoras}
$\quad \therefore \ 4x^2 + x^2 = 85^2$
$\quad \therefore \ 5x^2 = 7225$
$\quad \therefore \ x^2 = 1445$
$\quad \therefore \ x = \sqrt{1445} \quad$ {as $x > 0$}
$\quad \quad = \sqrt{289}\sqrt{5}$
$\quad \quad = 17\sqrt{5}$

b $\quad (x+3)^2 + (x-3)^2 = 12^2 \quad$ {Pythagoras}
$\quad \therefore \ x^2 + 6x + 9 + x^2 - 6x + 9 = 144$
$\quad \therefore \ 2x^2 = 126$
$\quad \therefore \ x^2 = 63$
$\quad \therefore \ x = \sqrt{63} \quad$ {as $x > 0$}
$\quad \quad = \sqrt{9}\sqrt{7}$
$\quad \quad = 3\sqrt{7}$

c The two triangles share a common hypotenuse.
$\quad \therefore \ x^2 + x^2 = 5^2 + 2^2 \quad$ {Pythagoras}
$\quad \therefore \ 2x^2 = 29$
$\quad \therefore \ x^2 = \dfrac{29}{2}$
$\quad \therefore \ x = \sqrt{\dfrac{29}{2}} \quad$ {as $x > 0$}

14 a $\quad x = 90 + 36 \quad$ {exterior angle}
$\quad \therefore \ x = 126$

b $\quad x + 65 + 35 = 180 \quad$ {angles of a triangle}
$\quad \therefore \ x = 80$

c $\quad x + x + 110 = 180 \quad$ {angles of isosceles triangle}
$\quad \therefore \ 2x + 110 = 180$
$\quad \therefore \ 2x = 70$
$\quad \therefore \ x = 35$

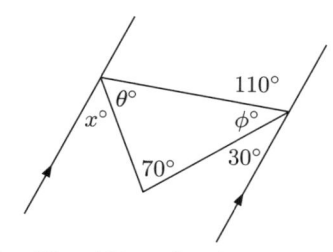

d

$\theta + \phi + 70 = 180 \quad$ {angles of a triangle}
$\therefore \ \theta + \phi = 110 \quad ....$ (1)
Also, $(x + \theta) + (\phi + 30) = 180 \quad$ {co-interior angles}
$\therefore \ x + (\theta + \phi) + 30 = 180$
$\therefore \ x + 110 + 30 = 180 \quad$ {using (1)}
$\therefore \ x = 40$

15 The figures are similar, $\quad \therefore \ \dfrac{x}{7} = \dfrac{8}{5}$
$\quad \therefore \ x = \dfrac{8}{5} \times 7$
$\quad \therefore \ x = \dfrac{56}{5}$ or 11.2

16

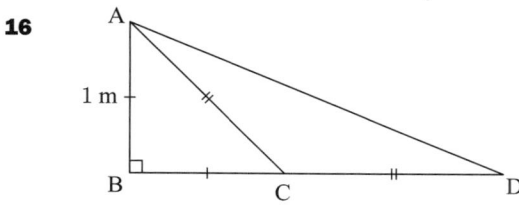

In $\triangle$ABC, $AC^2 = 1^2 + 1^2 \quad$ {Pythagoras}
$\quad \therefore \ AC^2 = 2$
$\quad \therefore \ AC = \sqrt{2} \quad$ {AC > 0}
So, $\quad CD = AC = \sqrt{2}$
In $\triangle$ABD, $AD^2 = 1^2 + BD^2 \quad$ {Pythagoras}
$\quad \therefore \ AD^2 = 1 + (1 + \sqrt{2})^2$
$\quad \therefore \ AD^2 \approx 6.828$
$\quad \therefore \ AD \approx 2.61 \quad$ {AD > 0}
So, AD is approximately 2.61 m long.

17 a $\quad 2x + (3x - 10) + (x + 40) = 180 \quad$ {angles of a triangle}
$\quad \therefore \ 6x + 30 = 180$
$\quad \therefore \ 6x = 150$
$\quad \therefore \ x = 25$

b Given $x = 25$, the angle at A is $\quad 2(25)° = 50°$
the angle at B is $\quad 3(25)° - 10° = 65°$
the angle at C is $\quad 25° + 40° = 65°$
So, the triangle is isosceles.

c The shortest side is opposite the smallest angle. The smallest angle is at A, so the shortest side is BC.

18 a Any two equilateral triangles are equiangular (both have angles of 60°, 60°, 60°).
$\quad \therefore \ $ all equilateral triangles are similar
$\quad \therefore \ $ the statement is true.

b Consider the following rhombuses:

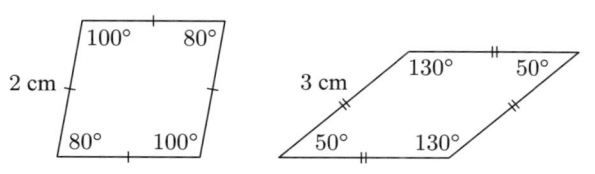

These rhombuses are not equiangular, so they are not similar.
$\quad \therefore \ $ the statement is false.

19 a $x + 71 + 71 = 180$ {angles of isosceles triangle}
  $\therefore\ x + 142 = 180$
  $\therefore\ x = 38$

  b $\theta + 112 = 180$
  {angles on a line}
  $\therefore\ \theta = 68$
  $\therefore\ \phi = 68$
  {isosceles triangle}
  $\therefore\ x + 68 + 68 = 180$
  {angles of a triangle}
  $\therefore\ x = 44$

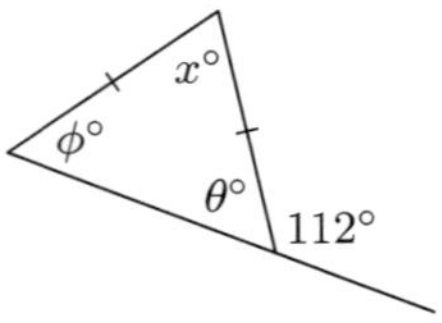

20 $(25k)^2 = 625k^2$
  and $(7k)^2 + (24k)^2 = 49k^2 + 576k^2 = 625k^2$
  $\therefore\ (7k)^2 + (24k)^2 = (25k)^2$ for all k
  So, $\{7k, 24k, 25k\}$ is a Pythagorean triple for all $k \in \mathbb{Z}^+$.

21 $(4x - 16) + (3x + 3) + (2x + 22) = 180$ {angles of a triangle}
  $\therefore\ 9x + 9 = 180$
  $\therefore\ 9x = 171$
  $\therefore\ x = 19$
  So, the angle at A is $4(19)° - 16° = 60°$
    the angle at B is $3(19)° + 3° = 60°$
    the angle at C is $2(19)° + 22° = 60°$
  $\therefore$ the triangle is equilateral.

22 $66.6^2 = 4435.56$
  and $54.2^2 + 38.7^2 = 4435.33$
  $\therefore\ 54.2^2 + 38.7^2 \approx 66.6^2$
  So, it appears that the corner of the frame is a right angle, and thus the frame is rectangular.

23 A hexagon has 6 sides.
  $\therefore$ the sum of the interior angles of a hexagon is
  $(6 - 2) \times 180°$
  $= 4 \times 180°$
  $= 720°$

24 a $\widehat{ABC} = \widehat{DEC} = 28°$ {given}
  $\widehat{ACB} = \widehat{DCE}$ {vertically opposite}
  $\therefore$ triangles ABC and DEC are similar.

  b $\alpha_1 = \alpha_2$
  {alternate angles}
  $\beta_1 = \beta_2$
  {vertically opposite}
  $\therefore$ triangles ABC and EDC are similar.

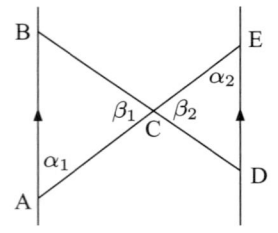

25 a

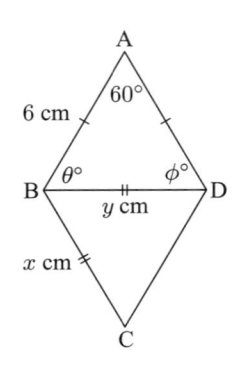

b $\theta = \phi$ {base angles of isosceles triangle}
  $\therefore\ \theta + \theta + 60 = 180$ {angles of a triangle}
  $\therefore\ 2\theta = 120$
  $\therefore\ \theta = 60$
  So, $\theta = \phi = 60$, hence triangle ABD is equilateral.
  $\therefore\ y = 6$
  $\therefore\ x = y = 6$

b

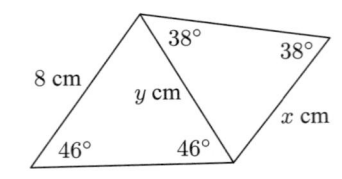

  $y = 8$ {isos. triangle with base angles $46°$}
  $\therefore\ x = y = 8$ {isos. triangle with base angles $38°$}

26

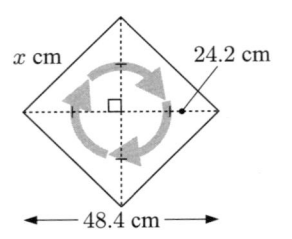

Let the sign have side length x cm.
  $\therefore\ x^2 = 24.2^2 + 24.2^2$ {Pythagoras}
  $\therefore\ x^2 = 1171.28$
  $\therefore\ x = \sqrt{1171.28}$ {as $x > 0$}
  $\therefore\ x \approx 34.2$
So, the sign has side lengths of 34.2 cm.

27 a The figure has 5 sides, so the sum of the interior angles is $(5 - 2) \times 180° = 540°$.
  $\therefore\ x + x + x + x + 90 = 540$
  $\therefore\ 4x + 90 = 540$
  $\therefore\ 4x = 450$
  $\therefore\ x = 112.5$

  b $2x + x + 81 + 90 = 360$ {angles of a quadrilateral}
  $\therefore\ 3x + 171 = 360$
  $\therefore\ 3x = 189$
  $\therefore\ x = 63$

28 a

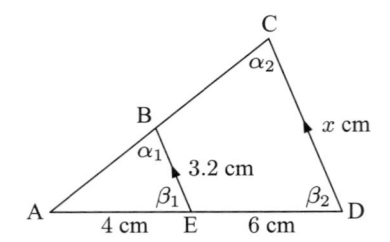

  $\alpha_1 = \alpha_2$ {corresponding angles}
  $\beta_1 = \beta_2$ {corresponding angles}
  So, $\triangle$s ABE and ACD are similar and
  $\dfrac{CD}{BE} = \dfrac{AD}{AE}$ {same ratio}
  $\therefore\ \dfrac{x}{3.2} = \dfrac{10}{4}$
  $\therefore\ x = \dfrac{10}{4} \times 3.2$
  $\therefore\ x = 8$

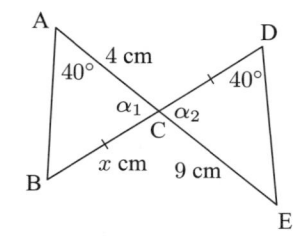

b

$\hat{BAC} = \hat{EDC}$ {given}
$\alpha_1 = \alpha_2$ {vertically opposite}
So, △s ABC and DEC are similar and
$$\frac{BC}{EC} = \frac{AC}{DC} \quad \text{{same ratio}}$$
$$\therefore \frac{x}{9} = \frac{4}{x}$$
$$\therefore x^2 = 36$$
$$\therefore x = 6 \quad \{\text{as } x > 0\}$$

29 a $\hat{ABC}$ measures $90°$ {angle in semi-circle}
$\therefore 4x + 5x + 90 = 180$ {angles of a triangle}
$\therefore 9x = 90$
$\therefore x = 10$

b $BP = AP = 12$ cm {tangents from external point}
$\hat{OBP} = 90°$ {radius-tangent}
$\therefore$ in △OBP, $OB^2 + BP^2 = OP^2$ {Pythagoras}
$\therefore x^2 + 12^2 = 15^2$
$\therefore x^2 + 144 = 225$
$\therefore x^2 = 81$
$\therefore x = 9 \quad \{\text{as } x > 0\}$

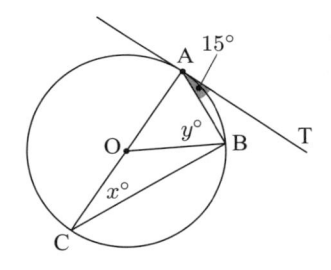

c

$\hat{OAT} = 90°$ {radius-tangent}
$\therefore \hat{OAB} = (90 - 15)° = 75°$
Also, $OA = OB$ {equal radii}
$\therefore y = 75$ {base angles of isos. triangle}
Now $\hat{ABC} = 90°$ {angle in a semi-circle}
$\therefore 75 + 90 + x = 180$ {angles in △ABC}
$\therefore x = 15$

30

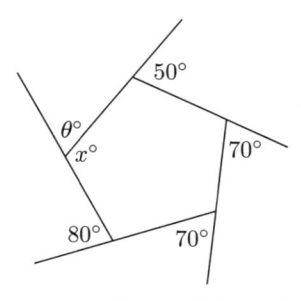

The sum of the exterior angles is $360°$.
$\therefore \theta + 50 + 70 + 70 + 80 = 360$
$\therefore \theta + 270 = 360$
$\therefore \theta = 90$
So, $x = 180 - 90 = 90$ {angles on a line}

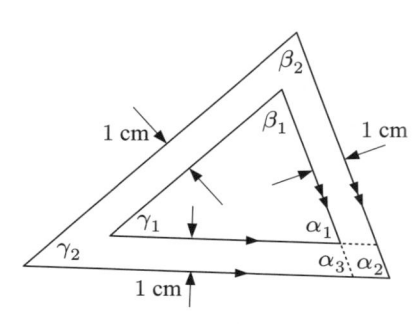

31

$\alpha_1 = \alpha_3$ {corresponding angles}
$\alpha_3 = \alpha_2$ {corresponding angles}
So, $\alpha_1 = \alpha_2$.
Likewise, $\beta_1 = \beta_2$ and $\gamma_1 = \gamma_2$.
$\therefore$ the triangles are equiangular and hence similar.

32 a The two shorter sides have lengths 8 cm and 15 cm.
Now $8^2 + 15^2 = 64 + 225 = 289$
and $17^2 = 289$
$\therefore 8^2 + 15^2 = 17^2$ and hence the triangle is right angled.

b The two shorter sides have lengths 5.0 m and 6.4 m.
Now $5.0^2 + 6.4^2 = 25 + 40.96 = 65.96$
but $8.0^2 = 64$
$\therefore 5.0^2 + 6.4^2 \neq 8.0^2$ and hence the triangle is not right angled.

33 a scale factor $k = \frac{15}{6}$
area of large triangle $= k^2 \times$ (area of small triangle)
$\therefore A = k^2 \times 16$
$\therefore A = \left(\frac{15}{6}\right)^2 \times 16$
$\therefore A = 100$

b large area $= k^2 \times$ small area
$\therefore 196 = k^2 \times 144$
$\therefore k^2 = \frac{196}{144}$
$\therefore k = \frac{14}{12} \quad \{k > 0\}$
Now, $x = k \times 36$
$\therefore x = \frac{14}{12} \times 36$
$\therefore x = 42$

34

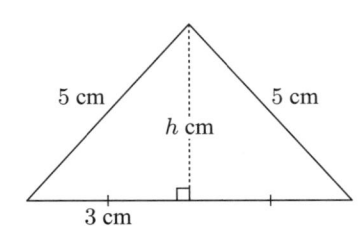

Since the triangle is isosceles, the altitude bisects the base.
$\therefore h^2 + 3^2 = 5^2$ {Pythagoras}
$\therefore h^2 + 9 = 25$
$\therefore h^2 = 16$
$\therefore h = 4 \quad \{h > 0\}$
$\therefore$ the triangle has area $= \frac{1}{2} \times 6 \times 4$
$= 12$ cm^2

35 **a** $A\hat{O}B = 2 \times 35° = 70°$ {angle at the centre}
Now $OA = OB$ {equal radii}
$\therefore O\hat{A}B = x°$ {base angles of isos. triangle}
$\therefore x + x + 70 = 180$ {angles in $\triangle OAB$}
$\therefore 2x = 110$
$\therefore x = 55$

b $x = 45$ {angles on same arc}

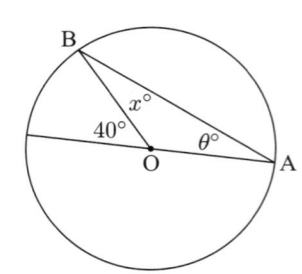

c

$\theta = \frac{1}{2} \times 40 = 20$ {angle at the centre}
$OA = OB$ {equal radii}
$\therefore x = \theta$ {base angles of isos. triangle}
$\therefore x = 20$

36 For other polygons, it is possible that one of these conditions is satisfied, but not the other.
For example,

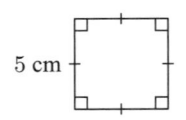

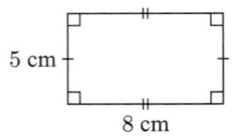

are equiangular, but the sides are not in proportion, so they are not similar.
Also,

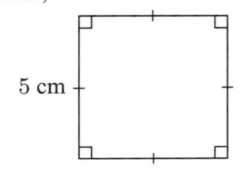

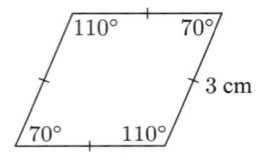

have their sides in proportion, but are not equiangular, so they are not similar.

37

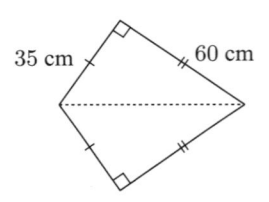

area of kite = area of 2 triangles
$= 2 \times (\frac{1}{2} \times 60 \times 35)$
$= 2100$ cm^2

38 A pentagon has 5 sides.
$\therefore$ the sum of the interior angles is $(5-2) \times 180° = 540°$.
$\therefore$ each angle of a regular pentagon has size $\frac{540°}{5} = 108°$.

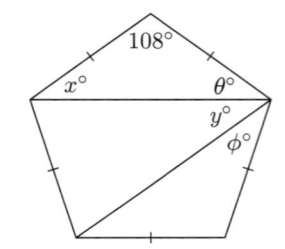

$x = \theta$ {base angles of isosceles triangle}
$\therefore x + x + 108 = 180$ {angles of a triangle}
$\therefore 2x = 72$
$\therefore x = 36$
So, $\theta = 36$, and by symmetry, $\phi = 36$ also.
$\therefore 36 + y + 36 = 108$ {angle of regular pentagon}
$\therefore y + 72 = 108$
$\therefore y = 36$

39

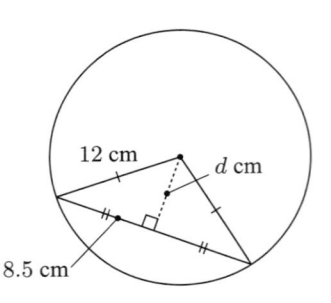

The perpendicular from the chord to the centre bisects the chord.
$\therefore d^2 + 8.5^2 = 12^2$ {Pythagoras}
$\therefore d^2 + 72.25 = 144$
$\therefore d^2 = 71.75$
$\therefore d \approx 8.47$ {$d > 0$}
So, the shortest distance from the centre of the circle to the chord is 8.47 cm.

40 Using the lengths of the bases, the scale factor $k = \frac{15}{9}$.
Now $h = k \times 15$
$\therefore h = \frac{15}{9} \times 15 = 25$
and large volume $= k^3 \times$ small volume
$\therefore 2000 = k^3 \times V$
$\therefore 2000 = \left(\frac{15}{9}\right)^3 \times V$
$\therefore V = 2000 \times \left(\frac{9}{15}\right)^3$
$\therefore V = 432$

41

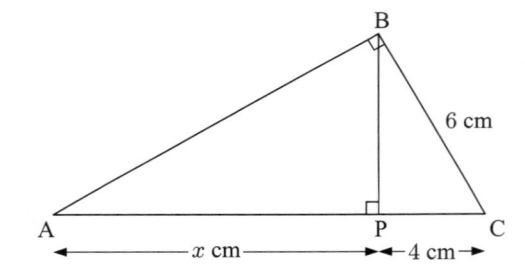

In $\triangle BPC$, $BP^2 + 4^2 = 6^2$ {Pythagoras}
$\therefore BP^2 + 16 = 36$
$\therefore BP^2 = 20$
In $\triangle ABP$, $AB^2 = AP^2 + BP^2$ {Pythagoras}
$\therefore AB^2 = x^2 + 20$
In $\triangle ABC$, $AC^2 = AB^2 + BC^2$ {Pythagoras}
$\therefore (x+4)^2 = (x^2 + 20) + 6^2$
$\therefore x^2 + 8x + 16 = x^2 + 56$
$\therefore 8x = 40$
$\therefore x = 5$

42 **a** Sum of opposite angles $= 90° + 90° = 180°$.
$\therefore$ PQRS is a cyclic quadrilateral
{supplementary opposite angles}

b $3x + 2x + 72 + 108 = 360$ {angles of a quadrilateral}
$$\therefore \ 5x + 180 = 360$$
$$\therefore \ 5x = 180$$
$$\therefore \ x = 36$$

So, the angle at P is $3(36)° = 108°$
and the angle at Q is $2(36)° = 72°$
$\therefore$ sum of opposite angles $= 108° + 72° = 180°$
$\therefore$ PQRS is a cyclic quadrilateral
{supplementary opposite angles}

43

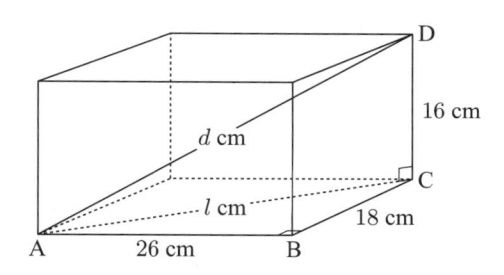

In $\triangle ABC$, $l^2 = 26^2 + 18^2$ {Pythagoras}
In $\triangle ACD$, $d^2 = l^2 + 16^2$ {Pythagoras}
$\therefore \ d^2 = 26^2 + 18^2 + 16^2$
$\therefore \ d^2 = 1256$
$\therefore \ d \approx 35.44$ $\{d > 0\}$

So, the diagonal of the parcel is 35.44 cm long.
$\therefore$ the cost to courier the parcel is
€$0.60 \times 35.44 \approx$ €21.26

44 a

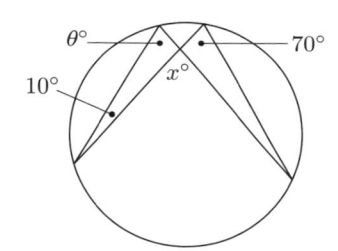

$\theta = 70$ {angles on same arc}
$x = \theta + 10$ {exterior angle of triangle}
$\therefore \ x = 70 + 10 = 80$

b $x = \frac{1}{2} \times 60 = 30$ {angle at the centre}
$y = x = 30$ {angles on same arc}

c $a = 120$ {angles on same arc}
$b = 2 \times 120 = 240$ {angle at the centre}

45

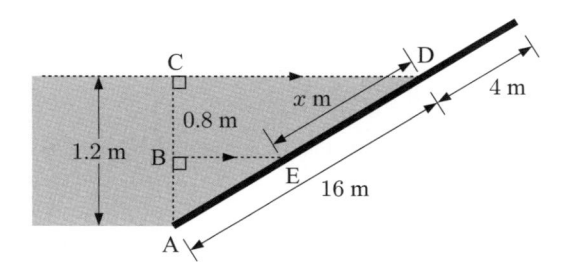

$A\hat{B}E = A\hat{C}D = 90°$, and angle A is common to both triangles.
$\therefore$ triangles ABE and ACD are similar, and

$\dfrac{AB}{AC} = \dfrac{AE}{AD}$ {same ratio}
$\therefore \ \dfrac{0.4}{1.2} = \dfrac{16 - x}{16}$
$\therefore \ 6.4 = 19.2 - 1.2x$
$\therefore \ 1.2x = 12.8$
$\therefore \ x \approx 10.7$

So, Chris needs to reverse $10.7 + 4 = 14.7$ m down the ramp.

46

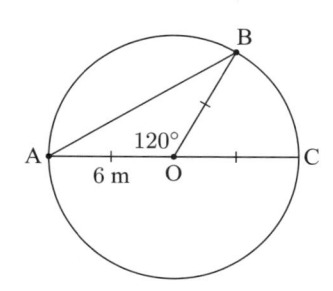

$B\hat{O}C = (180 - 120)° = 60°$ {angles on a line}
Also, $O\hat{B}C = O\hat{C}B = 60°$ {base angles of isos. $\triangle$}
$\therefore \ \triangle OBC$ is equilateral, so $BC = 6$ m.
Now $A\hat{B}C = 90°$ {angle in a semi-circle}
$\therefore$ in $\triangle ABC$, $AB^2 + BC^2 = AC^2$ {Pythagoras}
$\therefore \ AB^2 + 6^2 = 12^2$
$\therefore \ AB^2 + 36 = 144$
$\therefore \ AB^2 = 108$
$\therefore \ AB = \sqrt{108}$ $\{AB > 0\}$
$= \sqrt{36} \times \sqrt{3}$
$= 6\sqrt{3}$ m

47 a

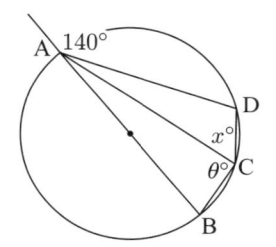

$\theta = 90$ {angle in a semi-circle}
ABCD is a cyclic quadrilateral.
$\therefore \ x + \theta = 140$ {exterior angle of cyclic quad.}
$\therefore \ x + 90 = 140$
$\therefore \ x = 50$

b

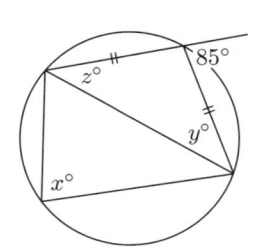

$x = 85$ {exterior angle of cyclic quadrilateral}
Now $z = y$ {base angles of isosceles triangle}
$\therefore \ y + y = 85$ {exterior angle of triangle}
$\therefore \ 2y = 85$
$\therefore \ y = 42.5$

c $x + 76 = 180$ {co-interior angles}
  $\therefore\ x = 104$
  Now $x + y = 180$ {opp. angles of cyclic quad.}
  $\therefore\ y = 180 - x = 76$

d $p = 2 \times 35 = 70$ {angle at the centre}
  $35 + q = 180$ {opposite angles of cyclic quad.}
  $\therefore\ q = 145$

e

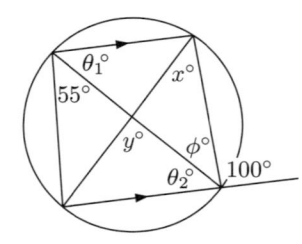

  $x = 55$ {angles on same arc}
  $55 + \theta_1 = 100$ {ext. angle of cyclic quad.}
  $\therefore\ \theta_1 = 45$
  $\therefore\ \theta_2 = \theta_1 = 45$ {alternate angles}
  $\therefore\ 45 + \phi + 100 = 180$ {angles on a line}
  $\therefore\ \phi = 35$
  Now $y = x + \phi$ {exterior angle of triangle}
  $\therefore\ y = 55 + 35 = 90$

48 a area $= \frac{1}{2} \times 3 \times 4 \times \sin 50° \approx 4.60$ cm^2

b $\widehat{BAC} = \widehat{DEC}$ {alternate angles}
  $\widehat{ACB} = \widehat{ECD}$ {vertically opposite}
  $\therefore\ \triangle$s ABC and EDC are similar, with scale factor
  $k = \dfrac{ED}{AB} = \dfrac{6}{4}$
  $\therefore\ $ area of CDE $= k^2 \times $ (area of ABC)
  $\approx \left(\dfrac{6}{4}\right)^2 \times 4.60 \approx 10.3$ cm^2

49

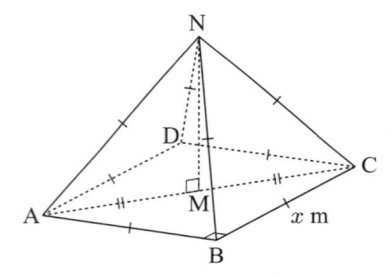

Let M be the point on the base directly beneath the apex N.
In $\triangle$ABC, $AC^2 = x^2 + x^2$ {Pythagoras}
  $\therefore\ AC^2 = 2x^2$
  $\therefore\ AC = x\sqrt{2}$ {AC > 0}
  $\therefore\ AM = \dfrac{x\sqrt{2}}{2} = \dfrac{x}{\sqrt{2}}$
In $\triangle$AMN, $NM^2 + AM^2 = AN^2$ {Pythagoras}
  $\therefore\ NM^2 + \left(\dfrac{x}{\sqrt{2}}\right)^2 = x^2$
  $\therefore\ NM^2 + \dfrac{x^2}{2} = x^2$
  $\therefore\ NM^2 = \dfrac{x^2}{2}$
  $\therefore\ NM = \dfrac{x}{\sqrt{2}}$ {NM > 0}

So, the height of the pyramid is $\dfrac{x}{\sqrt{2}}$ m.

50 a cross-section area $= \left(\dfrac{60 + 40}{2}\right) \times 30$
  $= 1500$ cm^2
  $\therefore\ $ volume of trough $= 1500 \times 200$ cm^3
  $= 300\,000$ cm^3
  $\therefore\ $ capacity of trough $= 300\,000$ ml
  $= 300$ litres

b

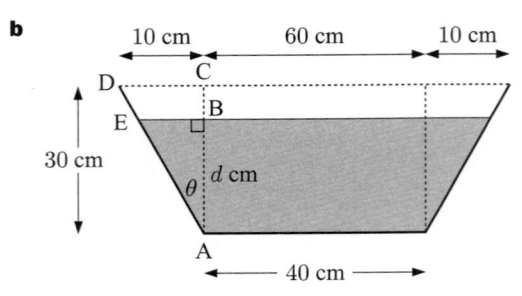

Suppose the trough is filled to a depth of d cm.
$\widehat{ABE} = \widehat{ACD} = 90°$, and θ is common to triangles ABE and ACD. So, the triangles are similar, and
  $\dfrac{BE}{CD} = \dfrac{AB}{AC}$ {same ratio}
  $\therefore\ \dfrac{BE}{10} = \dfrac{d}{30}$
  $\therefore\ BE = \dfrac{d}{3}$
$\therefore\ $ the cross-section area of the grain is
  $40 \times d + 2 \times \left(\dfrac{1}{2} \times d \times \dfrac{d}{3}\right)$
  $= 40d + \dfrac{1}{3}d^2$ cm^2
$\therefore\ $ the volume of the grain is
  $\left(40d + \dfrac{1}{3}d^2\right) \times 200$ cm^3
  $= 8000d + \dfrac{200}{3}d^2$ cm^3
Now, the volume of grain is 0.2 m^3 $= 200\,000$ cm^3
  $\therefore\ \dfrac{200}{3}d^2 + 8000d = 200\,000$
  $\therefore\ d^2 + 120d = 3000$ $\{\times \dfrac{3}{200}\}$
  $\therefore\ d^2 + 120d - 3000 = 0$
  $\therefore\ d \approx -141$ or 21.2 {technology}
But $d > 0$, so $d \approx 21.2$
$\therefore\ $ the grain is 21.2 cm deep.

51

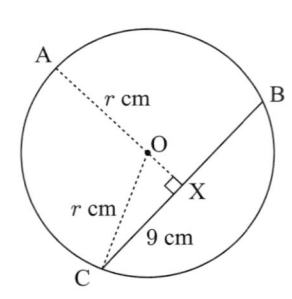

OX is perpendicular to CB, and so OX bisects CB.
  $\therefore\ CX = BX = 9$ cm
Now $OX = \frac{3}{8}AX$
  $\therefore\ OX = \frac{3}{8}(OX + r)$
  $\therefore\ OX = \frac{3}{8}OX + \frac{3}{8}r$
  $\therefore\ \frac{5}{8}OX = \frac{3}{8}r$
  $\therefore\ OX = \frac{3}{5}r$

$\therefore$ in $\triangle$OXC, $OC^2 = OX^2 + CX^2$ {Pythagoras}
$$\therefore r^2 = \left(\tfrac{3}{5}r\right)^2 + 9^2$$
$$\therefore r^2 = \tfrac{9}{25}r^2 + 81$$
$$\therefore \tfrac{16}{25}r^2 = 81$$
$$\therefore \tfrac{4}{5}r = 9 \quad \{r > 0\}$$
$$\therefore r = \tfrac{45}{4}$$
$\therefore$ the radius of the circle is 11.25 cm.

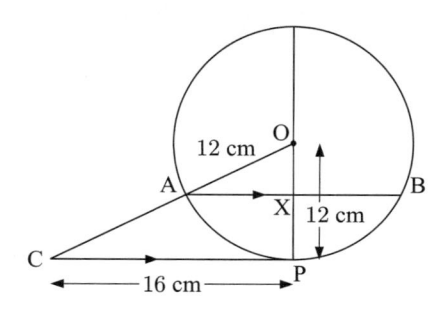

52

a $C\hat{P}O = 90°$ {radius-tangent}
$\therefore$ in $\triangle$CPO, $CO^2 = CP^2 + PO^2$ {Pythagoras}
$$\therefore CO^2 = 16^2 + 12^2$$
$$\therefore CO^2 = 400$$
$$\therefore CO = 20 \text{ cm} \quad \{CO > 0\}$$

b $A\hat{X}O = C\hat{P}O = 90°$ {corresponding angles}
$X\hat{A}O = P\hat{C}O$ {corresponding angles}
$\therefore$ triangles AXO and CPO are similar, and
$$\frac{XO}{PO} = \frac{AX}{CP} \quad \{\text{same ratio}\}$$
$$\therefore \frac{XO}{12} = \frac{AX}{16}$$
$$\therefore XO = \tfrac{3}{4}AX \quad .... (1)$$
In $\triangle$AXO, $AX^2 + XO^2 = AO^2$ {Pythagoras}
$$\therefore AX^2 + \left(\tfrac{3}{4}AX\right)^2 = 12^2$$
$$\therefore \tfrac{25}{16}AX^2 = 12^2$$
$$\therefore \tfrac{5}{4}AX = 12 \quad \{AX > 0\}$$
$$\therefore AX = 9.6$$
Now OX is perpendicular to AB, so OX bisects AB.
$\therefore$ AB $= 2 \times 9.6 = 19.2$ cm

c PX = PO − XO
$= 12 - \tfrac{3}{4}AX$ {using (1)}
$= 12 - \tfrac{3}{4} \times 9.6$
$= 4.8$
$\therefore$ the distance between the parallel line segments is 4.8 cm.

53 The solid with volume 30 cm³ is enlarged to the solid with volume 50 cm³.
$$\therefore 50 = k^3 \times 30$$
$$\therefore \tfrac{5}{3} = k^3$$
$$\therefore k = \sqrt[3]{\tfrac{5}{3}} \approx 1.1856$$
new surface area $= k^2 \times$ old surface area
$\approx 1.1856^2 \times 56$
≈ 78.7 cm²

54

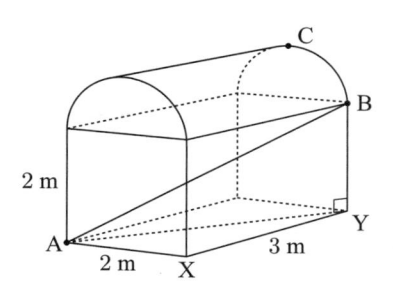

In $\triangle$AXY, $AY^2 = 2^2 + 3^2$ {Pythagoras}
In $\triangle$AYB, $AB^2 = AY^2 + BY^2$ {Pythagoras}
$$\therefore AB^2 = 2^2 + 3^2 + 2^2$$
$$\therefore AB^2 = 17$$
$$\therefore AB = \sqrt{17} \text{ m} \quad \{AB > 0\}$$

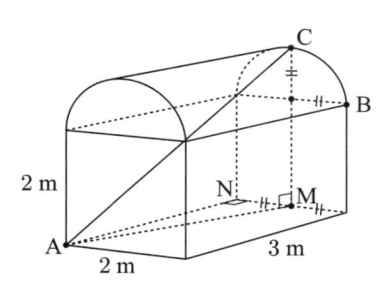

In $\triangle$ANM, $AM^2 = AN^2 + NM^2$ {Pythagoras}
$$\therefore AM^2 = 3^2 + 1^2$$
In $\triangle$AMC, $AC^2 = AM^2 + MC^2$ {Pythagoras}
$$\therefore AC^2 = 3^2 + 1^2 + 3^2$$
$$\therefore AC^2 = 19$$
$$\therefore AC = \sqrt{19} \text{ m} \quad \{AC > 0\}$$
$\sqrt{19} > \sqrt{17}$, so AC is further than AB.

55

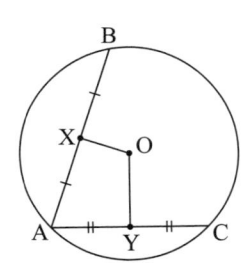

OX passes through the circle's centre and bisects AB, so OX is perpendicular to AB.
$$\therefore A\hat{X}O = 90°$$
Likewise, $A\hat{Y}O = 90°$
$\therefore A\hat{X}O + A\hat{Y}O = 90° + 90° = 180°$
$\therefore$ OXAY is a cyclic quadrilateral
{supplementary opposite angles}

56

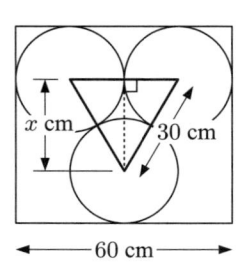

Two cans fit exactly along the 60 cm length of the box, so the diameter of each can is 30 cm.

∴ the centres of the cans form an equilateral triangle with side length 30 cm.
∴ $x^2 + 15^2 = 30^2$ {Pythagoras}
∴ $x^2 + 225 = 900$
∴ $x^2 = 675$
∴ $x \approx 26.0$ {$x > 0$}

So, the width of the box $\approx 15 + 26.0 + 15$
≈ 56.0 cm

57

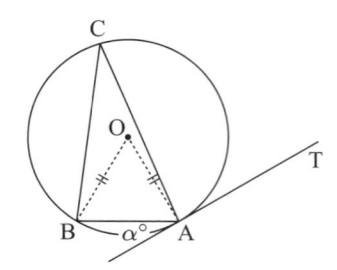

a $\quad O\hat{A}T = 90°$ {radius-tangent}
$\therefore O\hat{A}B = (90 - \alpha)°$
$\therefore O\hat{B}A = (90 - \alpha)°$ {base angles of isos. triangle}
$\therefore (90 - \alpha)° + (90 - \alpha)° + A\hat{O}B = 180°$
{angles of a triangle}
$\therefore 180° - 2\alpha° + A\hat{O}B = 180°$
$\therefore A\hat{O}B = 2\alpha°$

b $\quad A\hat{O}B = 2\alpha°$
$\therefore A\hat{C}B = \frac{1}{2} \times 2\alpha°$ {angle at the centre}
$= \alpha°$

c The angle between a tangent and a chord is equal to the angle subtended by the chord at the circle.

58

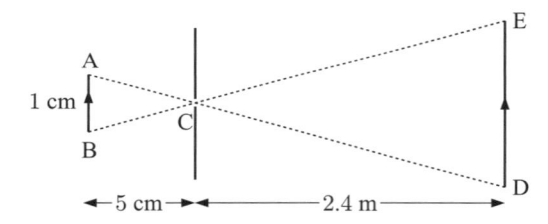

$A\hat{B}C = D\hat{E}C$ {alternate angles}
$B\hat{A}C = E\hat{D}C$ {alternate angles}
∴ △s ABC and DEC are similar.
△DEC is an enlargement of △ABC, with scale factor
$k = \frac{240}{5} = 48$ {2.4 m = 240 cm}
Now $DE = k \times AB$
$\therefore DE = 48 \times 1$
$= 48$ cm
∴ the beam of light on the screen is 48 cm wide.

59

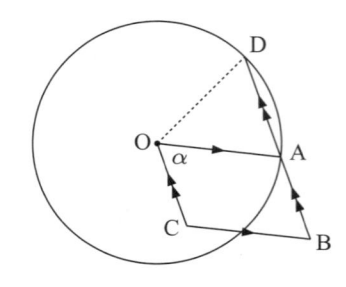

Let $A\hat{O}C = \alpha$
∴ $O\hat{A}D = \alpha$ {alternate angles}
∴ $O\hat{D}A = \alpha$ {base angles of isoc. △OAD}
Also, $O\hat{C}B = 180° - \alpha$ {co-interior angles}
∴ $O\hat{D}B + O\hat{C}B = \alpha + (180° - \alpha) = 180°$
∴ DOCB is a cyclic quadrilateral
{supplementary opposite angles}

SOLUTIONS TO TOPIC 5: TRANSFORMATIONS IN TWO DIMENSIONS

1 a 1 cm ≡ 10 m/s   **b** 1 cm ≡ 10 km/h

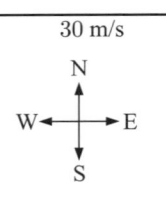

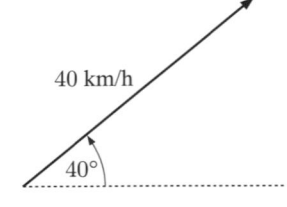

c 1 cm ≡ 10 N   **d** 1 cm ≡ 10 m

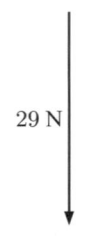

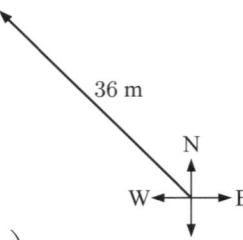

2 a $\overrightarrow{QO} = \begin{pmatrix} 0 - -4 \\ 0 - 5 \end{pmatrix} = \begin{pmatrix} 4 \\ -5 \end{pmatrix}$

b $\overrightarrow{RQ} = \begin{pmatrix} -4 - 3 \\ 5 - 1 \end{pmatrix} = \begin{pmatrix} -7 \\ 4 \end{pmatrix}$

c $\overrightarrow{PR} = \begin{pmatrix} 3 - 2 \\ 1 - -1 \end{pmatrix} = \begin{pmatrix} 1 \\ 2 \end{pmatrix}$

3 a $(4, 3) \xrightarrow{\binom{-2}{6}} (2, 9)$ **b** $(-4, 1) \xrightarrow{\binom{3}{2}} (-1, 3)$

4 a i $\overrightarrow{EF} = \mathbf{r}$ **ii** $\overrightarrow{GF} = -\mathbf{s}$ **iii** $\overrightarrow{EG} = -\mathbf{t}$
b $\mathbf{r} + \mathbf{s} + \mathbf{t} = 0$

5 a i   **ii**

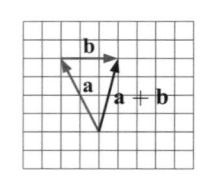

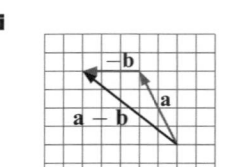

b i

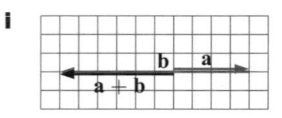

ii

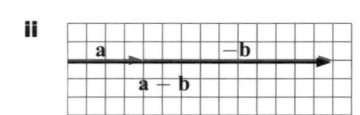

6

   a $(2, 3) \to (2, -3)$ **b** $(2, 3) \to (-3, -2)$
   c $(2, 3) \to (-2, 3)$ **d** $(2, 3) \to (2, 7)$

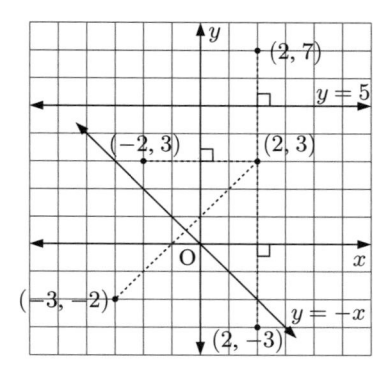

7 Let the point be (x, y).

$(x, y) \xrightarrow{\binom{3}{1}} (1, 5)$

$\therefore\ x + 3 = 1$ and $y + 1 = 5$
$\therefore\ x = -2$ and $y = 4$

So, the point is $(-2, 4)$.

8 **a** $\overrightarrow{WX} = \begin{pmatrix} 4-3 \\ -2-4 \end{pmatrix} = \begin{pmatrix} 1 \\ -6 \end{pmatrix}$

   b $\overrightarrow{XY} = \begin{pmatrix} -1-4 \\ 2--2 \end{pmatrix} = \begin{pmatrix} -5 \\ 4 \end{pmatrix}$

   c $\overrightarrow{ZV} = \begin{pmatrix} 1--2 \\ -4--3 \end{pmatrix} = \begin{pmatrix} 3 \\ -1 \end{pmatrix}$

   d $\overrightarrow{VX} = \begin{pmatrix} 4-1 \\ -2--4 \end{pmatrix} = \begin{pmatrix} 3 \\ 2 \end{pmatrix}$

   e $\overrightarrow{XV} = \begin{pmatrix} 1-4 \\ -4--2 \end{pmatrix} = \begin{pmatrix} -3 \\ -2 \end{pmatrix}$

   f $\overrightarrow{WZ} = \begin{pmatrix} -2-3 \\ -3-4 \end{pmatrix} = \begin{pmatrix} -5 \\ -7 \end{pmatrix}$

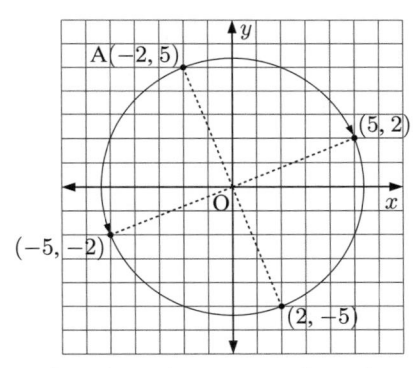

9

   a $(-2, 5) \to (2, -5)$ **b** $(-2, 5) \to (-5, -2)$
   c $(-2, 5) \to (5, 2)$

10 **a** $\overrightarrow{AB}$ and $\overrightarrow{CD}$ are opposite in direction.
   $\therefore$ the statement is false.

   b $\overrightarrow{BC}$ and $\overrightarrow{AD}$ are equal in length, and have the same direction.
   $\therefore$ the statement is true.

   c $|\overrightarrow{CB}| = |\overrightarrow{AB}| = |\mathbf{q}|$ {all sides have equal length}
   $\therefore$ the statement is true.

   d $\overrightarrow{BD} = -\mathbf{q} + \mathbf{p} \neq \mathbf{q}$
   $\therefore$ the statement is false.

   e $\overrightarrow{BA}$ and $\overrightarrow{DC}$ are parallel   {opp. sides of rhombus}
   $\therefore$ the statement is true.

11 **a** $\mathbf{a} + \mathbf{b} = \begin{pmatrix} 4 \\ 7 \end{pmatrix} + \begin{pmatrix} -2 \\ 1 \end{pmatrix} = \begin{pmatrix} 2 \\ 8 \end{pmatrix}$

   b $\mathbf{a} + \mathbf{a} + \mathbf{c} = \begin{pmatrix} 4 \\ 7 \end{pmatrix} + \begin{pmatrix} 4 \\ 7 \end{pmatrix} + \begin{pmatrix} -5 \\ -8 \end{pmatrix} = \begin{pmatrix} 3 \\ 6 \end{pmatrix}$

   c $\mathbf{c} - \mathbf{a} = \begin{pmatrix} -5 \\ -8 \end{pmatrix} - \begin{pmatrix} 4 \\ 7 \end{pmatrix} = \begin{pmatrix} -9 \\ -15 \end{pmatrix}$

   d $\mathbf{b} + \mathbf{c} = \begin{pmatrix} -2 \\ 1 \end{pmatrix} + \begin{pmatrix} -5 \\ -8 \end{pmatrix} = \begin{pmatrix} -7 \\ -7 \end{pmatrix}$

   $\therefore\ |\mathbf{b} + \mathbf{c}| = \sqrt{(-7)^2 + (-7)^2} = \sqrt{98}$
   $= \sqrt{49}\sqrt{2}$
   $= 7\sqrt{2}$ units

   e $\mathbf{b} - \mathbf{c} + \mathbf{a} = \begin{pmatrix} -2 \\ 1 \end{pmatrix} - \begin{pmatrix} -5 \\ -8 \end{pmatrix} + \begin{pmatrix} 4 \\ 7 \end{pmatrix} = \begin{pmatrix} 7 \\ 16 \end{pmatrix}$

   f $\mathbf{a} + \mathbf{b} - \mathbf{c} = \begin{pmatrix} 4 \\ 7 \end{pmatrix} + \begin{pmatrix} -2 \\ 1 \end{pmatrix} - \begin{pmatrix} -5 \\ -8 \end{pmatrix} = \begin{pmatrix} 7 \\ 16 \end{pmatrix}$

   $\therefore\ |\mathbf{a} + \mathbf{b} - \mathbf{c}| = \sqrt{7^2 + 16^2}$
   $= \sqrt{305}$ units

12 **a** Under a reduction with centre $O(0, 0)$ and scale factor $k = \frac{1}{2}$, $(1, 4) \to (\frac{1}{2} \times 1,\ \frac{1}{2} \times 4)$
   $\therefore\ (1, 4) \to (\frac{1}{2}, 2)$

   b

   $\overrightarrow{CP} = \begin{pmatrix} 5-3 \\ -6-4 \end{pmatrix} = \begin{pmatrix} 2 \\ -10 \end{pmatrix}$

   $\overrightarrow{CP'} = \frac{5}{2} \times \overrightarrow{CP} = \frac{5}{2} \begin{pmatrix} 2 \\ -10 \end{pmatrix} = \begin{pmatrix} 5 \\ -25 \end{pmatrix}$

   $\overrightarrow{OP'} = \overrightarrow{OC} + \overrightarrow{CP'}$
   $= \begin{pmatrix} 3 \\ 4 \end{pmatrix} + \begin{pmatrix} 5 \\ -25 \end{pmatrix}$
   $= \begin{pmatrix} 8 \\ -21 \end{pmatrix}$

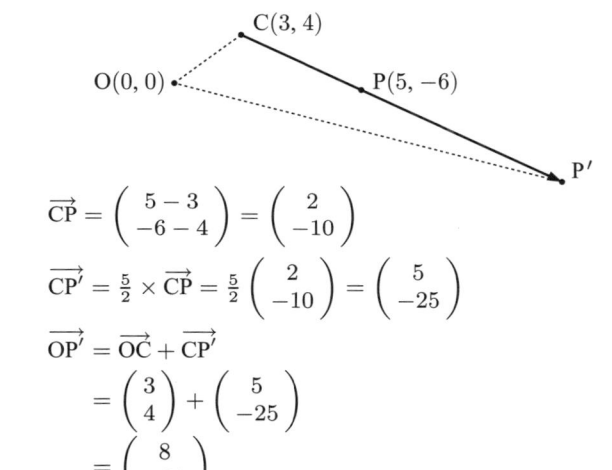

   So, the image of $(5, -6)$ is $(8, -21)$.

13 **a**

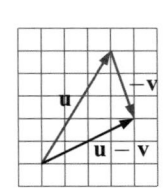

   b

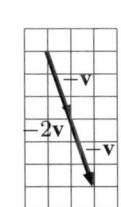

   $\mathbf{u} - \mathbf{v} = \begin{pmatrix} 4 \\ 2 \end{pmatrix}$ $-2\mathbf{v} = \begin{pmatrix} 2 \\ -6 \end{pmatrix}$

   c

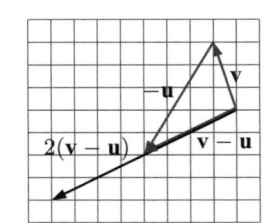

   $2(\mathbf{v} - \mathbf{u}) = \begin{pmatrix} -8 \\ -4 \end{pmatrix}$

14 a

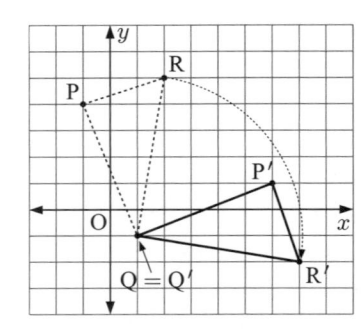

b P′(6, 1), Q′(1, −1), R′(7, −2)

15

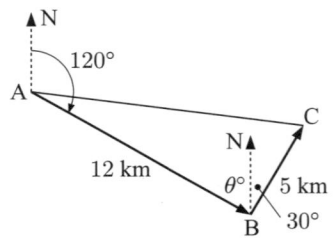

$\theta = 180 - 120 = 60$ {co-interior angles}
$\therefore \widehat{ABC} = 60° + 30° = 90°$
$\therefore$ in $\triangle ABC$, $AC^2 = AB^2 + BC^2$ {Pythagoras}
$\therefore AC^2 = 12^2 + 5^2$
$\therefore AC^2 = 169$
$\therefore AC = 13$ {AC > 0}

So, Mark is 13 km from his starting point.

16 a $\overrightarrow{BC} + \overrightarrow{CD} = \overrightarrow{BD}$
b $\overrightarrow{DA} + \overrightarrow{AB} - \overrightarrow{CB} = \overrightarrow{DB} + \overrightarrow{BC} = \overrightarrow{DC}$
c $\overrightarrow{BC} + \overrightarrow{CD} - \overrightarrow{AD} - \overrightarrow{BA} = \overrightarrow{BD} + \overrightarrow{DA} + \overrightarrow{AB}$
$= \overrightarrow{BB}$ or **0**
d $\overrightarrow{AB} - \overrightarrow{CB} - \overrightarrow{DC} = \overrightarrow{AB} + \overrightarrow{BC} + \overrightarrow{CD}$
$= \overrightarrow{AD}$

17 a, b

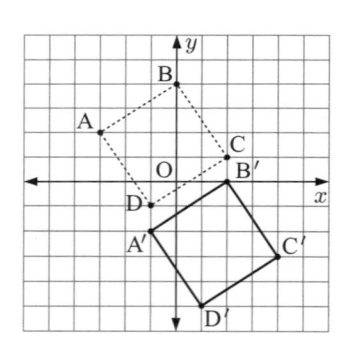

c A′(−1, −2), B′(2, 0), C′(4, −3), D′(1, −5)

d Each point has moved through $\sqrt{2^2 + (-4)^2}$
$= \sqrt{20}$
$= 2\sqrt{5}$ units

18 a $\left|\begin{pmatrix} 3 \\ 4 \end{pmatrix}\right|$
$= \sqrt{3^2 + 4^2}$
$= \sqrt{25}$
$= 5$ units

b $\left|\begin{pmatrix} 2 \\ -1 \end{pmatrix}\right|$
$= \sqrt{2^2 + (-1)^2}$
$= \sqrt{5}$ units

c $\left|\begin{pmatrix} 5 \\ 0 \end{pmatrix}\right|$
$= \sqrt{5^2 + 0^2}$
$= \sqrt{25}$
$= 5$ units

d $\left|\begin{pmatrix} -2 \\ 7 \end{pmatrix}\right|$
$= \sqrt{(-2)^2 + 7^2}$
$= \sqrt{53}$ units

e $\left|\begin{pmatrix} 0 \\ -9 \end{pmatrix}\right|$
$= \sqrt{0^2 + (-9)^2}$
$= \sqrt{81}$
$= 9$ units

f $\left|\begin{pmatrix} -2 \\ -6 \end{pmatrix}\right|$
$= \sqrt{(-2)^2 + (-6)^2}$
$= \sqrt{40}$
$= 2\sqrt{10}$ units

19

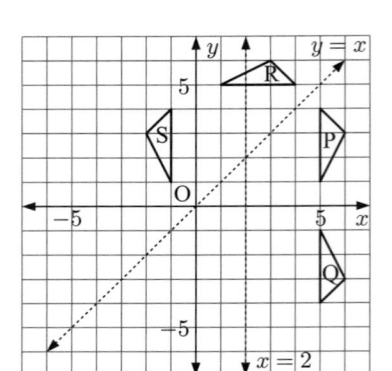

20 a $-3\mathbf{b}$
$= -3 \begin{pmatrix} -3 \\ -2 \end{pmatrix}$
$= \begin{pmatrix} 9 \\ 6 \end{pmatrix}$

b $2\mathbf{a} + \mathbf{b}$
$= 2 \begin{pmatrix} 2 \\ 1 \end{pmatrix} + \begin{pmatrix} -3 \\ -2 \end{pmatrix}$
$= \begin{pmatrix} 4 \\ 2 \end{pmatrix} + \begin{pmatrix} -3 \\ -2 \end{pmatrix} = \begin{pmatrix} 1 \\ 0 \end{pmatrix}$

c $\frac{1}{2}(5\mathbf{a} + 3\mathbf{b}) = \frac{1}{2}\left[5\begin{pmatrix} 2 \\ 1 \end{pmatrix} + 3\begin{pmatrix} -3 \\ -2 \end{pmatrix}\right]$
$= \frac{1}{2}\left[\begin{pmatrix} 10 \\ 5 \end{pmatrix} + \begin{pmatrix} -9 \\ -6 \end{pmatrix}\right]$
$= \frac{1}{2}\begin{pmatrix} 1 \\ -1 \end{pmatrix} = \begin{pmatrix} \frac{1}{2} \\ -\frac{1}{2} \end{pmatrix}$

d $\frac{1}{4}\mathbf{a} - \mathbf{b} = \frac{1}{4}\begin{pmatrix} 2 \\ 1 \end{pmatrix} - \begin{pmatrix} -3 \\ -2 \end{pmatrix}$
$= \begin{pmatrix} \frac{1}{2} \\ \frac{1}{4} \end{pmatrix} + \begin{pmatrix} 3 \\ 2 \end{pmatrix} = \begin{pmatrix} \frac{7}{2} \\ \frac{9}{4} \end{pmatrix}$

21 a $k = 2$ **b** $k = \frac{1}{2}$ **c** $k = 3$

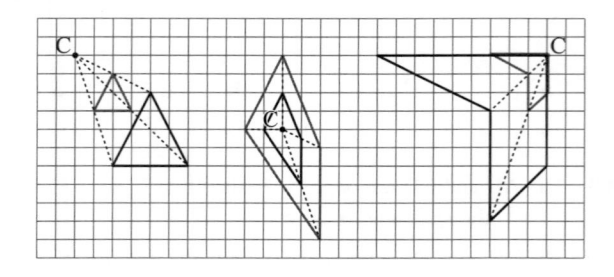

22 a i $\overrightarrow{AB} = \begin{pmatrix} -5 - 1 \\ -6 - 1 \end{pmatrix} = \begin{pmatrix} -6 \\ -7 \end{pmatrix}$
ii $AB = \sqrt{(-6)^2 + (-7)^2}$
$= \sqrt{85}$ units

b i $\overrightarrow{AB} = \begin{pmatrix} 7 - 7 \\ -9 - 4 \end{pmatrix} = \begin{pmatrix} 0 \\ -13 \end{pmatrix}$

ii $AB = \sqrt{0^2 + (-13)^2}$
$= \sqrt{169}$
$= 13$ units

c i $\vec{AB} = \begin{pmatrix} 3 - -2 \\ 6 - 5 \end{pmatrix} = \begin{pmatrix} 5 \\ 1 \end{pmatrix}$

ii $AB = \sqrt{5^2 + 1^2}$
$= \sqrt{26}$ units

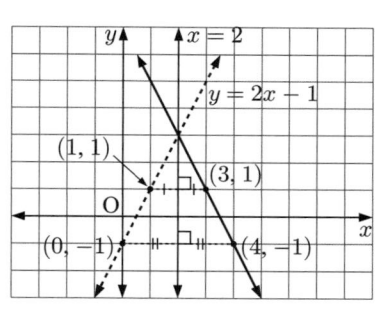

23

Two points which lie on the line $y = 2x - 1$ are $(1, 1)$ and $(0, -1)$.
When reflected in the line $x = 2$, these points are mapped to $(3, 1)$ and $(4, -1)$ respectively.
$\therefore (3, 1)$ and $(4, -1)$ lie on the image line.
$\therefore$ the image line has gradient $\frac{-1-1}{4-3} = -2$
so its equation is $2x + y = 2(3) + 1$
$\therefore y = -2x + 7$

24 a $s - 2t + u = 0$
$\therefore u = 2t - s$
$= 2\begin{pmatrix} -3 \\ 4 \end{pmatrix} - \begin{pmatrix} 4 \\ 6 \end{pmatrix}$
$= \begin{pmatrix} -6 \\ 8 \end{pmatrix} - \begin{pmatrix} 4 \\ 6 \end{pmatrix}$
$= \begin{pmatrix} -10 \\ 2 \end{pmatrix}$

b $u + \frac{1}{2}s = 3t$
$\therefore u = 3t - \frac{1}{2}s$
$= 3\begin{pmatrix} -3 \\ 4 \end{pmatrix} - \frac{1}{2}\begin{pmatrix} 4 \\ 6 \end{pmatrix}$
$= \begin{pmatrix} -9 \\ 12 \end{pmatrix} - \begin{pmatrix} 2 \\ 3 \end{pmatrix}$
$= \begin{pmatrix} -11 \\ 9 \end{pmatrix}$

25 a

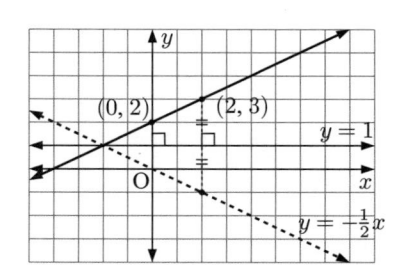

$(0, 2)$ and $(2, 3)$ lie on the image line.
$\therefore$ the image line has gradient $\frac{3-2}{2-0} = \frac{1}{2}$ and
y-intercept 2, so its equation is $y = \frac{1}{2}x + 2$.

b

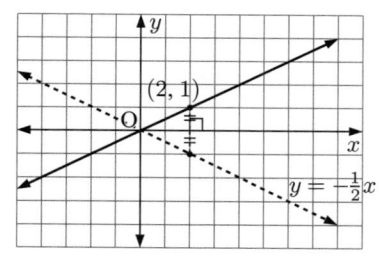

$(0, 0)$ and $(2, 1)$ lie on the image line.
$\therefore$ the image line has gradient $\frac{1-0}{2-0} = \frac{1}{2}$ and
y-intercept 0, so its equation is $y = \frac{1}{2}x$.

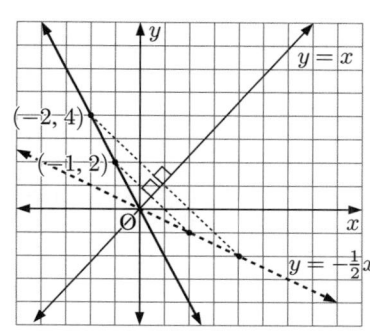

c

$(-2, 4)$ and $(-1, 2)$ lie on the image line.
$\therefore$ the image line has gradient $\frac{2-4}{-1--2} = -2$
so its equation is $2x + y = 2(-2) + 4$
or $y = -2x$.

26 a $\mathbf{a} = -\frac{1}{2}\mathbf{b}$
$\therefore \mathbf{a}$ is parallel to $\mathbf{b}$, and $|\mathbf{a}| = \left|-\frac{1}{2}\right| |\mathbf{b}| = \frac{1}{2}|\mathbf{b}|$
$\therefore \mathbf{a}$ is a half as long as $\mathbf{b}$, and has the opposite direction.

b $\mathbf{a} = 3\mathbf{b}$
$\therefore \mathbf{a}$ is parallel to $\mathbf{b}$, and $|\mathbf{a}| = |3||\mathbf{b}| = 3|\mathbf{b}|$
$\therefore \mathbf{a}$ is 3 times longer than $\mathbf{b}$, and they have the same direction.

27 a

$k = 2$

b

$k = \frac{1}{2}$

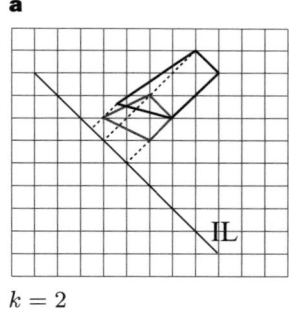

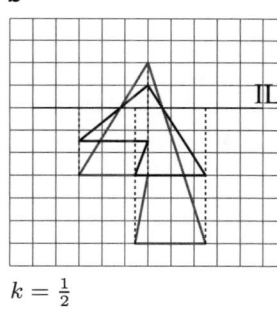

28 a TR consists of a 90° clockwise rotation about $(0, 0)$, followed by a translation of $\begin{pmatrix} 3 \\ -1 \end{pmatrix}$.

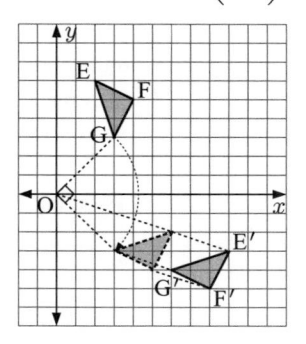

b RT consists of a translation of $\begin{pmatrix} 3 \\ -1 \end{pmatrix}$, followed by a 90° clockwise rotation about $(0, 0)$.

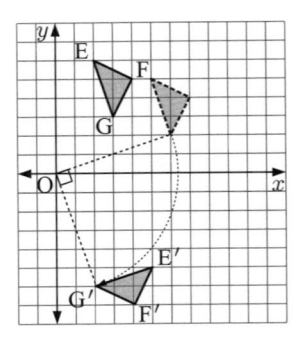

29 $\left| \begin{pmatrix} -5 \\ q \end{pmatrix} \right| = 13 \quad \therefore \sqrt{(-5)^2 + q^2} = 13$
$\therefore 25 + q^2 = 169$
$\therefore q^2 = 144$
$\therefore q = \pm 12$

30 a

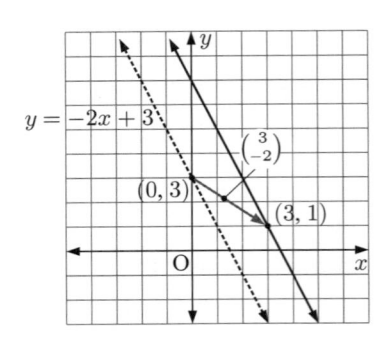

The image line is parallel to the object line.
$\therefore$ the image of $y = -2x+3$ has the form $y = -2x+c$.
$(0, 3)$ lies on the line $y = -2x+3$, and since
$(0, 3) \xrightarrow{\begin{pmatrix} 3 \\ -2 \end{pmatrix}} (3, 1)$, $(3, 1)$ lies on the image.
$\therefore 1 = -2(3) + c$
$\therefore c = 7$
$\therefore$ the equation of the image is $y = -2x + 7$.

b

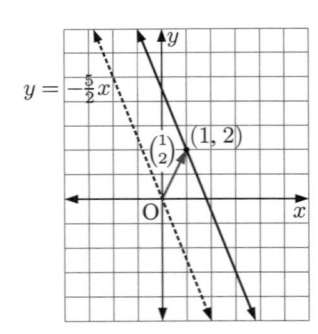

The image line is parallel to the object line.
$\therefore$ the image of $y = -\frac{5}{2}x$ has the form $y = -\frac{5}{2}x + c$.
$(0, 0)$ lies on the line $y = -\frac{5}{2}x$, and since
$(0, 0) \xrightarrow{\begin{pmatrix} 1 \\ 2 \end{pmatrix}} (1, 2)$, $(1, 2)$ lies on the image.
$\therefore 2 = -\frac{5}{2}(1) + c$
$\therefore c = \frac{9}{2}$
$\therefore$ the equation of the image is $y = -\frac{5}{2}x + \frac{9}{2}$.

31 Let V have coordinates (a, b).
Since STUV is a parallelogram, $\overrightarrow{TU} = \overrightarrow{SV}$
$\therefore \begin{pmatrix} 3 - 0 \\ 6 - 4 \end{pmatrix} = \begin{pmatrix} a - 5 \\ b - 1 \end{pmatrix}$
$\therefore a - 5 = 3$ and $b - 1 = 2$
$\therefore a = 8$ and $b = 3$
So, the remaining vertex is $V(8, 3)$.

32 a Under a stretch with invariant y-axis and scale factor $k = \frac{1}{3}$, $(2, 6) \to (\frac{1}{3} \times 2, 6)$
$\therefore (2, 6) \to (\frac{2}{3}, 6)$

b

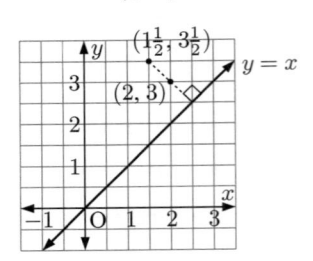

Under a stretch with invariant line $y = x$ and scale factor $k = 2$, $(2, 3) \to (1\frac{1}{2}, 3\frac{1}{2})$.

33 a $\overrightarrow{AB} = \begin{pmatrix} -4 \\ 2 \end{pmatrix}$, $\overrightarrow{BC} = \begin{pmatrix} 2 \\ 5 \end{pmatrix}$, $\overrightarrow{CD} = \begin{pmatrix} 4 \\ -1 \end{pmatrix}$,
$\overrightarrow{DE} = \begin{pmatrix} 3 \\ 2 \end{pmatrix}$, $\overrightarrow{EA} = \begin{pmatrix} -5 \\ -8 \end{pmatrix}$

b $\overrightarrow{AE} = \begin{pmatrix} 5 \\ 8 \end{pmatrix}$

c i The sum of all the vectors is
$\overrightarrow{AB} + \overrightarrow{BC} + \overrightarrow{CD} + \overrightarrow{DE} + \overrightarrow{EA} = \overrightarrow{AA} = \mathbf{0}$
This occurs because the lap starts and finishes at the same place.
ii Ingmar needs to find the length of each leg individually, then add the results.
iii $|\overrightarrow{AB}| = \sqrt{(-4)^2 + 2^2} = \sqrt{20}$ units
$|\overrightarrow{BC}| = \sqrt{2^2 + 5^2} = \sqrt{29}$ units
$|\overrightarrow{CD}| = \sqrt{4^2 + (-1)^2} = \sqrt{17}$ units
$|\overrightarrow{DE}| = \sqrt{3^2 + 2^2} = \sqrt{13}$ units
$|\overrightarrow{EA}| = \sqrt{(-5)^2 + (-8)^2} = \sqrt{89}$ units
$\therefore$ the length of one lap
$= \sqrt{20} + \sqrt{29} + \sqrt{17} + \sqrt{13} + \sqrt{89}$
≈ 27.0 units

34

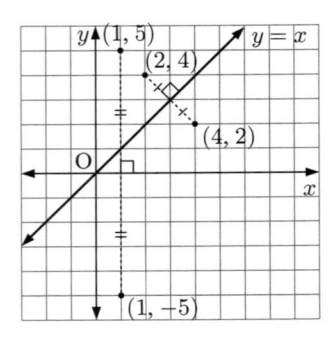

$(1, -5) \to (1, 5) \to (2, 4) \to (4, 2)$
So, the image of $(1, -5)$ is $(4, 2)$.

35 a $\overrightarrow{AB} = \begin{pmatrix} 4-10 \\ -1-1 \end{pmatrix} = \begin{pmatrix} -6 \\ -2 \end{pmatrix}$

$\overrightarrow{AC} = \begin{pmatrix} -2-10 \\ -3-1 \end{pmatrix} = \begin{pmatrix} -12 \\ -4 \end{pmatrix} = 2\begin{pmatrix} -6 \\ -2 \end{pmatrix}$

$\therefore \overrightarrow{AC} = 2\overrightarrow{AB}$, so $\overrightarrow{AC}$ is parallel to $\overrightarrow{AB}$.

$\therefore$ A, B and C are collinear.

b $k = 2$

36 Every point of the image triangle A'B'C' is twice as far from the line $x = 1$ as the corresponding point of the object triangle ABC.

$\therefore$ a stretch with invariant line $x = 1$ and scale factor 2 has occurred.

37 Let **c** be the velocity of the current, **x** be the velocity of the kayak in still conditions, and **k** be the actual velocity of the kayak.

$\therefore |\mathbf{c}| = 3.75$ and $|\mathbf{k}| = 9$.

a $\tan\theta = \frac{3.75}{9}$

$\therefore \theta = \tan^{-1}\left(\frac{3.75}{9}\right) \approx 22.6°$

$\therefore$ the kayaker is facing 22.6° west of due south.

b $|\mathbf{x}|^2 = 9^2 + 3.75^2$ {Pythagoras}

$\therefore |\mathbf{x}|^2 = 95.0625$

$\therefore |\mathbf{x}| = 9.75$ $\{|\mathbf{x}| > 0\}$

If the water was still, the kayaker would be paddling at 9.75 km/h.

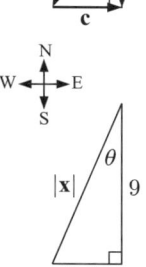

38 a M_1M_2 consists of a reflection in the line $y = -x$, followed by a reflection in the line $x = 2$.

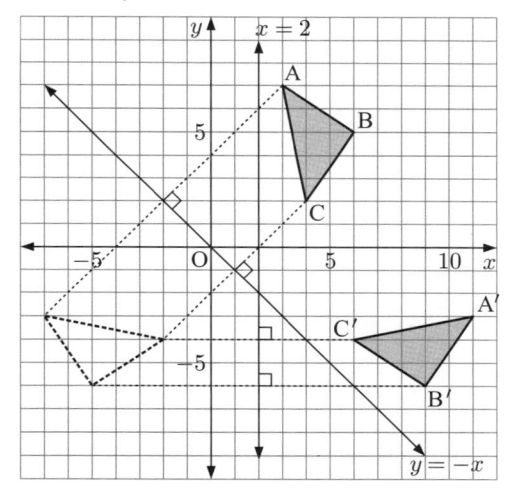

b M_2M_1 consists of a reflection in the line $x = 2$, followed by a reflection in the line $y = -x$.

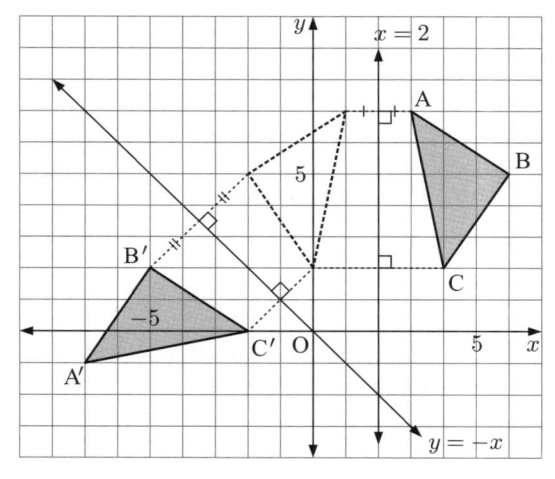

39 Since PQRS is a parallelogram, $\overrightarrow{SR} = \mathbf{a}$ and $\overrightarrow{QR} = \mathbf{b}$.

a $\overrightarrow{PR} = \overrightarrow{PQ} + \overrightarrow{QR}$
$= \mathbf{a} + \mathbf{b}$

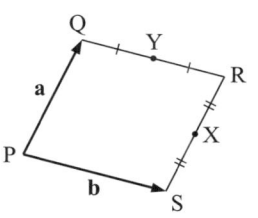

b $\overrightarrow{PX} = \overrightarrow{PS} + \overrightarrow{SX}$
$= \overrightarrow{PS} + \frac{1}{2}\overrightarrow{SR}$
$= \mathbf{b} + \frac{1}{2}\mathbf{a}$

c $\overrightarrow{QX} = \overrightarrow{QR} + \overrightarrow{RX}$
$= \overrightarrow{QR} - \overrightarrow{XR}$
$= \mathbf{b} - \frac{1}{2}\mathbf{a}$

d $\overrightarrow{PY} = \overrightarrow{PQ} + \overrightarrow{QY}$
$= \mathbf{a} + \frac{1}{2}\mathbf{b}$

e $\overrightarrow{XY} = \overrightarrow{XR} + \overrightarrow{RY}$
$= \overrightarrow{XR} - \overrightarrow{YR}$
$= \frac{1}{2}\mathbf{a} - \frac{1}{2}\mathbf{b}$

40 $x + 2y = -2$ has x-intercept -2 (when $y = 0$)
and y-intercept -1 (when $x = 0$)

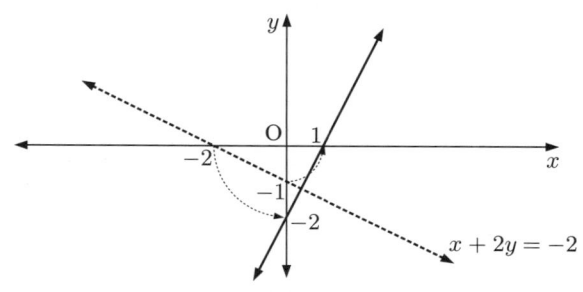

When these intercepts are rotated 90° anticlockwise about O, the image line has x-intercept 1 and y-intercept -2.

$\therefore$ (0, -2) and (1, 0) lie on the image line.

$\therefore$ the image line has gradient $\frac{0 - -2}{1 - 0} = 2$

so its equation is $2x - y = 2(1) - 0$
or $y = 2x - 2$

41 $\overrightarrow{CE} = \overrightarrow{CP} + \overrightarrow{PE}$
$= \overrightarrow{PE} - \overrightarrow{PC}$

$\overrightarrow{CD} = \frac{5}{12}\overrightarrow{CE}$
$= \frac{5}{12}\left(\overrightarrow{PE} - \overrightarrow{PC}\right)$

$\overrightarrow{PD} = \overrightarrow{PC} + \overrightarrow{CD}$
$= \overrightarrow{PC} + \frac{5}{12}\left(\overrightarrow{PE} - \overrightarrow{PC}\right)$
$= \frac{7}{12}\overrightarrow{PC} + \frac{5}{12}\overrightarrow{PE}$

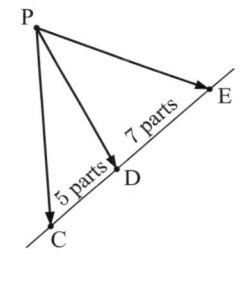

42 a a translation of $\begin{pmatrix} -4 \\ 7 \end{pmatrix}$

b a rotation about point P, clockwise through 120°

c a stretch with invariant line $y = x$ and scale factor $\frac{1}{3}$

d a reflection in the line $x = -4$

e an enlargement with centre (1, 2) and scale factor 5

43 a Since ABCD is a rhombus,
$\overrightarrow{DC} = \overrightarrow{AB} = \mathbf{a}$ and
$\overrightarrow{BC} = \overrightarrow{AD} = \mathbf{b}$.

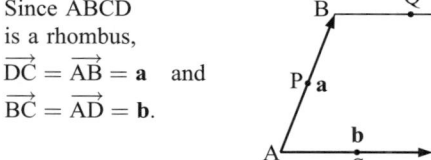

i $\overrightarrow{PS} = \overrightarrow{PA} + \overrightarrow{AS}$
$= -\frac{1}{2}\mathbf{a} + \frac{1}{2}\mathbf{b}$

ii $\overrightarrow{QR} = \overrightarrow{QC} + \overrightarrow{CR}$
$= \frac{1}{2}\mathbf{b} - \frac{1}{2}\mathbf{a}$

iii $\overrightarrow{PQ} = \overrightarrow{PB} + \overrightarrow{BQ}$
$= \frac{1}{2}\mathbf{a} + \frac{1}{2}\mathbf{b}$

iv $\overrightarrow{SR} = \overrightarrow{SD} + \overrightarrow{DR}$
$= \frac{1}{2}\mathbf{b} + \frac{1}{2}\mathbf{a}$

b **i** $\vec{PR} = \vec{PB} + \vec{BC} + \vec{CR}$
$= \frac{1}{2}\mathbf{a} + \mathbf{b} - \frac{1}{2}\mathbf{a}$
$= \mathbf{b}$
$\therefore |\vec{PR}| = |\mathbf{b}|$
$= |\mathbf{a}|$ {ABCD rhombus $\therefore |\mathbf{a}| = |\mathbf{b}|$}

ii $\vec{SQ} = \vec{SD} + \vec{DC} + \vec{CQ}$
$= \frac{1}{2}\mathbf{b} + \mathbf{a} - \frac{1}{2}\mathbf{b}$
$= \mathbf{a}$
$\therefore |\vec{SQ}| = |\mathbf{a}|$

c From **a** we have $\vec{PS} = \vec{QR}$ and $\vec{PQ} = \vec{SR}$
$\therefore$ we deduce PQRS is a parallelogram.
Also, from **b**, the diagonals are equal in length.
$\therefore$ PQRS is a rectangle.

44 a T_5 maps triangle 0 to triangle 5
T_7 maps triangle 5 to triangle 2
$\therefore$ T_5 then T_7 is equivalent to T_2.

b T_1 maps triangle 0 to triangle 1
T_4 maps triangle 1 to triangle 5
$\therefore$ T_1 then T_4 is equivalent to T_5.

c T_6 maps triangle 0 to triangle 6
T_6 maps triangle 6 to triangle 4
$\therefore$ T_6 then T_6 is equivalent to T_4.

d T_3 maps triangle 0 to triangle 3
T_2 maps triangle 3 to triangle 5
$\therefore$ T_3 then T_2 is equivalent to T_5.

SOLUTIONS TO TOPIC 6: MENSURATION

1 a 72 mm
$= (72 \div 10)$ cm
$= 7.2$ cm

b 5.8 m
$= (5.8 \times 100)$ cm
$= (5.8 \times 100 \times 10)$ mm
$= 5800$ mm

c 9.75 km
$= (9.75 \times 1000)$ m
$= 9750$ m

d 28 000 000 cm
$= (28\,000\,000 \div 100)$ m
$= (28\,000\,000 \div 100 \div 1000)$ km
$= 280$ km

2 Distance between light poles $= \dfrac{2.4 \text{ km}}{80} = \dfrac{2400 \text{ m}}{80} = 30$ m

3 a Perimeter
$= 2 \times 15 + 12$
$= 42$ cm

b Perimeter
$= 2 \times 3.5 + 2 \times 2$
$= 11$ m

c Perimeter
$= 3 \times 1.5 + 2.5 + 4 + 3$
$= 14$ m

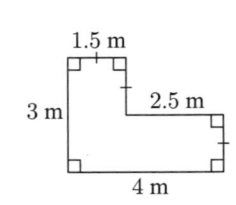

4 a 44 mm^2
$= (44 \div 100)$ cm^2
$= 0.44$ cm^2

b 0.059 ha
$= (0.059 \times 10\,000)$ m^2
$= (0.059 \times 10\,000 \times 10\,000)$ cm^2
$= 5\,900\,000$ cm^2

c 21.85 ha
$= (21.85 \div 100)$ km^2
$= 0.2185$ km^2

d 0.000 006 2 km^2
$= (0.000\,006\,2 \times 1\,000\,000)$ m^2
$= (0.000\,006\,2 \times 1\,000\,000 \times 1\,000\,000)$ mm^2
$= 6\,200\,000$ mm^2

e 360 m^2
$= (360 \times 10\,000)$ cm^2
$= 3\,600\,000$ cm^2

f 39 500 m^2
$= (39\,500 \div 10\,000)$ ha
$= 3.95$ ha

5 The rectangle has perimeter $= 2 \times 3.2 + 2 \times 2.4$
$= 11.2$ m
$\therefore$ the perimeter of the square is also 11.2 m, and hence the length of its sides is $\dfrac{11.2 \text{ m}}{4} = 2.8$ m.

6 1.36 m$^2 = 13\,600$ cm^2
$\therefore$ the number of boxes on the pallet $= \dfrac{13\,600}{85} = 160$

7 a $P = 4z$

b $P = a + 2b$

c $P = 3p + 5 \times p + 2 \times q$
$= 8p + 2q$

8 a $\pi r^2 = 36.4$
$\therefore r^2 = \dfrac{36.4}{\pi}$
$\therefore r = \sqrt{\dfrac{36.4}{\pi}}$ $\{r > 0\}$
$\therefore r \approx 3.4039$
The radius of the circle is 3.40 m.

b $C = 2\pi r$
$\approx 2\pi \times 3.4039$
≈ 21.4
The circumference of the circle is 21.4 m.

9 a Area $= 9 \times 8 - 3 \times 5$
$= 57$ cm^2

b Area $=$ base $\times$ height
$= 9 \times 6$
$= 54$ m^2

10 a The cube has 6 identical faces, each with
area $= 16 \times 16 = 256$ cm^2.
$\therefore$ the surface area of the cube $= 6 \times 256 = 1536$ cm^2.

b The net of the prism is:

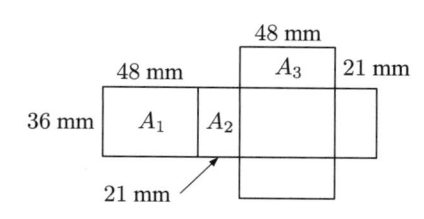

$A_1 = 36 \times 48 = 1728 \text{ mm}^2$ {bottom and top}
$A_2 = 36 \times 21 = 756 \text{ mm}^2$ {sides}
$A_3 = 48 \times 21 = 1008 \text{ mm}^2$ {front and back}

$\therefore$ total surface area $= 2 \times A_1 + 2 \times A_2 + 2 \times A_3$
$= 2 \times 1728 + 2 \times 756 + 2 \times 1008$
$= 6984 \text{ mm}^2$

11 a 3.71 litres
$= (3.71 \times 100)$ cl
$= 371$ cl

b 58 215 ml
$= (58\,215 \div 1000)$ litres
$= 58.215$ litres

12

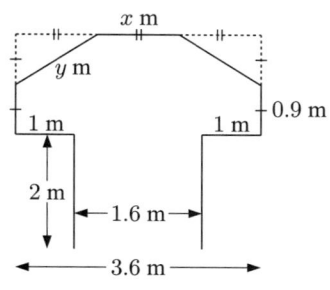

$x = \frac{3.6}{3} = 1.2$
$\therefore y^2 = 1.2^2 + 0.9^2$ {Pythagoras}
$\therefore y^2 = 2.25$
$\therefore y = 1.5$ {$y > 0$}

$\therefore$ guard rail length
$= 1.2 + 2 \times 1.5 + 2 \times 0.9 + 2 \times 1 + 2 \times 2$
$= 12$ m

13

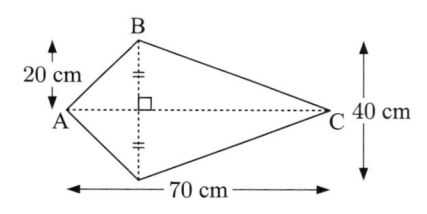

area of kite $= 2 \times$ (area of $\triangle ABC$)
$= 2 \left(\frac{1}{2} \times 70 \times 20 \right)$
$= 1400 \text{ cm}^2$

14 a The figure has:
- 3 rectangular faces 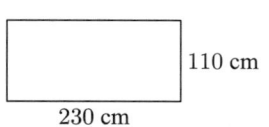
- 2 triangular faces

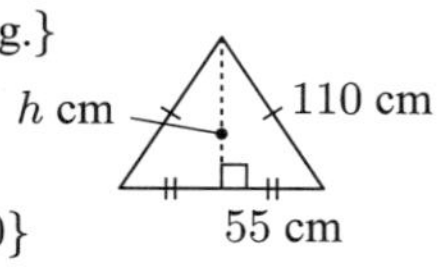

$h^2 + 55^2 = 110^2$ {Pythag.}
$\therefore h^2 + 3025 = 12\,100$
$\therefore h^2 = 9075$
$\therefore h \approx 95.3$ {$h > 0$}

$\therefore$ total surface area
$\approx 3 \times (230 \times 110) + 2 \times \left(\frac{1}{2} \times 110 \times 95.3 \right)$
$\approx 86\,400 \text{ cm}^2$

b The figure has:
- 1 square base 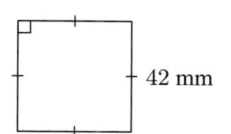
- 4 triangular faces

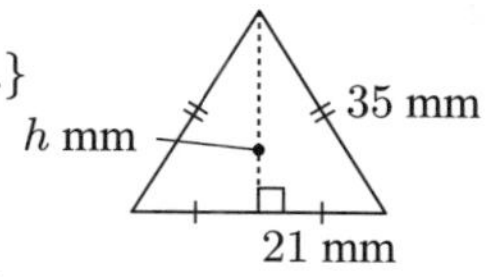

$h^2 + 21^2 = 35^2$ {Pythag.}
$\therefore h^2 + 441 = 1225$
$\therefore h^2 = 784$
$\therefore h = 28$ {$h > 0$}

$\therefore$ total surface area
$= 42 \times 42 + 4 \times \left(\frac{1}{2} \times 42 \times 28 \right)$
$= 4116 \text{ mm}^2$

15 a Distance around semi-circle $= \frac{1}{2} \times \pi \times 10 \approx 15.7$ mm
$\therefore$ total perimeter $\approx 15.7 + 3 \times 10 \approx 45.7$ mm
Area $= 10 \times 10 + \frac{1}{2} \times \pi \times 5^2 \approx 139 \text{ mm}^2$

b Perimeter $= 2.2 + 2.2 +$ length of arc
$= 4.4 + \left(\frac{80}{360} \right) \times 2 \times \pi \times 2.2$
≈ 7.47 m

Area $= \left(\frac{80}{360} \right) \times \pi \times 2.2^2 \approx 3.38 \text{ m}^2$

16 a 7.25 m^3
$= (7.25 \times 1\,000\,000) \text{ cm}^3$
$= 7\,250\,000 \text{ cm}^3$

b $2\,900\,000\,000 \text{ mm}^3$
$= (2\,900\,000\,000 \div 1000) \text{ cm}^3$
$= (2\,900\,000\,000 \div 1000 \div 1\,000\,000) \text{ m}^3$
$= 2.9 \text{ m}^3$

c 2500 cm^3
$= (2500 \times 1000) \text{ mm}^3$
$= 2\,500\,000 \text{ mm}^3$

17 Volume of milk used each week
$= 235 \times 75$ ml
$= 17\,625$ ml
$= 17.625$ l

18

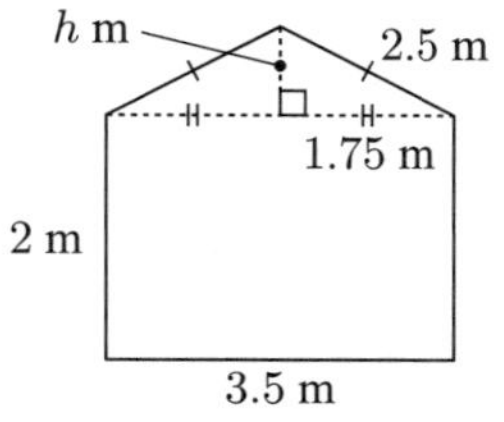

$h^2 + 1.75^2 = 2.5^2$ {Pythagoras}
$\therefore h^2 + 3.0625 = 6.25$
$\therefore h^2 = 3.1875$
$\therefore h \approx 1.785$ {$h > 0$}

$\therefore$ area of end $\approx 2 \times 3.5 + \frac{1}{2} \times 3.5 \times 1.785$
$\approx 10.124 \text{ m}^2$

$\therefore$ total surface area
$\approx 2 \times 10.124 + 2 \times (6 \times 2) + 2 \times (6 \times 2.5)$
$\approx 74.2 \text{ m}^2$

So, 74.2 m^2 of sheet metal is required.

19 Area $= \left(\dfrac{\theta}{360}\right) \times \pi r^2$
$= \dfrac{250}{360} \times \pi \times 4^2$
$\approx 34.9 \text{ cm}^2$

20

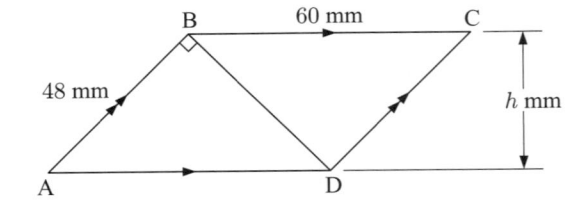

a ABCD is a parallelogram
$\therefore$ AD $= 60$ mm, DC $= 48$ mm.
$\therefore$ in $\triangle$ABD, BD$^2 + 48^2 = 60^2$ {Pythagoras}
$\therefore$ BD $= \sqrt{60^2 - 48^2}$ {BD > 0}
$= 36$ mm
Now $\widehat{BDC} = \widehat{ABD} = 90°$ {alternate angles}
$\therefore$ area of parallelogram $=$ area $\triangle$ABD $+$ area $\triangle$BCD
$= \tfrac{1}{2} \times 48 \times 36 + \tfrac{1}{2} \times 48 \times 36$
$= 1728$ mm^2

b area of parallelogram $= 1728$ mm^2
$\therefore 60 \times h = 1728$
$h = 28.8$

21 a The sphere has radius 30 cm.
$\therefore$ surface area $= 4 \times \pi \times 30^2$
$\approx 11\,300 \text{ cm}^2$

b surface area $= 2\pi r^2 + 2\pi rh$
$= 2 \times \pi \times 20^2 + 2 \times \pi \times 20 \times 380$
$\{3.8 \text{ m} = 380 \text{ cm}\}$
$\approx 50\,300 \text{ cm}^2$

22 a

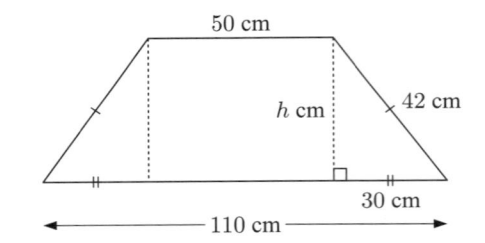

$h^2 + 30^2 = 42^2$ {Pythagoras}
$\therefore h = \sqrt{42^2 - 30^2}$ $\{h > 0\}$
$\therefore h \approx 29.39$
$\therefore$ area of top $\approx \left(\dfrac{50 + 110}{2}\right) \times 29.39$
$\approx 2350 \text{ cm}^2$

b area of four sides
$= (110 \times 55) + 2 \times (42 \times 55) + (50 \times 55)$
$= 13\,420 \text{ cm}^2$

23 $1000 \text{ cm}^3 = 1\,000\,000 \text{ mm}^3$
$\therefore$ number of resistors $= \dfrac{1\,000\,000}{40} = 25\,000$

24 $500 \text{ cm}^3 \equiv 500$ ml
$\equiv 0.5$ l

25

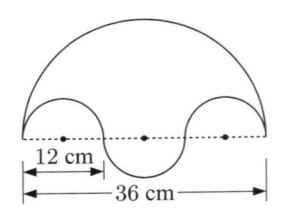

Perimeter of figure
$=$ (perimeter of large semi-circle) $+$
$\quad 3 \times$ (perimeter of small semi-circle)
$= \tfrac{1}{2} \times \pi \times 36 + 3 \times \left(\tfrac{1}{2} \times \pi \times 12\right)$
≈ 113 cm

Area of figure
$=$ (area of large semi-circle) $-$ (area of 1 small semi-circle)
$= \tfrac{1}{2} \times \pi \times 18^2 - \tfrac{1}{2} \times \pi \times 6^2$
$\approx 452 \text{ cm}^2$

26 The beach balls have radius 18 cm.
$\therefore$ surface area of ball $= 4 \times \pi \times 18^2$
$= 1296\pi \text{ cm}^2$
$\therefore$ surface area of 200 balls $= 200 \times 1296\pi$
$\approx 814\,000 \text{ cm}^2$
$\approx 81.4 \text{ m}^2$

So, 81.4 m^2 of rubber is needed.

27 a Volume
$=$ length $\times$ width $\times$ depth
$= 3.5 \times 4.2 \times 2.5$
$\approx 36.8 \text{ m}^3$

b Volume
$= \pi r^2 \times h$
$= \pi \times 24^2 \times 86$
$\approx 156\,000 \text{ mm}^3$

c Volume
$=$ area of end $\times$ height
$= 3.6 \times 25$
$= 90 \text{ cm}^3$

d Volume
$= \tfrac{4}{3}\pi r^3$
$= \tfrac{4}{3} \times \pi \times 2.5^3$
$\approx 65.4 \text{ cm}^3$

e Volume
$= \tfrac{1}{3}\pi r^2 h$
$= \tfrac{1}{3} \times \pi \times 2.5^2 \times 8$
$\approx 52.4 \text{ m}^3$

28 a Surface area $= 2\pi rh + \pi r^2$
$= 2 \times \pi \times 6 \times 8 + \pi \times 6^2$
$\approx 415 \text{ cm}^2$

b $l^2 = 14^2 + 28^2$ {Pythagoras}
$\therefore l = \sqrt{14^2 + 28^2}$ $\{l > 0\}$
$\therefore l \approx 31.3$
$\therefore$ surface area $= \pi rl$
$\approx \pi \times 14 \times 31.3$
$\approx 1380 \text{ mm}^2$

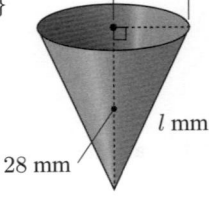

29

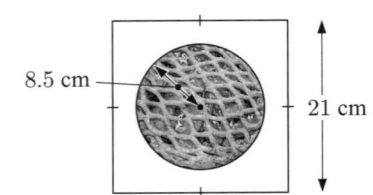

area of pie $= \pi \times 8.5^2 = 72.25\pi \text{ cm}^2$
area of plate $= 21 \times 21 = 441 \text{ cm}^2$
Now $\dfrac{72.25\pi}{441} \approx 0.515 \approx 51.5\%$
So, the pie covers 51.5% of the plate.

30 Volume of flask $= \pi r^2 h$
$= \pi \times 3.42^2 \times 16.33$
$\approx 600 \text{ cm}^3$
$\therefore$ capacity of flask $\approx 600 \text{ ml}$

31 Volume of cookies $= \pi r^2 h$
$= \pi \times 2.5^2 \times 1$
$= 6.25 \pi \text{ cm}^3$
Volume of dough $= 20 \times 15 \times 8$
$= 2400 \text{ cm}^3$
$\dfrac{2400}{6.25\pi} \approx 122$, so 122 cookies can be made from the block.

32 The pattern can be divided into 25 cm by 25 cm squares as shown:

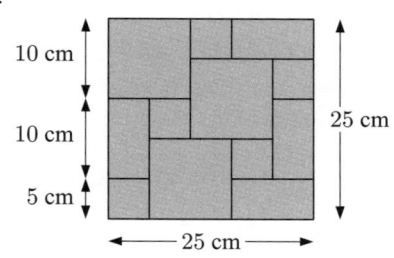

The square has area $25 \times 25 = 625 \text{ cm}^2$.
The small tiles are 5 cm by 5 cm, and there are 5 of them in the square.
So, the small tiles occupy a total of $5 \times (5 \times 5) = 125 \text{ cm}^2$.
$\therefore$ the proportion of the area covered by the smallest tile is $\dfrac{125}{625} = 20\%$.

33 **a** Total surface area
$= 2 \times (10 \times 10) \;+\; 4 \times (10 \times 20)$
$\quad\quad\;\uparrow \quad\quad\quad\quad\quad\quad \uparrow$
front and back   top, bottom and sides
$= 1000 \text{ cm}^2$

b For the front and back:

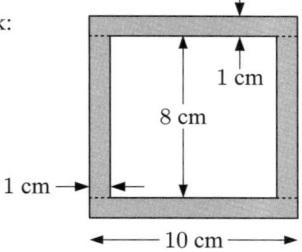

painted area $= 2 \times (10 \times 1) + 2 \times (8 \times 1) = 36 \text{ cm}^2$
For the top, bottom and sides:

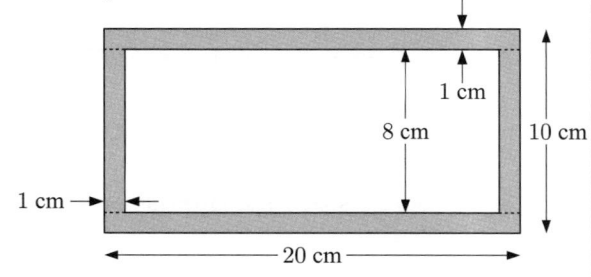

painted area $= 2 \times (20 \times 1) + 2 \times (8 \times 1) = 56 \text{ cm}^2$
$\therefore$ total painted area $= 2 \times 36 + 4 \times 56 = 296 \text{ cm}^2$

c Unpainted area $= 1000 - 296 = 704 \text{ cm}^2$

34 **a** $A =$ area of 5 semi-circles
$= 5 \times \left(\tfrac{1}{2}\pi r^2\right)$
$= \tfrac{5}{2}\pi r^2$

b $A =$ sector area $+$ area of $\triangle$
$= \tfrac{270}{360} \times \pi q^2 + \tfrac{1}{2} \times q \times q$
$= q^2 \left(\tfrac{3}{4}\pi + \tfrac{1}{2}\right)$

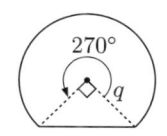

c $A =$ area of trapezium $-$ area of semi-circle
$= \left(\dfrac{2a+b}{2}\right) \times c - \tfrac{1}{2}\pi a^2$
$= \dfrac{(2a+b)c - \pi a^2}{2}$

d

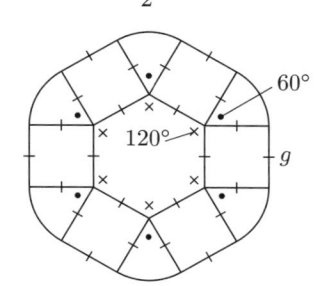

Area of each square $= g^2$
Area of each circle sector
$= \tfrac{60}{360} \times \pi g^2$
The inner hexagon is made up of 6 equilateral triangles of length g.

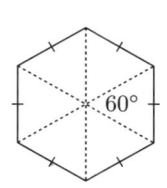

$\therefore$ area of hexagon $= 6 \times \left(\tfrac{1}{2} \times g \times g \times \sin 60°\right)$
$= 3g^2 \times \tfrac{\sqrt{3}}{2}$
$= \tfrac{3\sqrt{3}}{2}g^2$

$\therefore\; A = 6 \times g^2 + 6 \times \left(\tfrac{60}{360}\right)\pi g^2 + \tfrac{3\sqrt{3}}{2}g^2$
$= \left(6 + \pi + \tfrac{3\sqrt{3}}{2}\right)g^2$

35 Surface area $= \pi r l + \pi r^2$
$= \pi \times 7.5 \times 34 + \pi \times 7.5^2$
$\approx 978 \text{ mm}^2$

36 **a** Total mass $= 1.08$ tonnes $= 1080$ kg
$\therefore$ mass of each post $= \tfrac{1080}{60}$ kg $= 18$ kg

b Volume $= \pi r^2 h$
$= \pi \times 0.08^2 \times 1.8 \quad \{8 \text{ cm} = 0.08 \text{ m}\}$
$\approx 0.0362 \text{ m}^3$

37 **a** $A =$ area of hollow cylinder $+$ area of sphere
$= 2\pi \left(\dfrac{d}{2}\right) l + 4\pi \left(\dfrac{d}{2}\right)^2$
$= \pi d l + \pi d^2$
$= \pi d (l + d)$

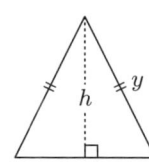

b In the triangular faces,
$h^2 + x^2 = y^2$
$\therefore\; h = \sqrt{y^2 - x^2} \quad \{h > 0\}$
$\therefore$ area of each triangular face
$= \tfrac{1}{2} \times 2x \times \sqrt{y^2 - x^2}$
$= x\sqrt{y^2 - x^2}$
$\therefore\; A = \underbrace{4 \times x\sqrt{y^2 - x^2}}_{\text{triangular faces}} + \underbrace{4(2x \times x)}_{\text{sides}} + \underbrace{(2x \times 2x)}_{\text{base}}$
$= 4x\sqrt{y^2 - x^2} + 8x^2 + 4x^2$
$= 4x\left(\sqrt{y^2 - x^2} + 3x\right)$

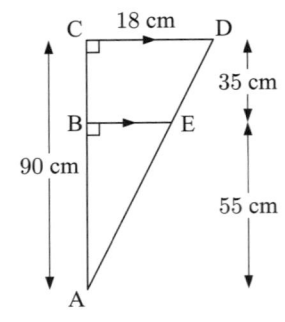

38

$A\hat{E}B = A\hat{D}C$ {corresponding angles}
$A\hat{B}E = A\hat{C}D = 90°$
$\therefore$ $\triangle$s ABE and ACD are similar, and
$$\frac{BE}{CD} = \frac{AB}{AC} \quad \{\text{same ratio}\}$$
$\therefore \frac{BE}{18} = \frac{55}{90}$
$\therefore$ BE = 11
$\therefore$ volume of bucket
$=$ volume of large cone $-$ volume of small cone
$= \frac{1}{3}\pi(18)^2 \times 90 - \frac{1}{3}\pi(11)^2 \times 55$
$\approx 23\,567\,\text{cm}^3$
$\therefore$ the bucket has capacity 23 567 ml.
In 3 hours or 180 minutes, the bucket loses
$180 \times 1.2 = 216$ ml of water.
$\therefore$ the amount of water remaining $\approx (23\,567 - 216)$ ml
$\approx 23\,351$ ml
≈ 23.4 litres

39 **a** $V = \frac{1}{3} \times$ area of base $\times$ height
$= \frac{1}{3}abh$

b $V =$ volume of hemisphere
$= \frac{1}{2} \times \left(\frac{4}{3}\pi p^3\right)$
$= \frac{2}{3}\pi p^3$

c $V =$ area of trapezium $\times$ length
$= \left(\frac{3a + 5a}{2} \times h\right) \times b$
$= 4abh$

40 Let the radius of the wedge be r cm.
Now volume $= 460$ cm^3
$\therefore \frac{1}{4}\pi r^2 \times 6.1 = 460$
$\therefore r^2 = \frac{1840}{6.1\pi}$
$\therefore r = \sqrt{\frac{1840}{6.1\pi}} \quad \{r > 0\}$
$\therefore r \approx 9.80$
$\therefore$ the radius of the wedge is 9.80 cm.

41 Area of end $= \frac{1}{2} \times \pi \left(\frac{a}{2}\right)^2 + 2a \times a$
$+ 2a \times 3a$
$= \frac{\pi a^2}{8} + 2a^2 + 6a^2$
$= a^2\left(\frac{\pi}{8} + 8\right)$
$\therefore V = a^2\left(\frac{\pi}{8} + 8\right) \times l$
$= a^2l\left(\frac{\pi}{8} + 8\right)$

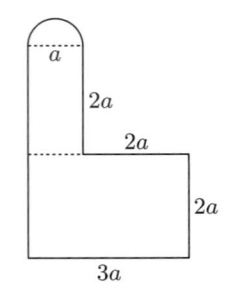

42 Volume of each handle $= \pi \times 3^2 \times 4$
$= 36\pi$ cm^3
Volume of shaft $= \pi \times 1.5^2 \times 12$
$= 27\pi$ cm^3
$\therefore$ total volume of door handle $= 2 \times 36\pi \times 27\pi$
$= 99\pi$
≈ 311 cm^3

43 **a** 55 litres $= 55\,000$ ml
$\therefore$ the water has volume 55 000 cm^3.
Suppose the water rises to a height of h cm.
$\therefore \pi \times 20^2 \times h = 55\,000$
$\therefore h = \frac{55\,000}{400\pi}$
$\therefore h \approx 43.77$
$\therefore$ the water is $50 - 43.77 = 6.23$ cm from the top.

b Volume of space remaining in aquarium
$\approx \pi \times 20^2 \times 6.23$
≈ 7831.853 cm^3
$\approx 7\,831\,853$ mm^3
Volume of each marble $= \frac{4}{3}\pi \times 6^3$
≈ 904.8 mm^3
So, $\frac{7\,831\,853}{904.8} \approx 8660$ marbles can be added before the aquarium overflows.

SOLUTIONS TO TOPIC 7: COORDINATE GEOMETRY

1 **a**

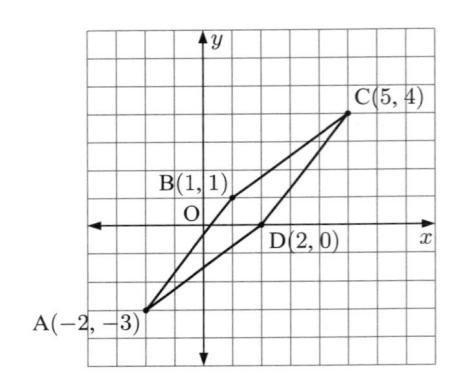

b $\overrightarrow{AB} = \begin{pmatrix} 3 \\ 4 \end{pmatrix}$, and $AB = \sqrt{3^2 + 4^2} = 5$ units

$\overrightarrow{BC} = \begin{pmatrix} 4 \\ 3 \end{pmatrix}$, and $BC = \sqrt{4^2 + 3^2} = 5$ units

$\overrightarrow{CD} = \begin{pmatrix} -3 \\ -4 \end{pmatrix}$, and $CD = \sqrt{(-3)^2 + (-4)^2} = 5$ units

$\overrightarrow{DA} = \begin{pmatrix} -4 \\ -3 \end{pmatrix}$, and $DA = \sqrt{(-4)^2 + (-3)^2} = 5$ units

All four sides are equal in length, so the points form a rhombus.

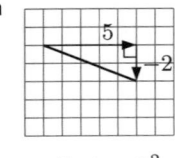

2 **a** gradient $= \frac{-2}{5}$
$= -\frac{2}{5}$

b gradient $= \frac{4}{2}$
$= 2$

c gradient $= \frac{4}{0}$
which is undefined

d gradient $= \frac{5}{6}$

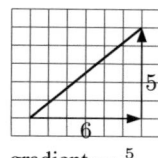

e gradient $= \frac{0}{4} = 0$

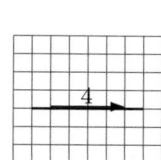

f gradient $= \frac{-1}{4} = -\frac{1}{4}$

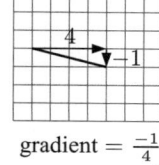

3 a $AB = \sqrt{(1--3)^2 + (4--1)^2}$
$= \sqrt{4^2 + 5^2}$
$= \sqrt{41}$ units

b $BC = \sqrt{(6-1)^2 + (-2-4)^2}$
$= \sqrt{5^2 + (-6)^2}$
$= \sqrt{61}$ units

c $AC = \sqrt{(6--3)^2 + (-2--1)^2}$
$= \sqrt{9^2 + (-1)^2}$
$= \sqrt{82}$ units

Hence, triangle ABC is scalene.

4 a $y = 5$ **b** $x = 3$

c $m = -2, \ c = 3 \ \therefore \ y = -2x + 3$

d The points $(4, 0)$ and $(0, -2)$ lie on the line.
$\therefore$ the line has gradient $\frac{-2-0}{0-4} = \frac{1}{2}$, and $c = -2$,
so its equation is $y = \frac{1}{2}x - 2$.

5 a $AB = \sqrt{(2-0)^2 + (1--3)^2}$
$= \sqrt{2^2 + 4^2}$
$= \sqrt{20}$ units
$BC = \sqrt{(6-2)^2 + (-1-1)^2}$
$= \sqrt{4^2 + (-2)^2}$
$= \sqrt{20}$ units
$AC = \sqrt{(6-0)^2 + (-1--3)^2}$
$= \sqrt{6^2 + 2^2}$
$= \sqrt{40}$ units

So $AB = BC$, and $AB^2 + BC^2 = 20 + 20$
$= 40$
$= AC^2$

So, triangle ABC is a right-angled isosceles triangle. The right angle is at B.

b $AB = \sqrt{(5-1)^2 + (4-1)^2}$
$= \sqrt{4^2 + 3^2}$
$= 5$ units
$BC = \sqrt{(-2-5)^2 + (-3-4)^2}$
$= \sqrt{(-7)^2 + (-7)^2}$
$= \sqrt{98}$ units
$AC = \sqrt{(-2-1)^2 + (-3-1)^2}$
$= \sqrt{(-3)^2 + (-4)^2}$
$= 5$ units
$AB = AC$, so triangle ABC is isosceles.

6 a 2 **b** $-\frac{1}{4}$ **c** $\frac{1}{3}$

7 a x-coordinate of M y-coordinate of M
$= \frac{5+0}{2}$ $= \frac{5+0}{2}$
$= \frac{5}{2}$ $= \frac{5}{2}$
So, M is $\left(\frac{5}{2}, \frac{5}{2}\right)$.

b $AP = OP$
$\therefore \ \sqrt{(x-5)^2 + (2-5)^2} = \sqrt{(x-0)^2 + (2-0)^2}$
$\therefore \ (x-5)^2 + (-3)^2 = x^2 + 2^2$
$\therefore \ x^2 - 10x + 25 + 9 = x^2 + 4$
$\therefore \ -10x = -30$
$\therefore \ x = 3$

c A point B on the line segment AD has the form $B(x, x), \ 0 \leqslant x \leqslant 5$.
$AB = 3\sqrt{2}$
$\therefore \ \sqrt{(x-5)^2 + (x-5)^2} = 3\sqrt{2}$
$\therefore \ (x-5)^2 + (x-5)^2 = 18$
$\therefore \ x^2 - 10x + 25 + x^2 - 10x + 25 = 18$
$\therefore \ 2x^2 - 20x + 32 = 0$
$\therefore \ x^2 - 10x + 16 = 0$
$\therefore \ (x-2)(x-8) = 0$
$\therefore \ x = 2 \text{ or } 8$
But $0 \leqslant x \leqslant 5$, so $x = 2$
$\therefore$ B is $(2, 2)$.

8 a gradient
$= \frac{-3-5}{3-1}$
$= \frac{-8}{2}$
$= -4$

b gradient
$= \frac{-9-1}{-5-1}$
$= \frac{-10}{-6}$
$= \frac{5}{3}$

c gradient
$= \frac{2-2}{15-4}$
$= \frac{0}{11}$
$= 0$

d gradient
$= \frac{-4-0}{5-5}$
$= \frac{-4}{0}$
which is undefined

9 a The midpoint of BC is
$M\left(\frac{4+3}{2}, \frac{5+-1}{2}\right)$ or $M\left(\frac{7}{2}, 2\right)$.
$\therefore \ AM = \sqrt{\left(\frac{7}{2}-1\right)^2 + (2-3)^2}$
$= \sqrt{\left(\frac{5}{2}\right)^2 + (-1)^2}$
$= \sqrt{\frac{29}{4}}$
$= \frac{\sqrt{29}}{2}$ units

b The diameter of the circle is $\frac{\sqrt{29}}{2}$ units
$\therefore$ the radius of the circle is $\frac{\sqrt{29}}{4}$ units.
$\therefore \ \text{area} = \pi r^2 = \pi \left(\frac{\sqrt{29}}{4}\right)^2$
$= \frac{29\pi}{16}$
≈ 5.69 units^2

10 a

x	-3	-2	-1	0	1	2	3
y	0	0.5	1	1.5	2	2.5	3

b

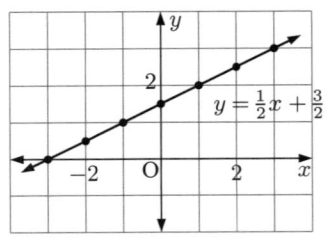

Graph showing line $y = \frac{1}{2}x + \frac{3}{2}$.

c Using the points $(1, 2)$ and $(3, 3)$

gradient $= \dfrac{3-2}{3-1} = \dfrac{1}{2}$

From the table, the x-intercept is -3 and the y-intercept is 1.5.

11 a gradient of RS $= -2$

{parallel lines have equal gradient}

$\therefore \dfrac{t-3}{-4--1} = -2$

$\therefore t - 3 = 6$

$\therefore t = 9$

b gradient of AB $= -\dfrac{3}{2}$

{perpendicular to line of gradient $\frac{2}{3}$}

$\therefore \dfrac{3-5}{2-t} = -\dfrac{3}{2}$

$\therefore \dfrac{-2}{2-t} \times \left(\dfrac{2}{2}\right) = \dfrac{-3}{2} \times \left(\dfrac{2-t}{2-t}\right)$

$\therefore -4 = -3(2-t)$ {equating numerators}

$\therefore -4 = -6 + 3t$

$\therefore 2 = 3t$

$\therefore t = \dfrac{2}{3}$

12 a The gradient $m = \dfrac{5-3}{3-0} = \dfrac{2}{3}$.

The y-intercept $c = 3$.

$\therefore$ the equation is $y = \dfrac{2}{3}x + 3$.

b $(0, 4)$ and $(6, 0)$ lie on the line.

$\therefore$ the gradient $m = \dfrac{0-4}{6-0} = \dfrac{-4}{6} = -\dfrac{2}{3}$.

The y-intercept $c = 4$.

$\therefore$ the equation is $y = -\dfrac{2}{3}x + 4$.

c $(1, 4)$ and $(3, 0)$ lie on the line.

$\therefore$ the gradient $m = \dfrac{0-4}{3-1} = \dfrac{-4}{2} = -2$.

$\therefore$ the equation is $y = -2x + c$.

But when $x = 1$, $y = 4$

$\therefore 4 = -2(1) + c$

$\therefore c = 6$

So, the equation is $y = -2x + 6$.

d $(0, -2)$ and $(3, 0)$ lie on the line.

$\therefore$ the gradient $m = \dfrac{0--2}{3-0} = \dfrac{2}{3}$.

The y-intercept $c = -2$.

$\therefore$ the equation is $y = \dfrac{2}{3}x - 2$.

13 gradient of DE $= \dfrac{6-5}{1--2} = \dfrac{1}{3}$

gradient of EF $= \dfrac{3-6}{-6-1} = \dfrac{-3}{-7} = \dfrac{3}{7}$

$\therefore$ DE is not parallel to EF, so D, E, and F do not lie in a straight line.

14 a Let C be (a, b).

$\therefore \dfrac{a+1}{2} = 3$ and $\therefore \dfrac{b+2}{2} = 6$

$\therefore a + 1 = 6$ and $\therefore b + 2 = 12$

$\therefore a = 5$ and $\therefore b = 10$

$\therefore$ C is $(5, 10)$.

b AB $= \sqrt{(9-1)^2 + (3-2)^2}$

$= \sqrt{8^2 + 1^2}$

$= \sqrt{65}$ units

BC $= \sqrt{(5-9)^2 + (10-3)^2}$

$= \sqrt{(-4)^2 + 7^2}$

$= \sqrt{65}$ units

AC $= \sqrt{(5-1)^2 + (10-2)^2}$

$= \sqrt{4^2 + 8^2}$

$= \sqrt{80}$ units

AB $=$ BC, so triangle ABC is isosceles.

15 a $m = -1$ and $c = 7$,

so the equation is $y = -x + 7$.

b $m = 2$, so the equation is $y = 2x + c$.

But when $x = -2$, $y = -5$

$\therefore -5 = 2(-2) + c$

$\therefore c = -1$

So, the equation is $y = 2x - 1$.

c The gradient $m = \dfrac{3--5}{-1-2} = \dfrac{8}{-3} = -\dfrac{8}{3}$

$\therefore$ the equation is $y = -\dfrac{8}{3}x + c$.

But when $x = -1$, $y = 3$

$\therefore 3 = -\dfrac{8}{3}(-1) + c$

$\therefore 3 = \dfrac{8}{3} + c$

$\therefore c = \dfrac{1}{3}$

So, the equation is $y = -\dfrac{8}{3}x + \dfrac{1}{3}$.

16 a The gradient $m = \dfrac{80-35}{20-0} = \dfrac{45}{20} = \dfrac{9}{4}$

The y-intercept $c = 35$.

The variable on the vertical axis is n, and the variable on the horizontal axis is t.

So, the equation is $n = \dfrac{9}{4}t + 35$.

b The gradient $m = \dfrac{26-86}{18-6} = \dfrac{-60}{12} = -5$

$\therefore$ the equation is $y = -5x + c$.

But when $x = 6$, $y = 86$

$\therefore 86 = -5(6) + c$

$\therefore c = 116$

The variable on the vertical axis is P, and the variable on the horizontal axis is t.

So, the equation is $P = -5t + 116$.

17 **a** The intercept on the vertical axis is 105, which means there is a supply charge of $105.

  b Using the points $(0, 105)$ and $(14, 140)$,
  gradient $= \dfrac{140 - 105}{14 - 0} = \dfrac{35}{14} = 2.5$
  This means that the cost per kilolitre of water is $2.50.

  c The amount A charged for using k kilolitres of water is $A = 2.5k + 105$ dollars.
  When $k = 7$, $A = 2.5(7) + 105 = 122.5$
  So, the amount paid for using 7 kl of water is $122.50.

  d When $A = 120$, $120 = 2.5k + 105$
  $\therefore\ 15 = 2.5k$
  $\therefore\ k = 6$
  So, a household which was billed $120 used 6 kl of water.

  e The new amount charged $A' = 1.4k + 115$ dollars.
  When $k = 7$, $A' = 1.4(7) + 115 = 124.8$
  Under the new cost structure the household is charged $124.80.
  So, using **c**, the household is worse off.

18 The line $2x - y = 5$ has gradient $\dfrac{2}{1} = 2$.

  $\therefore\ $ the required line has gradient $-\dfrac{1}{2}$, and passes through $(-1, 2)$.

  $\therefore\ $ its equation is $x + 2y = (-1) + 2(2)$
  $\therefore\ x + 2y = 3$

19 Since J, K, and L lie in a straight line,
  gradient of JK = gradient of KL
  $\therefore\ \dfrac{1 - -2}{-2 - z} = \dfrac{-1 - 1}{3 - -2}$
  $\therefore\ \dfrac{3}{-2 - z} = \dfrac{-2}{5}$
  $\therefore\ 15 = -2(-2 - z)$
  $\therefore\ 15 = 4 + 2z$
  $\therefore\ 11 = 2z$
  $\therefore\ z = \dfrac{11}{2}$

20 Substituting $(a, 4)$ into $2x - 3y = a$ gives
  $2(a) - 3(4) = a$
  $\therefore\ 2a - 12 = a$
  $\therefore\ a = 12$

21 **a** $x \geqslant 0$ and $y \leqslant 0$ **b** $y > 5$

  c The top line has gradient 1 and y-intercept 2, so its equation is $y = x + 2$.
  The bottom line has gradient 1 and y-intercept -1, so its equation is $y = x - 1$.
  So, the unshaded region is $y \leqslant x + 2$ and $y > x - 1$.

  d The diagonal line has gradient 1 and y-intercept 3, so its equation is $y = x + 3$.
  So, the unshaded region is $y \leqslant x + 3$, $x \leqslant 0$, $y \geqslant 0$.

22 **a** $y = -\dfrac{1}{3}x + 2$
  $\therefore\ \dfrac{1}{3}x + y = 2$
  $\therefore\ x + 3y = 6$

  b $2x + 3y = -4$
  $\therefore\ 3y = -2x - 4$
  $\therefore\ y = -\dfrac{2}{3}x - \dfrac{4}{3}$

23 **a** The gradient $= \dfrac{13 - 7}{5 - -3} = \dfrac{6}{8} = \dfrac{3}{4}$.
  $\therefore\ $ its equation is $3x - 4y = 3(-3) - 4(7)$
  $\therefore\ 3x - 4y = -37$

  b $\left(\dfrac{1}{2}, \dfrac{9}{2}\right)$ and $(2, 0)$ lie on the line.
  $\therefore\ $ the gradient $= \dfrac{0 - \frac{9}{2}}{2 - \frac{1}{2}} = \dfrac{-\frac{9}{2}}{\frac{3}{2}} = -3$.
  $\therefore\ $ its equation is $3x + y = 3(2) + 0$
  $\therefore\ 3x + y = 6$

24 **a** The midpoint of AB is $M\left(\dfrac{1+7}{2}, \dfrac{1+5}{2}\right)$ or $M(4, 3)$.

  b gradient $= \dfrac{5 - 1}{7 - 1} = \dfrac{4}{6} = \dfrac{2}{3}$

  c The line through M and perpendicular to AB has gradient $-\dfrac{3}{2}$, and passes through $(4, 3)$
  $\therefore\ $ its equation is $3x + 2y = 3(4) + 2(3)$
  $\therefore\ 3x + 2y = 18$

  d AC = BC
  $\therefore\ \sqrt{(8-1)^2 + (y-1)^2} = \sqrt{(8-7)^2 + (y-5)^2}$
  $\therefore\ 7^2 + (y-1)^2 = 1^2 + (y-5)^2$
  $\therefore\ 49 + y^2 - 2y + 1 = 1 + y^2 - 10y + 25$
  $\therefore\ 8y = -24$
  $\therefore\ y = -3$
  $\therefore\ $ the y-coordinate of C is -3.

  e Substituting $(8, -3)$ into $3x + 2y = 18$ gives
  $3(8) + 2(-3) = 18$, which is true
  $\therefore\ $ C lies on the line through M which is perpendicular to AB.

  f The points on the perpendicular bisector of AB are equidistant from A and B.

25 **a** gradient of line $A = \dfrac{20 - 0}{5 - 0} = 4$
  The cost of sending parcels using normal mail is $4 per kg.
  gradient of line $B = \dfrac{45 - 0}{5 - 0} = 9$
  The cost of sending parcels using express mail is $9 per kg.

  b Sending a 3 kg parcel costs $4 \times 3 = \$12$ using normal mail and $9 \times 3 = \$27$ using express mail.
  $\therefore\ $ it costs $\$27 - \$12 = \$15$ more to send the parcel using express mail.

26 **a** A line parallel to $3x - y = 5$ has the form $3x - y = k$.
  But when $x = 0$, $y = 0$
  $\therefore\ 3(0) - 0 = k$
  $\therefore\ k = 0$
  $\therefore\ $ the equation is $3x - y = 0$.

  b The line $x + 4y = -3$ has gradient $-\dfrac{1}{4}$.
  $\therefore\ $ the required line has gradient 4, and passes through $(3, 2)$.
  $\therefore\ $ it has equation $4x - y = 4(3) - 2$
  $\therefore\ 4x - y = 10$

27 a, b, 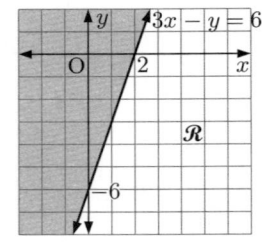 c, 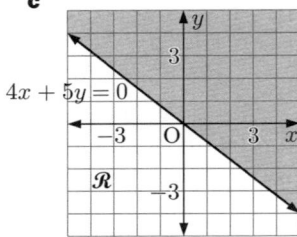 d, 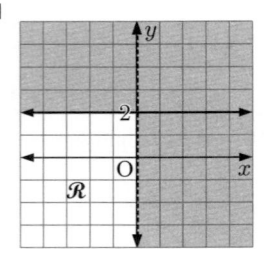 e, 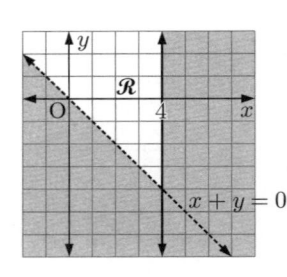 f [graphs showing shaded regions $\mathcal{R}$ for various linear inequalities]

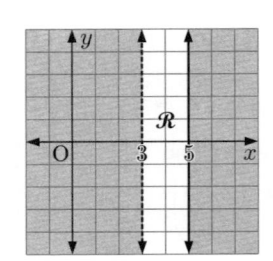

b: $3x - y = 6$
c: $4x + 5y = 0$
e: $x + y = 0$
f: $y = x$

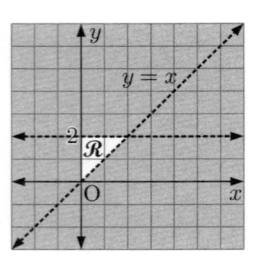

28 The line $y = 3x - 7$ has gradient 3, and the line $2x + ay = -1$ has gradient $-\dfrac{2}{a}$.

  a If the lines are parallel, then
  $$3 = -\dfrac{2}{a}$$
  $$\therefore \ a = -\tfrac{2}{3}$$

  b If the lines are perpendicular, then
  $$-\dfrac{2}{a} = -\dfrac{1}{3}$$
  $$\therefore \ -2 = -\tfrac{1}{3}a$$
  $$\therefore \ a = 6$$

29 **a** The y-intercept is 2.
  When $x = 1$, $y = 3 + 2 = 5$.
  $\therefore \ (0, 2)$ and $(1, 5)$ lie on the line.

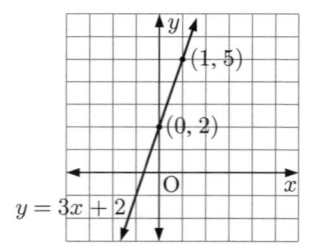

  [graph of $y = 3x + 2$ showing points $(0, 2)$ and $(1, 5)$]

  b When $x = 0$, $-y = 8$
  $$\therefore \ y = -8$$
  When $y = 0$, $4x = 8$
  $$\therefore \ x = 2$$

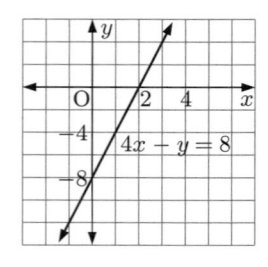

  [graph of $4x - y = 8$]

c The y-intercept is 6.
  When $x = 2$, $y = -1 + 6 = 5$.
  $\therefore \ (0, 6)$ and $(2, 5)$ lie on the line.

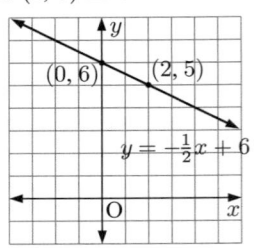

  [graph of $y = -\tfrac{1}{2}x + 6$ through $(0, 6)$ and $(2, 5)$]

d When $x = 0$, $3y = 9$
  $\therefore \ y = 3$
  When $y = 0$, $x = 9$

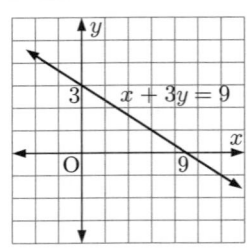

  [graph of $x + 3y = 9$]

30 **a** **i** $OA = \sqrt{(1-0)^2 + (3-0)^2}$
  $= \sqrt{1^2 + 3^2}$
  $= \sqrt{10}$ units

  ii $AB = \sqrt{(4-1)^2 + (4-3)^2}$
  $= \sqrt{3^2 + 1^2}$
  $= \sqrt{10}$ units

  iii $BC = \sqrt{(6-4)^2 + (-2-4)^2}$
  $= \sqrt{2^2 + (-6)^2}$
  $= \sqrt{40}$ units

  iv $OC = \sqrt{(6-0)^2 + (-2-0)^2}$
  $= \sqrt{6^2 + (-2)^2}$
  $= \sqrt{40}$ units

  b $OA = AB$, $OC = BC$, and AC is common to triangles OAC and BAC.
  $\therefore$ triangles OAC and BAC are congruent {SSS}

  c OABC contains two pairs of equal adjacent sides.
  $\therefore$ OABC is a kite.

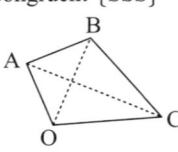

  [diagram of kite OABC with diagonals]

  d **i** gradient of OB $= \dfrac{4-0}{4-0} = \tfrac{4}{4} = 1$

  ii gradient of AC $= \dfrac{-2-3}{6-1} = \tfrac{-5}{5} = -1$

  e Since 1 and -1 are negative reciprocals, OB is perpendicular to AC.
  We have shown that the diagonals of a kite are perpendicular.

  f The midpoint of OB is $\left(\dfrac{0+4}{2}, \dfrac{0+4}{2}\right)$ or $(2, 2)$.
  This is the point where the diagonals of the kite intersect.

31 **a** The line $y = \tfrac{1}{3}x + 4$ has gradient $\tfrac{1}{3}$ and y-intercept 4.

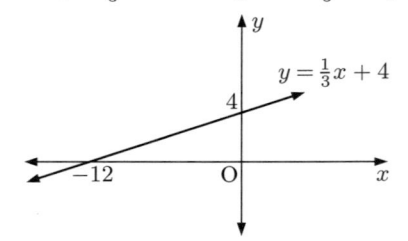

  [graph of $y = \tfrac{1}{3}x + 4$ through $(-12, 0)$ and $(0, 4)$]

b

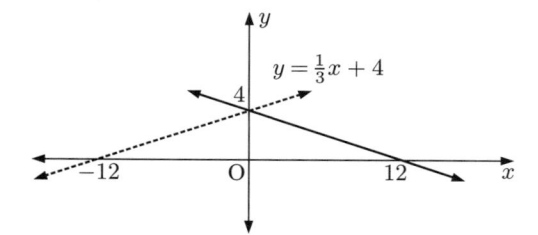

The reflected line has gradient $-\frac{1}{3}$ and y-intercept 4.

∴ its equation is $y = -\frac{1}{3}x + 4$.

c

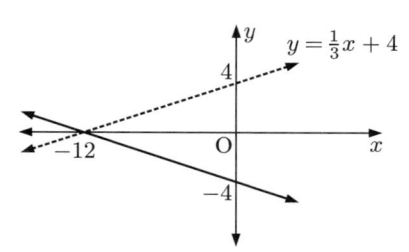

The reflected line has gradient $-\frac{1}{3}$ and y-intercept -4.

∴ its equation is $y = -\frac{1}{3}x - 4$.

d

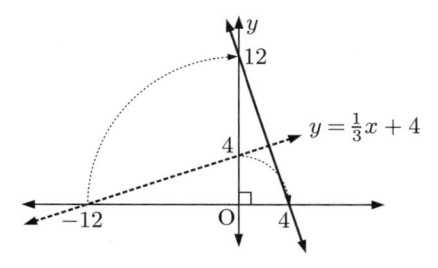

$(0, 12)$ and $(4, 0)$ lie on the rotated line.

∴ the line has gradient $= \dfrac{0-12}{4-0} = \dfrac{-12}{4} = -3$ and y-intercept 12.

∴ the equation is $y = -3x + 12$.

32 a The dotted line passes through $(-1, 3)$ and $(0, 4)$.

∴ the line has gradient $= \dfrac{4-3}{0--1} = 1$ and y-intercept 4.

∴ the equation is $y = x + 4$.

The solid line passes through $(-2, 0)$ and $(-1, 3)$.

∴ the line has gradient $= \dfrac{3-0}{-1--2} = 3$

∴ the line has equation $3x - y = 3(-2) - 0$

∴ $y = 3x + 6$

∴ $\mathcal{R}$ is represented by the inequalities $y < x + 4$, $y \leqslant 3x + 6$, $x \leqslant 0$, $y \geqslant 0$.

b

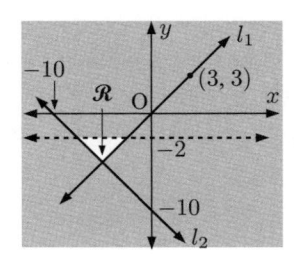

Line l_1 passes through $O(0, 0)$ and $(3, 3)$.

∴ l_1 has equation $y = x$.

Line l_2 passes through $(-10, 0)$ and $(0, -10)$.

∴ the line has gradient $= \dfrac{-10-0}{0--10} = \dfrac{-10}{10} = -1$

∴ l_2 has equation $x + y = (-10) + 0$

∴ $x + y = -10$

∴ $\mathcal{R}$ is represented by the inequalities $y \geqslant x$, $x + y \geqslant -10$, $y < -2$.

33 a

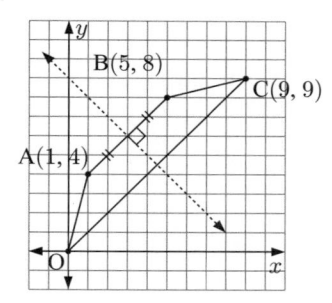

b gradient of AB $= \dfrac{8-4}{5-1} = \dfrac{4}{4} = 1$

gradient of OC $= \dfrac{9-0}{9-0} = \dfrac{9}{9} = 1$

∴ AB and OC are parallel, so OABC is a trapezium.

c The line symmetry is perpendicular to AB, and passes through the midpoint of AB.

From **b**, the gradient of AB is 1, and the midpoint of AB is $\left(\dfrac{1+5}{2}, \dfrac{4+8}{2}\right)$ or $(3, 6)$.

So, the line of symmetry has gradient -1 and passes through $(3, 6)$.

∴ its equation is $x + y = (3) + (6)$

∴ $x + y = 9$

34 a

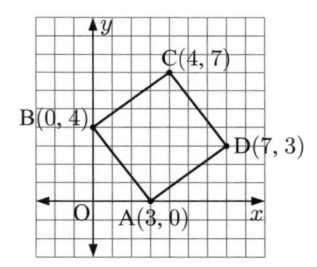

b AB $= \sqrt{(0-3)^2 + (4-0)^2} = \sqrt{(-3)^2 + 4^2}$
$= 5$ units

BC $= \sqrt{(4-0)^2 + (7-4)^2} = \sqrt{4^2 + 3^2}$
$= 5$ units

CD $= \sqrt{(7-4)^2 + (3-7)^2} = \sqrt{3^2 + (-4)^2}$
$= 5$ units

DA $= \sqrt{(3-7)^2 + (0-3)^2} = \sqrt{(-4)^2 + (-3)^2}$
$= 5$ units

So, all the sides are equal in length.

Now, gradient of AB $= \dfrac{4-0}{0-3} = \dfrac{4}{-3} = -\dfrac{4}{3}$ and

gradient of BC $= \dfrac{7-4}{4-0} = \dfrac{3}{4}$

Since $-\dfrac{4}{3}$ and $\dfrac{3}{4}$ are negative reciprocals, AB is perpendicular to BC.

∴ ABCD is a square.

c

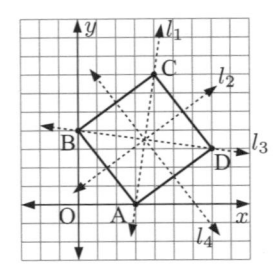

ABCD has 4 lines of symmetry.

l_1 passes through A(3, 0) and C(4, 7).

∴ it has gradient $\frac{7-0}{4-3} = 7$

and equation $7x - y = 7(3) - (0)$

∴ $7x - y = 21$

l_2 is parallel to BC, and passes through the midpoint of C(4, 7) and D(7, 3)

∴ l_2 has gradient $\frac{3}{4}$, and passes through $\left(\frac{11}{2}, 5\right)$.

∴ l_2 has equation $3x - 4y = 3\left(\frac{11}{2}\right) - 4(5)$

∴ $3x - 4y = -\frac{7}{2}$

l_3 passes through B(0, 4) and D(7, 3).

∴ it has gradient $\frac{3-4}{7-0} = -\frac{1}{7}$

and equation $x + 7y = (0) + 7(4)$

∴ $x + 7y = 28$

l_4 is parallel to AB, and passes through the midpoint of A(3, 0) and D(7, 3).

∴ l_4 has gradient $-\frac{4}{3}$, and passes through $\left(5, \frac{3}{2}\right)$.

∴ l_4 has equation $4x + 3y = 4(5) + 3\left(\frac{3}{2}\right)$

∴ $4x + 3y = \frac{49}{2}$

35 a

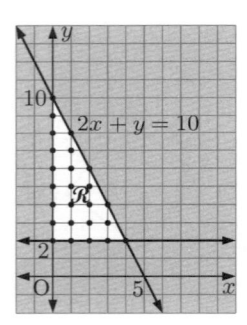

b 25 points in $\mathcal{R}$ have integer coordinates.

c The points (0, 7), (1, 6), (2, 5), and (3, 4) satisfy $x + y = 7$.

∴ 4 of the points obey the rule $x + y = 7$.

36 a If AB ∥ CD then
gradient of AB = gradient of CD

∴ $\frac{3-1}{6-2} = \frac{k-5}{1-7}$

∴ $\frac{2}{4} = \frac{k-5}{-6}$

∴ $-3 = k - 5$

∴ $k = 2$

b i gradient of AD = $\frac{2-1}{1-2} = \frac{1}{-1} = -1$

gradient of DE = $\frac{-1-2}{4-1} = \frac{-3}{3} = -1$

∴ AD is parallel to DE, so A, D, and E are collinear.

ii gradient of BC = $\frac{5-3}{7-6} = 2$

gradient of CE = $\frac{-1-5}{4-7} = \frac{-6}{-3} = 2$

∴ BC is parallel to CE, so B, C, and E are collinear.

c i AE = $\sqrt{(4-2)^2 + (-1-1)^2} = \sqrt{2^2 + (-2)^2}$
    = $\sqrt{8}$ units

ii DE = $\sqrt{(4-1)^2 + (-1-2)^2} = \sqrt{3^2 + (-3)^2}$
    = $\sqrt{18}$ units

iii BE = $\sqrt{(4-6)^2 + (-1-3)^2} = \sqrt{(-2)^2 + (-4)^2}$
    = $\sqrt{20}$ units

iv CE = $\sqrt{(4-7)^2 + (-1-5)^2} = \sqrt{(-3)^2 + (-6)^2}$
    = $\sqrt{45}$ units

d AB ∥ CD

∴ $\alpha_1 = \alpha_2$ and $\beta_1 = \beta_2$
{corresponding angles}

θ is common to both triangles.

∴ triangles ABE and DCE are equiangular and hence similar.

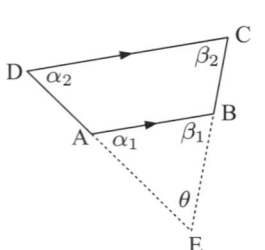

Check: $\frac{AE}{DE} = \frac{\sqrt{8}}{\sqrt{18}} = \frac{2\sqrt{2}}{3\sqrt{2}} = \frac{2}{3}$

and $\frac{BE}{CE} = \frac{\sqrt{20}}{\sqrt{45}} = \frac{2\sqrt{5}}{3\sqrt{5}} = \frac{2}{3}$ ✓

37 a

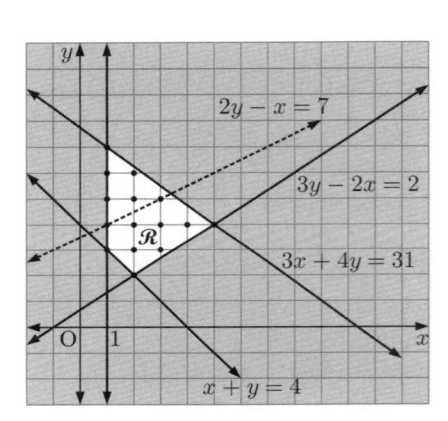

b The points in $\mathcal{R}$ with integer coordinates which lie on the line $2y - x = 7$ are (1, 4) and (3, 5).

SOLUTIONS TO TOPIC 8: TRIGONOMETRY

1 a The hypotenuse is the longest side, which is QR.

b The side opposite α is PQ.

c The side adjacent to α is PR.

2 $BC^2 = 3^2 + 7^2$ {Pythagoras}

∴ $BC^2 = 58$

∴ $BC = \sqrt{58}$ {BC > 0}

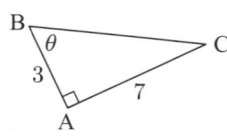

So, the hypotenuse has length $\sqrt{58}$ units.

a $\sin\theta = \frac{OPP}{HYP} = \frac{7}{\sqrt{58}}$ **b** $\cos\theta = \frac{ADJ}{HYP} = \frac{3}{\sqrt{58}}$

c $\tan\theta = \frac{OPP}{ADJ} = \frac{7}{3}$

3 a $\sin 25° = \dfrac{x}{12}$ $\left\{\sin\theta = \dfrac{\text{OPP}}{\text{HYP}}\right\}$

$\therefore \sin 25° \times 12 = x$

$\therefore x \approx 5.07$

b $\tan 40° = \dfrac{3}{x}$ $\left\{\tan\theta = \dfrac{\text{OPP}}{\text{ADJ}}\right\}$

$\therefore x = \dfrac{3}{\tan 40°}$

$\therefore x \approx 3.58$

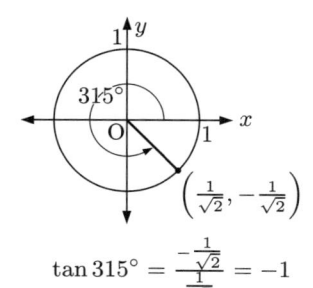

4 a $\left(-\dfrac{1}{\sqrt{2}}, \dfrac{1}{\sqrt{2}}\right)$, 135°

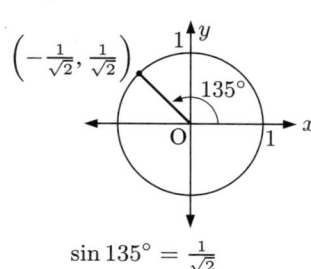

$\sin 135° = \dfrac{1}{\sqrt{2}}$

b 315°, $\left(\dfrac{1}{\sqrt{2}}, -\dfrac{1}{\sqrt{2}}\right)$

$\tan 315° = \dfrac{-\frac{1}{\sqrt{2}}}{\frac{1}{\sqrt{2}}} = -1$

c 240°, $\left(-\dfrac{1}{2}, -\dfrac{\sqrt{3}}{2}\right)$

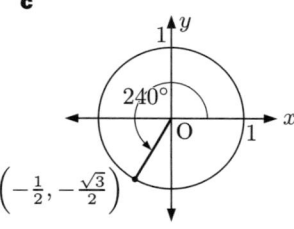

$\cos 240° = -\dfrac{1}{2}$

5 a $\cos 21° = \dfrac{16.12}{x}$ $\left\{\cos\theta = \dfrac{\text{ADJ}}{\text{HYP}}\right\}$

$\therefore x = \dfrac{16.12}{\cos 21°}$

$\therefore x \approx 17.27$

b $\sin 50° = \dfrac{5.21}{x}$ $\left\{\sin\theta = \dfrac{\text{OPP}}{\text{HYP}}\right\}$

$\therefore x = \dfrac{5.21}{\sin 50°}$

$\therefore x \approx 6.80$

6 $\cos 45° = \dfrac{3}{x}$

$\therefore x = \dfrac{3}{\cos 45°}$

$\therefore x \approx 4.24$

$\therefore$ length of metal required
$\approx 2 + 4.24 + 2$
≈ 8.24 cm

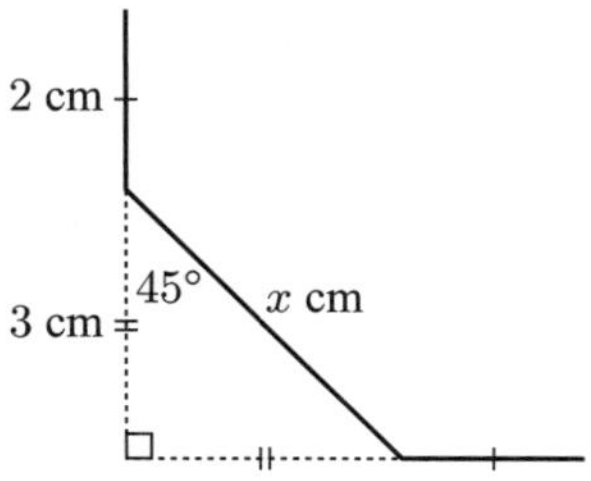

7 a Area $= \frac{1}{2} \times 6 \times 8 \times \sin 50° \approx 18.4$ cm^2

b Area $= \frac{1}{2} \times 10 \times 10 \times \sin 27° \approx 22.7$ cm^2

c Area $= \frac{1}{2} \times 1.7 \times 2.5 \times \sin 39° \approx 1.34$ km^2

8 a $\cos\theta = \dfrac{5}{11}$ $\left\{\cos\theta = \dfrac{\text{ADJ}}{\text{HYP}}\right\}$

$\therefore \theta = \cos^{-1}\left(\dfrac{5}{11}\right)$

$\therefore \theta \approx 63.0°$

b $\tan\theta = \dfrac{6}{5}$ $\left\{\tan\theta = \dfrac{\text{OPP}}{\text{ADJ}}\right\}$

$\therefore \theta = \tan^{-1}\left(\dfrac{6}{5}\right)$

$\therefore \theta \approx 50.2°$

9

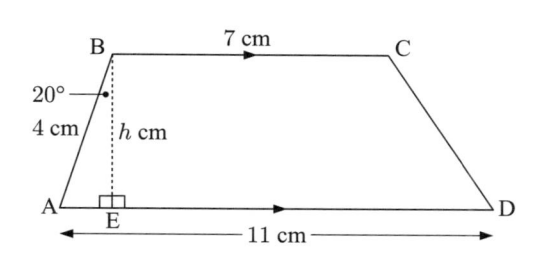

We draw BE perpendicular to BC and AD.

In $\triangle$ABE, $\cos 20° = \dfrac{h}{4}$

$\therefore \cos 20° \times 4 = h$

$\therefore h \approx 3.759$

$\therefore$ area of trapezium $\approx \left(\dfrac{7+11}{2}\right) \times 3.759 \approx 33.8$ cm^2

10 a $\cos\theta = x$-coordinate of P $= 0.5299$

b $\sin\theta = y$-coordinate of P $= 0.8480$

c $\tan\theta = \dfrac{\sin\theta}{\cos\theta} = \dfrac{0.8480}{0.5299} \approx 1.60$

11 area $= 175$ cm^2

$\therefore \frac{1}{2} \times x \times 31 \times \sin 72° = 175$

$\therefore x = \dfrac{175 \times 2}{31 \times \sin 72°}$

$\therefore x \approx 11.9$

12

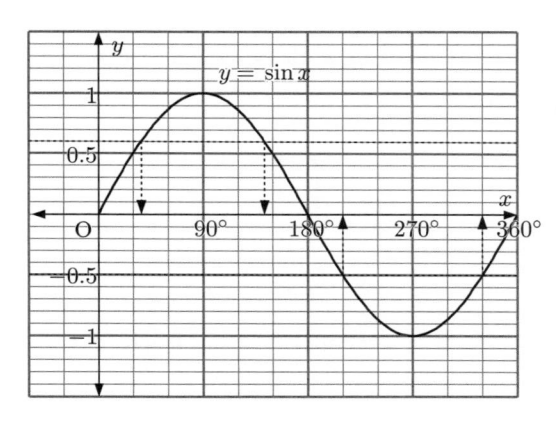

a The solutions to $\sin x = 0.6$ are $x \approx 37°, 143°$

b The solutions to $\sin x = -0.5$ are $x = 210°, 330°$

13 a $\sin 30° = \dfrac{x}{2}$ $\left\{\sin\theta = \dfrac{\text{OPP}}{\text{HYP}}\right\}$

$\therefore \dfrac{1}{2} = \dfrac{x}{2}$

$\therefore x = 1$

b $\sin 45° = \dfrac{3}{x}$

$\therefore \dfrac{1}{\sqrt{2}} = \dfrac{3}{x}$

$\therefore x = 3\sqrt{2}$

c $\sin 60° = \dfrac{x}{4}$

$\therefore \dfrac{\sqrt{3}}{2} = \dfrac{x}{4}$

$\therefore x = 2\sqrt{3}$

14 a $\dfrac{x}{\sin 125°} = \dfrac{5.7}{\sin 26°}$ {sine rule}

$\therefore x = \dfrac{5.7 \times \sin 125°}{\sin 26°}$

$\therefore x \approx 10.7$

b $\dfrac{x}{\sin 37°} = \dfrac{12.1}{\sin 51°}$ {sine rule}

$\therefore\ x = \dfrac{12.1 \times \sin 37°}{\sin 51°}$

$\therefore\ x \approx 9.37$

c $\dfrac{x}{\sin 49°} = \dfrac{3.5}{\sin 86°}$ {sine rule}

$\therefore\ x = \dfrac{3.5 \times \sin 49°}{\sin 86°}$

$\therefore\ x \approx 2.65$

15

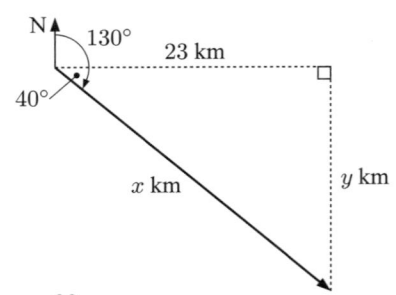

a $\cos 40° = \dfrac{23}{x}$

$\therefore\ x = \dfrac{23}{\cos 40°}$

$\therefore\ x \approx 30.0$

The hovercraft has travelled 30.0 km.

b $\tan 40° = \dfrac{y}{23}$

$\therefore\ y = 23 \times \tan 40°$

$\therefore\ y \approx 19.3$

The hovercraft is 19.3 km south of its starting point.

16

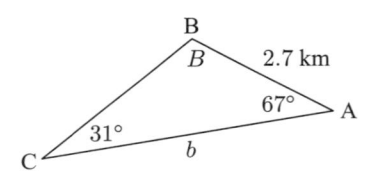

$31° + 67° + B = 180°$ {angles of a triangle}

$\therefore\ B = 82°$

$\therefore\ \dfrac{b}{\sin 82°} = \dfrac{2.7}{\sin 31°}$ {sine rule}

$\therefore\ b = \dfrac{2.7 \times \sin 82°}{\sin 31°}$

$\therefore\ b \approx 5.19$ km

17 a

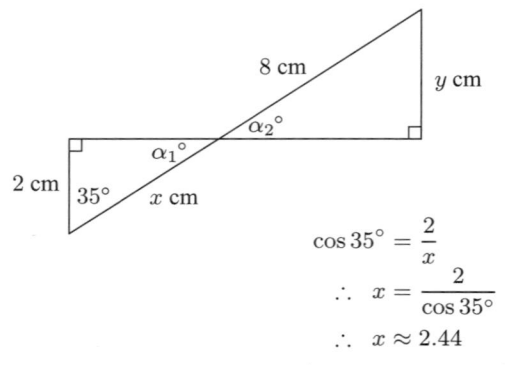

$\cos 35° = \dfrac{2}{x}$

$\therefore\ x = \dfrac{2}{\cos 35°}$

$\therefore\ x \approx 2.44$

Now $90 + 35 + \alpha_1 = 180$ {angles of a triangle}

$\therefore\ \alpha_1 = 55$

$\therefore\ \alpha_2 = 55$ {vertically opposite angles}

So, $\sin 55° = \dfrac{y}{8}$

$\therefore\ y = 8 \times \sin 55°$

$\therefore\ y \approx 6.55$

b

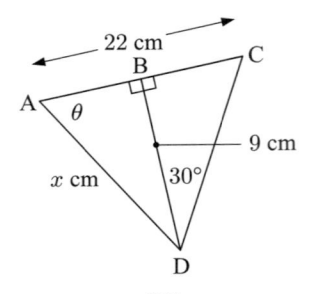

In $\triangle BCD$, $\tan 30° = \dfrac{BC}{9}$

$\therefore\ BC = 9 \times \tan 30°$

≈ 5.196 cm

$\therefore\ AB \approx 22 - 5.196 \approx 16.804$ cm

$\therefore$ in $\triangle ABD$, $\tan \theta \approx \dfrac{9}{16.804}$

$\therefore\ \theta \approx \tan^{-1}\left(\dfrac{9}{16.804}\right)$

$\approx 28.2°$

and $x^2 \approx 9^2 + 16.804^2$ {Pythagoras}

$\therefore\ x \approx \sqrt{9^2 + 16.804^2}$ $\{x > 0\}$

$\therefore\ x \approx 19.1$

18

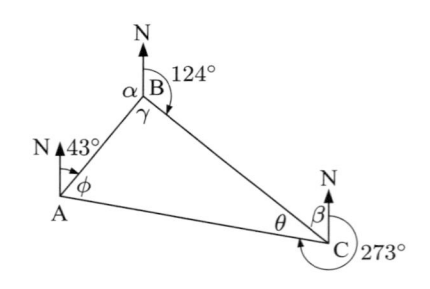

$\alpha = 180° - 43° = 137°$ {co-interior angles}

$\therefore\ \gamma = 360° - 124° - 137° = 99°$ {angles at a point}

$\beta = 180° - 124° = 56°$ {co-interior angles}

$\therefore\ \theta = 360° - 273° - 56° = 31°$ {angles at a point}

$\therefore\ \phi = 180° - 99° - 31° = 50°$ {angles of a triangle}

a bearing of B from A $= 043°$

b bearing of C from B $= 124°$

c bearing of B from C $= 273° + \theta = 304°$

d bearing of C from A $= 43° + \phi = 093°$

e bearing of A from B $= 124° + \gamma = 223°$

f bearing of A from C $= 273°$

19 a By the cosine rule:

$x^2 = 8^2 + 9^2 - 2 \times 8 \times 9 \times \cos 77°$

$\therefore\ x = \sqrt{8^2 + 9^2 - 2 \times 8 \times 9 \times \cos 77°}$ $\{x > 0\}$

$\therefore\ x \approx 10.6$

b $\cos x° = \dfrac{5^2 + 14^2 - 17^2}{2 \times 5 \times 14}$ {cosine rule}

$\therefore\ x° = \cos^{-1}\left(\dfrac{5^2 + 14^2 - 17^2}{2 \times 5 \times 14}\right)$

$\therefore\ x \approx 119$

c By the cosine rule:
$$210^2 = x^2 + x^2 - 2 \times x \times x \times \cos 33°$$
$$\{2.1\,\text{m} = 210\,\text{cm}\}$$
$$\therefore\ 210^2 = 2x^2 - 2x^2 \times \cos 33°$$
$$\therefore\ 210^2 = 2x^2(1 - \cos 33°)$$
$$\therefore\ x^2 = \frac{210^2}{2(1 - \cos 33°)}$$
$$\therefore\ x = \sqrt{\frac{210^2}{2(1 - \cos 33°)}} \quad \{x > 0\}$$
$$\therefore\ x \approx 370$$

20 $\tan 60° \sin 60° + \cos 60° = \sqrt{3} \times \frac{\sqrt{3}}{2} + \frac{1}{2}$
$$= \frac{3}{2} + \frac{1}{2} = 2$$

21

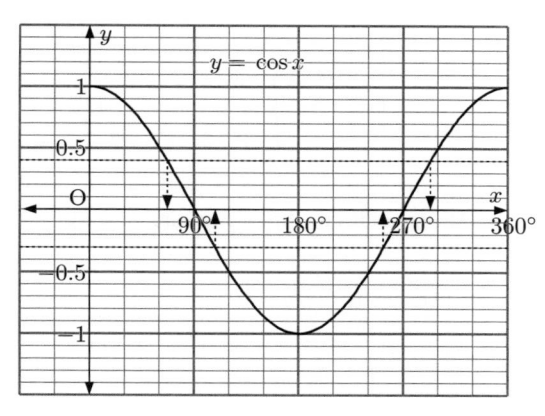

a The solutions to $\cos x = -0.3$ are $x \approx 107°, 253°$
b The solutions to $\cos x = 0.4$ are $x \approx 66°, 294°$

22 a Tangents from an external point are equal in length.
b

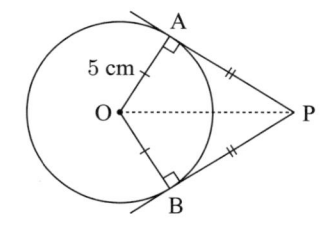

$$\begin{array}{ll}\text{OA} = \text{OB} & \{\text{equal radii}\} \\ \widehat{\text{OAP}} = \widehat{\text{OBP}} = 90° & \{\text{radius-tangent}\} \\ \text{AP} = \text{BP} & \{\text{from } \textbf{a}\}\end{array}$$

$\therefore$ triangles OAP and OBP are congruent. {SAS}
$\therefore\ \widehat{\text{AOP}} = \widehat{\text{BOP}} = 75°$
$\therefore$ in $\triangle\text{AOP}$, $\tan 75° = \dfrac{\text{AP}}{5}$
$$\therefore\ \text{AP} = 5 \times \tan 75°$$
$$\therefore\ \text{AP} \approx 18.66$$
$\therefore$ area of AOBP $= 2 \times (\text{area of } \triangle\text{AOP})$
$$\approx 2 \times \left(\tfrac{1}{2} \times 18.66 \times 5\right)$$
$$\approx 93.3\,\text{cm}^2$$

23 a The required angle is $\widehat{\text{AHE}}$.
$\tan \theta = \frac{2}{5}$
$\therefore\ \theta = \tan^{-1}\left(\frac{2}{5}\right)$
$\therefore\ \theta \approx 21.8°$
$\therefore$ the angle is about $21.8°$.

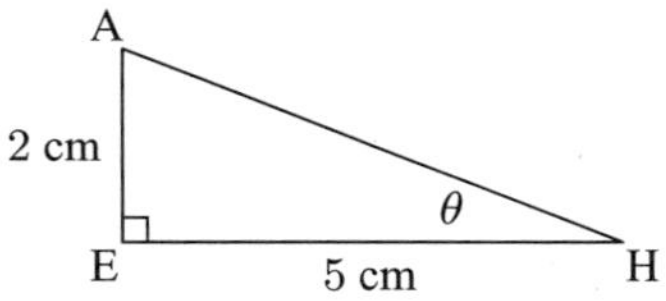

b The required angle is $\widehat{\text{CEG}}$.

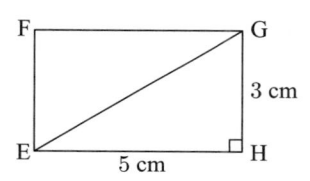

$$\text{EG}^2 = 5^2 + 3^2 \quad \{\text{Pythagoras}\}$$
$$\therefore\ \text{EG}^2 = 34$$
$$\therefore\ \text{EG} = \sqrt{34}\,\text{cm}$$
$$\therefore\ \tan \theta = \tfrac{2}{\sqrt{34}}$$
$$\therefore\ \theta = \tan^{-1}\left(\tfrac{2}{\sqrt{34}}\right)$$
$$\therefore\ \theta \approx 18.9°$$
$\therefore$ the angle is about $18.9°$. 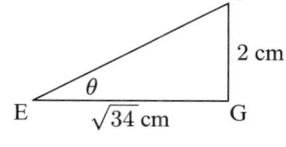

24 $\cos R = \dfrac{11^2 + 22^2 - 14^2}{2 \times 11 \times 22}$ {cosine rule}
$$\therefore\ R = \cos^{-1}\left(\dfrac{11^2 + 22^2 - 14^2}{2 \times 11 \times 22}\right)$$
$$\therefore\ R \approx 32.324° \approx 32.3°$$
$\cos S = \dfrac{11^2 + 14^2 - 22^2}{2 \times 11 \times 14}$ {cosine rule}
$$\therefore\ S = \cos^{-1}\left(\dfrac{11^2 + 14^2 - 22^2}{2 \times 11 \times 14}\right)$$
$$\therefore\ S \approx 122.83° \approx 122.8°$$
$\therefore\ T \approx 180° - 122.83° - 32.324°$ {angles of a triangle}
$\therefore\ T \approx 24.8$

25

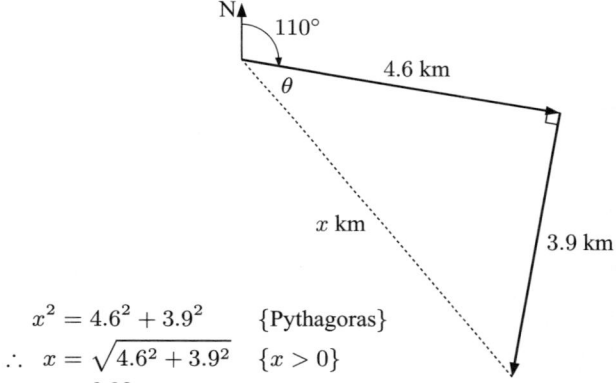

$$x^2 = 4.6^2 + 3.9^2 \quad \{\text{Pythagoras}\}$$
$$\therefore\ x = \sqrt{4.6^2 + 3.9^2} \quad \{x > 0\}$$
$$\therefore\ x \approx 6.03$$
Now $\tan \theta = \frac{3.9}{4.6}$
$$\therefore\ \theta = \tan^{-1}\left(\tfrac{3.9}{4.6}\right)$$
$$\therefore\ \theta \approx 40.3°$$
so $110° + \theta \approx 150.3°$
$\therefore$ Robert is 6.03 km from his starting point on a bearing of $150.3°$.

26 a

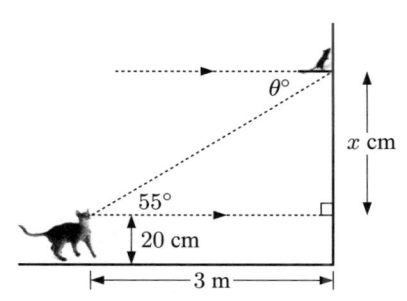

$\theta = 55$ {alternate angles}
$\therefore$ the angle of depression from the mouse to the cat is $55°$.

b $\tan 55° = \dfrac{x}{300}$ {3 m = 300 cm}

$\therefore\ x = 300 \times \tan 55°$

$\therefore\ x \approx 428$

$\therefore$ the shelf is $428 + 20 = 448$ cm above the floor.

27 a $\widehat{AOP} = 180° - 120° = 60°$ {angles on a line}
$OA = OP$ {equal radii}
$\widehat{OAP} = \widehat{OPA} = 60°$ {angles of isos. $\triangle$}
$\therefore\ \triangle APO$ is equilateral.

b i $ON = \tfrac{1}{2}$ {altitude PN bisects the base}

ii In $\triangle PNO$, $PN^2 + ON^2 = OP^2$ {Pythagoras}

$\therefore\ PN^2 + \left(\tfrac{1}{2}\right)^2 = 1^2$

$\therefore\ PN^2 + \tfrac{1}{4} = 1$

$\therefore\ PN^2 = \tfrac{3}{4}$

$\therefore\ PN = \tfrac{\sqrt{3}}{2}$ {PN > 0}

c P is at $\left(-\tfrac{1}{2}, \tfrac{\sqrt{3}}{2}\right)$.

d i $\cos 120° = x$-coordinate of P $= -\tfrac{1}{2}$

ii $\sin 120° = y$-coordinate of P $= \tfrac{\sqrt{3}}{2}$

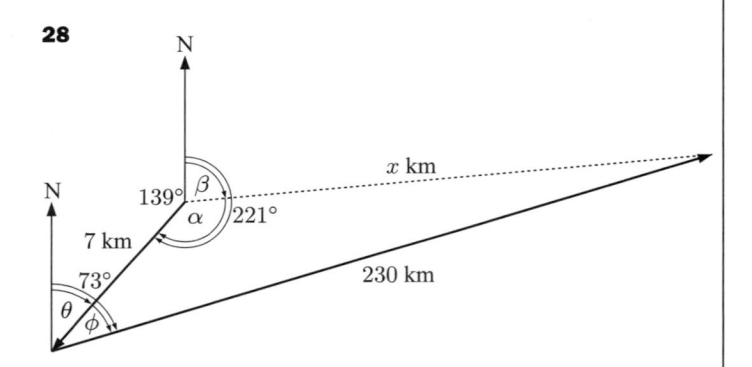

iii $\tan 120° = \dfrac{\sin 120°}{\cos 120°}$

$= \dfrac{\tfrac{\sqrt{3}}{2}}{-\tfrac{1}{2}}$

$= -\sqrt{3}$

28

a $\theta = 180° - 139° = 41°$ {co-interior angles}

$\therefore\ \phi = 73° - 41° = 32°$

By the cosine rule:

$x^2 = 7^2 + 230^2 - 2 \times 7 \times 230 \times \cos 32°$

$\therefore\ x = \sqrt{7^2 + 230^2 - 2 \times 7 \times 230 \times \cos 32°}$

$\therefore\ x \approx 224.09$

The aeroplane is about 224 km from Adelaide.

b $\cos \alpha \approx \dfrac{7^2 + 224.09^2 - 230^2}{2 \times 7 \times 224.09}$ {cosine rule}

$\therefore\ \alpha \approx \cos^{-1}\left(\dfrac{7^2 + 224.09^2 - 230^2}{2 \times 7 \times 224.09}\right)$

$\therefore\ \alpha \approx 147.1°$

$\therefore\ \beta = 221° - \alpha \approx 73.9°$

The aeroplane is on a bearing of 073.9° from Adelaide.

29

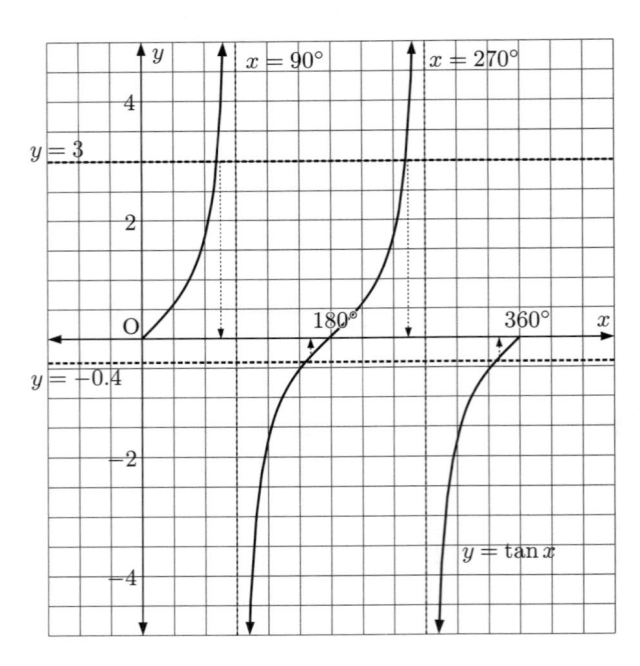

a The solutions to $\tan x = -0.4$ are $x \approx 158°, 338°$

b The solutions to $\tan x = 3$ are $x \approx 72°, 252°$

30 Using Pythagoras:

a In $\triangle ABC$, $AC^2 = \left(3\sqrt{2}\right)^2 + \left(3\sqrt{2}\right)^2$

$\therefore\ AC^2 = 18 + 18$

$\therefore\ AC^2 = 36$

$\therefore\ AC = 6$ cm {AC > 0}

b $GF = \tfrac{1}{2} \times 6 = 3$ cm

In $\triangle CFG$, $CG^2 = 3^2 + 4^2$

$\therefore\ CG^2 = 25$

$\therefore\ CG = 5$ cm {CG > 0}

c In $\triangle EFG$, $EG^2 + 3^2 = \left(3\sqrt{2}\right)^2$

$\therefore\ EG^2 + 9 = 18$

$\therefore\ EG^2 = 9$

$\therefore\ EG = 3$ cm {EG > 0}

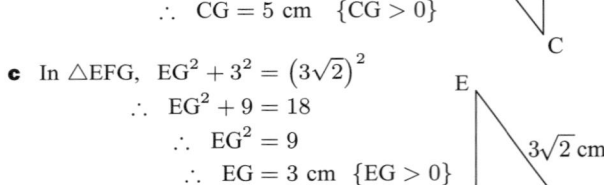

In $\triangle CEG$, $CE^2 = 5^2 + 3^2$

$\therefore\ CE^2 = 34$

$\therefore\ CE = \sqrt{34}$ cm {CE > 0}

d $\tan \widehat{ECG} = \tfrac{3}{5}$

$\therefore\ \widehat{ECG} = \tan^{-1}\left(\tfrac{3}{5}\right)$

$\therefore\ \widehat{ECG} \approx 31.0°$

31

Area = 5 m²

$\therefore\ \tfrac{1}{2} \times 3.1 \times 4.5 \times \sin \theta = 5$

$\therefore\ 6.975 \sin \theta = 5$

$\therefore\ \sin \theta = \tfrac{5}{6.975}$

Now $\sin^{-1}\left(\tfrac{5}{6.975}\right) \approx 45.8°$

$\therefore\ \theta \approx 45.8°$ or $134.2°$

32

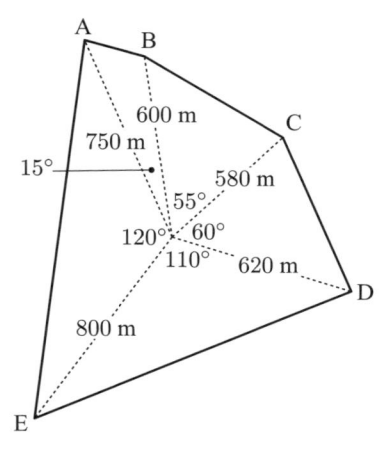

Using the cosine rule,
$$AB^2 = 750^2 + 600^2 - 2 \times 750 \times 600 \times \cos 15°$$
$$\therefore AB = \sqrt{750^2 + 600^2 - 2 \times 750 \times 600 \times \cos 15°}$$
$$\approx 230.58 \text{ m}$$

Likewise,
$$BC = \sqrt{600^2 + 580^2 - 2 \times 600 \times 580 \times \cos 55°}$$
$$\approx 545.15 \text{ m}$$
$$CD = \sqrt{580^2 + 620^2 - 2 \times 580 \times 620 \times \cos 60°}$$
$$\approx 601.00 \text{ m}$$
$$DE = \sqrt{620^2 + 800^2 - 2 \times 620 \times 800 \times \cos 110°}$$
$$\approx 1167.77 \text{ m}$$
$$EA = \sqrt{800^2 + 750^2 - 2 \times 800 \times 750 \times \cos 120°}$$
$$\approx 1342.57 \text{ m}$$

$\therefore$ perimeter $\approx 230.58 + 545.15 + 601.00 + 1167.77$
$$+ 1342.57$$
$$\approx 3890 \text{ m}$$

33

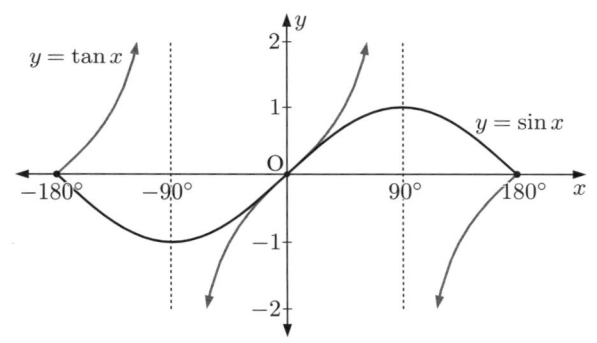

Since $\tan x = \dfrac{\sin x}{\cos x}$, $\tan x = 0$ whenever $\sin x = 0$.

34 a The required angle is $P\widehat{A}O$.
$\tan \theta = \dfrac{8}{3}$
$\therefore \theta = \tan^{-1}\left(\dfrac{8}{3}\right) \approx 69.4°$
$\therefore$ the angle is about $69.4°$.

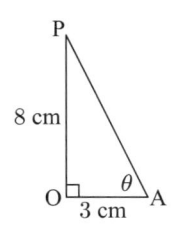

b 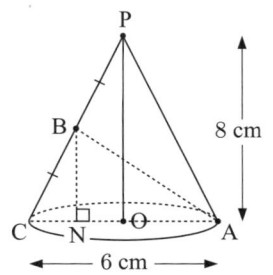 The required angle is $B\widehat{A}N$.

$\triangle$s CBN and CPO are similar since $C\widehat{N}B = C\widehat{O}P = 90°$, and the angle at C is common to both triangles.

$\therefore \dfrac{BN}{PO} = \dfrac{CB}{CP}$ and $\dfrac{CN}{CO} = \dfrac{CB}{CP}$

$\therefore \dfrac{BN}{8} = \dfrac{1}{2}$ and $\dfrac{CN}{3} = \dfrac{1}{2}$

$\therefore BN = 4$ and $CN = 1.5$

$\therefore AN = 6 - 1.5 = 4.5 \text{ cm}$

So, $\tan \theta = \dfrac{4}{4.5}$
$\therefore \theta = \tan^{-1}\left(\dfrac{4}{4.5}\right)$
$\therefore \theta \approx 41.6°$

$\therefore$ the angle is about $41.6°$.

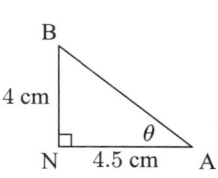

35 a $\cos \theta = \dfrac{9^2 + 10^2 - 11^2}{2 \times 9 \times 10}$
$\therefore \theta = \cos^{-1}\left(\dfrac{9^2 + 10^2 - 11^2}{2 \times 9 \times 10}\right)$
$\approx 70.53°$

So, area $\approx \dfrac{1}{2} \times 9 \times 10 \times \sin 70.53°$
$\approx 42.4 \text{ cm}^2$

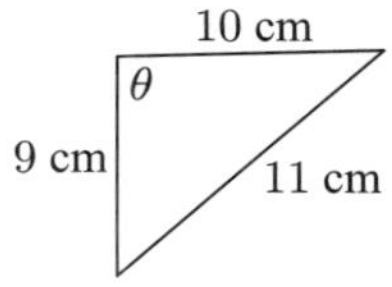

b $\cos \theta = \dfrac{6^2 + 10^2 - 15^2}{2 \times 6 \times 10}$
$\therefore \theta = \cos^{-1}\left(\dfrac{6^2 + 10^2 - 15^2}{2 \times 6 \times 10}\right)$
$\approx 137.87°$

So, area $= \dfrac{1}{2} \times 6 \times 10 \times \sin 137.87°$
$\approx 20.1 \text{ cm}^2$

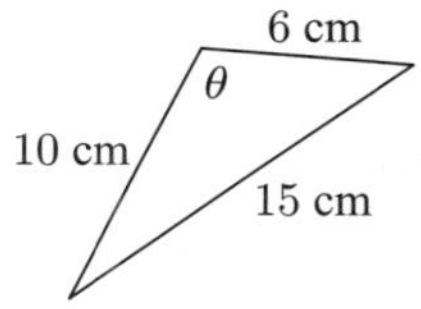

36 a

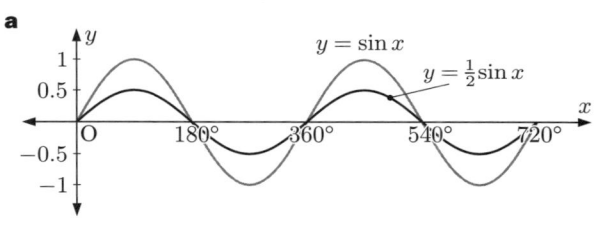

b

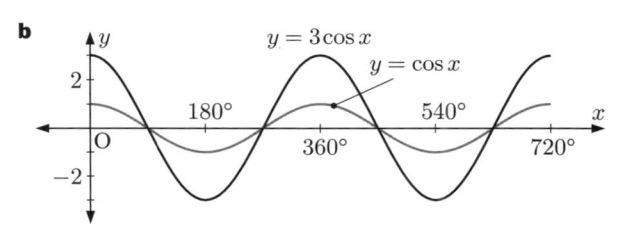

37 a i $\sin 34° = \dfrac{44}{x}$
$\therefore x = \dfrac{44}{\sin 34°}$
$\therefore x \approx 78.68$

The distance from the commentator to the ball is about 78.7 m.

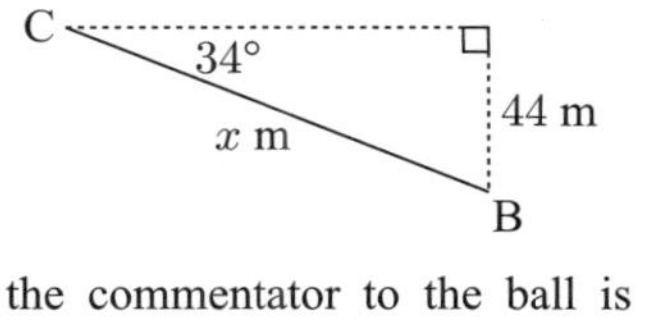

ii $\sin 26° = \dfrac{44}{x}$
$\therefore x = \dfrac{44}{\sin 26°}$
$\therefore x \approx 100.37$

The distance from the commentator to the goal mouth is about 100.4 m.

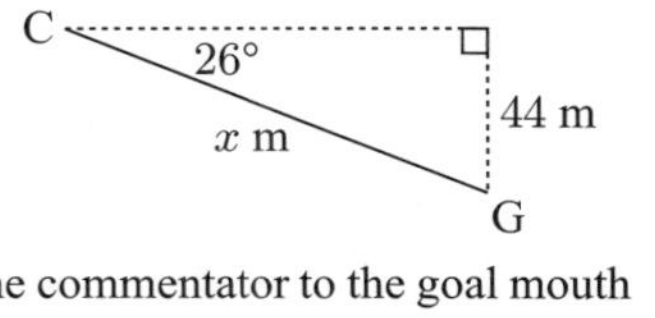

iii $x^2 + 78.68^2 \approx 100.37^2$
$\therefore\ x \approx \sqrt{100.37^2 - 78.68^2}$
$\therefore\ x \approx 62.3$

The distance from the ball to the goal mouth is about 62.3 m.

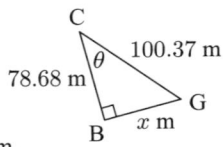

b $\cos \theta \approx \frac{78.68}{100.37}$
$\therefore\ \theta \approx \cos^{-1}\left(\frac{78.68}{100.37}\right)$
$\therefore\ \theta \approx 38.4°$

38 a

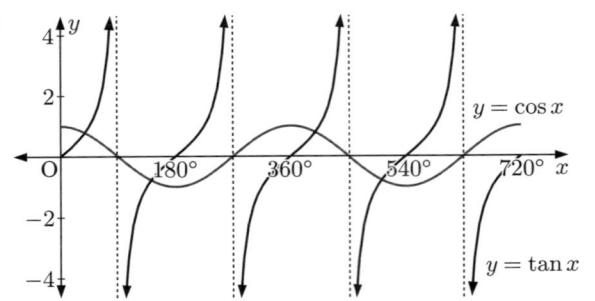

b Since $\tan x = \frac{\sin x}{\cos x}$, the asymptotes of $y = \tan x$ occur when $\cos x = 0$.

39 a The required angle is $E\hat{D}C$.
$\tan \theta = \frac{14}{26}$
$\therefore\ \theta = \tan^{-1}\left(\frac{14}{26}\right)$
$\approx 28.3°$
$\therefore\$ the angle is about $28.3°$.

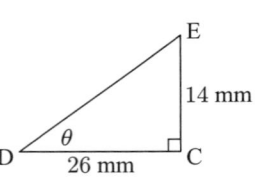

b 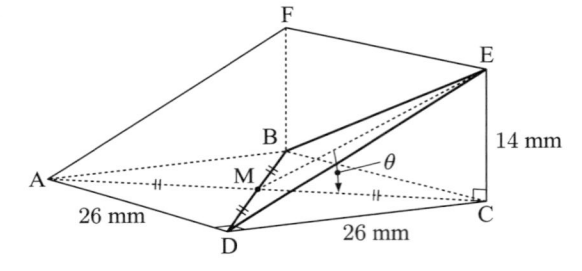 The required angle is $E\hat{M}C$, where M is the centre of the square base ABCD.

In $\triangle ADC$, $AC^2 = 26^2 + 26^2$ {Pythagoras}
$\therefore\ AC^2 = 1352$
$\therefore\ AC = \sqrt{1352}$
$\therefore\ AC = 26\sqrt{2}$ mm
So, $MC = 13\sqrt{2}$ mm

$\tan \theta = \frac{14}{13\sqrt{2}}$
$\therefore\ \theta = \tan^{-1}\left(\frac{14}{13\sqrt{2}}\right) \approx 37.3°$

The angle is about $37.3°$.

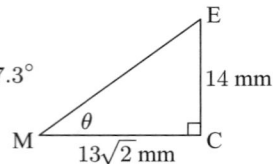

40 a

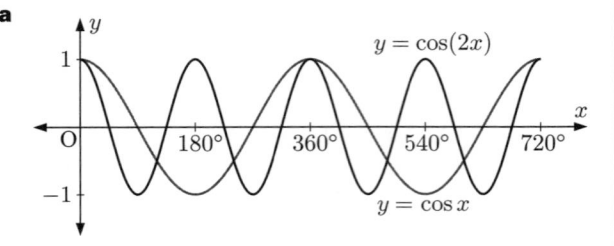

b

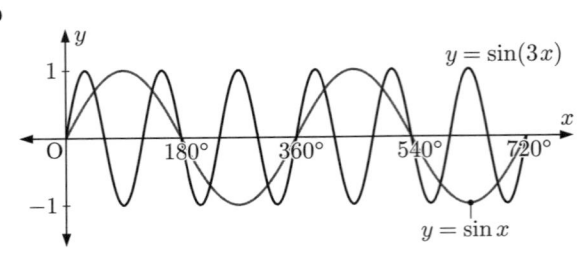

41

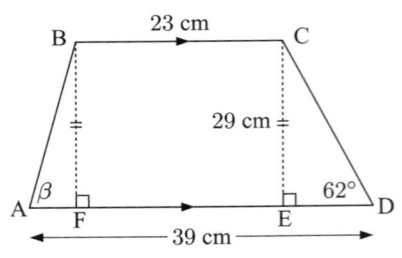

In $\triangle CDE$, $\tan 62° = \frac{29}{ED}$
$\therefore\ ED = \frac{29}{\tan 62°} \approx 15.42$ cm
$\therefore\ AF \approx 39 - 15.42 - 23 \approx 0.58$ cm
$\therefore\$ in $\triangle ABF$, $\tan \beta = \frac{29}{0.58}$
$\therefore\ \beta = \tan^{-1}\left(\frac{29}{0.58}\right)$
$\therefore\ \beta \approx 88.9°$

SOLUTIONS TO TOPIC 9: SETS

1 a i $U = \{1, 2, 3, 4, 5, 6, 7, 8, 9, 10\}$
$\therefore\ n(U) = 10$
ii $n(P) = 4$ **iii** $n(Q) = 4$

b $P' = \{1, 2, 7, 8, 9, 10\}$

c i false {3 and 5 are in P, but not in Q}
ii true **iii** false **iv** true

d

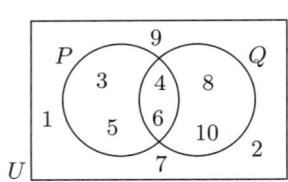

2 a false **b** true $\{(\sqrt{5})^2 = 5\}$
c false **d** true

3 a **b** finite

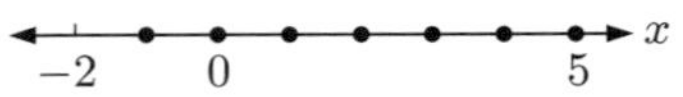

4 a $A = \{10, 12, 14, 15, 16, 18, 20\}$
$B = \{10, 12, 14, 16, 18, 20\}$
b i $n(A) = 7$ **ii** $n(B) = 6$
c i false **ii** false

5 a **b**

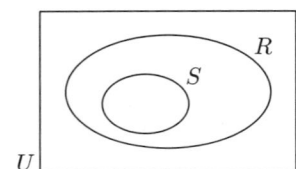

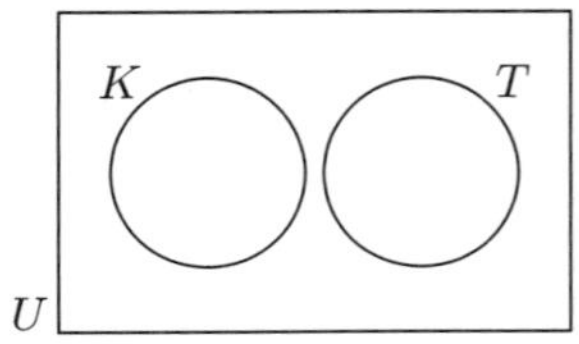

6 **a** false   **b** true   **c** false

7 **a** 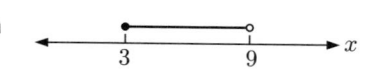
Number line with closed dot at 3, open dot at 9.
   b infinite

8
Venn diagram showing $\mathbb{N} \subset \mathbb{Z} \subset \mathbb{Q} \subset \mathbb{R}$.

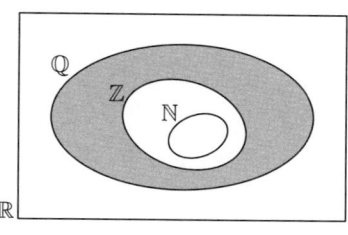

9 **a**
  i $A = \{2, 3, 5, 6, 8\}$
  ii $A \cap B = \{2, 5, 8\}$
  iii $A' \cap B = \{4, 7\}$

b
  i $A \cup B = \{2, 3, 4, 5, 6, 7, 8\}$
    $\therefore\ n(A \cup B) = 7$
  ii $A \cap B' = \{3, 6\}$
    $\therefore\ n(A \cap B') = 2$
  iii $A' \cup B' = \{1, 3, 4, 6, 7, 9\}$
    $\therefore\ n(A' \cup B') = 6$

10 **a** Venn diagram with $A \setminus B$ shaded.   **b** Venn diagram with complement of A shaded.

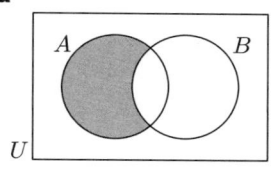

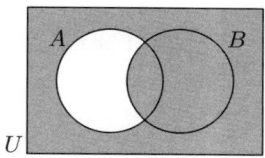

11 **a** $\{x \mid 0 < x \leqslant 2,\ x \in \mathbb{R}\}$
b $\{x \mid x \leqslant -3 \text{ or } x > 3,\ x \in \mathbb{R}\}$
c $\{x \mid -2 \leqslant x \leqslant 3,\ x \in \mathbb{Z}\}$

12 **a** $n(U) = 11 + 9 + 3 + 2 = 25$
b $n(P \cap D) = 9$ **c** $n(P \cap D') = 11$
d $n(P \cup D) = 11 + 9 + 3 = 23$

13 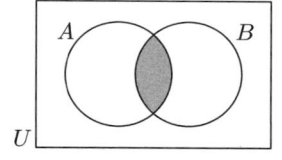
Venn diagram: The shaded region is $A \cap B$. 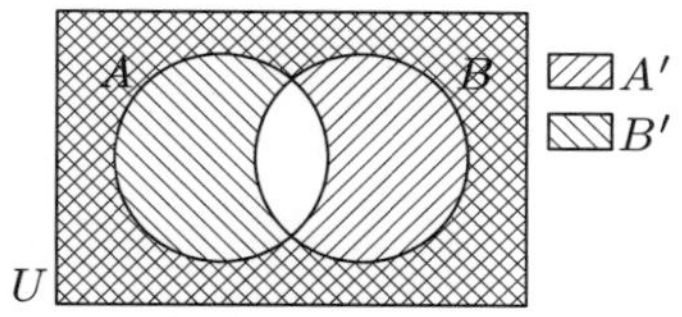
Venn diagram: The unshaded region is $(A' \cup B')'$.
$A \cap B$ and $(A' \cup B')'$ are represented by the same regions, so $A \cap B = (A' \cup B')'$.

14 **a**
Venn diagram with J and B: regions (a), (b), (c), (d).

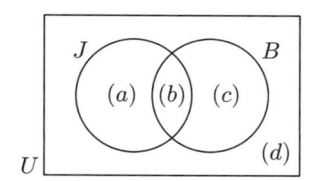

$d = 4$ {4 like neither}
$\therefore\ a + b + c = 13$
$\quad a + b = 11$ {11 like jazz}
$\quad b + c = 8$ {8 like blues}
So, $a = 5,\ b = 6,\ c = 2$.

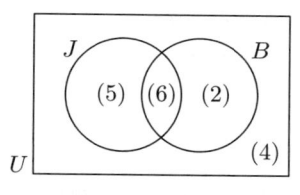

Venn diagram with J and B: (5), (6), (2), (4).

b **i** $n(J \cap B) = 6$ **ii** $n(J \cap B') = 5$

15 **a** infinite   {for example, $\frac{1}{2}, \frac{1}{3}, \frac{1}{4}, \frac{1}{5}, \ldots$ are all elements of the set}
b We cannot place an infinite number of discrete points on a number line.

16 We let A represent the DVDs watched by Angela, and B represent the DVDs watched by Ben.

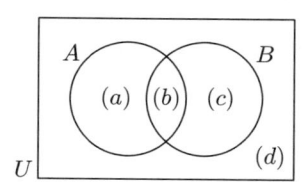

Venn diagram with A and B: regions (a), (b), (c), (d).

$d = 9$ {9 have been watched by neither}
$\therefore\ a + b + c = 66$
$\quad a + b = 58$ {Angela has watched 58}
$\quad b + c = 29$ {Ben has watched 29}
So, $a = 37,\ b = 21,\ c = 8$.

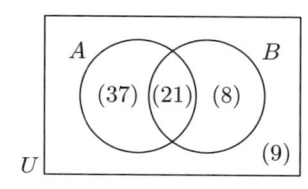

Venn diagram: (37), (21), (8), (9).

a $n(A \cap B) = 21$ **b** $n(B \cap A') = 8$

17 **a** In X or Y, but not both, or $(X \cap Y') \cup (X' \cap Y)$.
b In only A, or in both B and C, or $(A \cap B' \cap C') \cup (B \cap C)$.

18
Three-circle Venn diagram with C, L, R and regions $(a), (b), (c), (d), (e), (f), (g), (h)$.

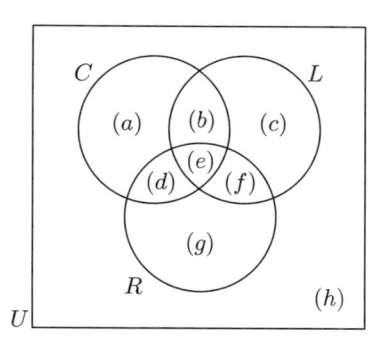

$e = 5$ {5 like all three}
$b + e = 12$ {12 are in C and L}
$d + e = 9$ {9 are in C and R}
$f + e = 6$ {6 are in L and R}
So, $b = 7,\ d = 4,\ f = 1$.
Also, $\quad a + b + d + e = 31$ {31 are in C}
$\therefore\ a + 7 + 4 + 5 = 31$ $\therefore\ a = 15$
$\quad b + c + e + f = 19$ {19 are in L}
$\therefore\ 7 + c + 5 + 1 = 19$ $\therefore\ c = 6$
$\quad d + e + f + g = 20$ {20 are in R}
$\therefore\ 4 + 5 + 1 + g = 20$ $\therefore\ g = 10$
$\quad h = 12$ {12 like none}

∴ number of students surveyed
$$= a+b+c+d+e+f+g+h$$
$$= 15+7+6+4+5+1+10+12$$
$$= 60$$

SOLUTIONS TO TOPIC 10: PROBABILITY

1 a The event has a 50-50 chance of occurring.
   b The event is very likely to occur.
   c The event is unlikely to occur.

2 a extremely unlikely   **b** very likely

3 a The company carried $2812 + 465 + 38 + 11 = 3326$ parcels last year.

  b
Description	Frequency	Relative Frequency
Delivered on time	2812	0.845
Delivered late	465	0.140
Returned to sender	38	0.011
Lost	11	0.003
	3326	

  c i P(lose a parcel) ≈ 0.003
   ii P(deliver a parcel) $\approx 0.845 + 0.140$
   ≈ 0.985

4 a
	Allowed	Not Allowed	Total
Smokers	86	19	105
Non-smokers	24	215	239
Total	110	234	344

  b i There are 105 smokers out of 344 people surveyed.
   ∴ P(smoker) $= \frac{105}{344} \approx 0.305$
   ii 234 of the 344 people want smoking banned.
   ∴ P(wants smoking banned) $= \frac{234}{344} \approx 0.680$

  c Of the 239 non-smokers, 24 want smoking allowed.
   ∴ P(Amy wants smoking allowed)
   $= \frac{24}{239} \approx 0.100$

5 a These events are independent. The light globe in the garage blowing will not affect the milk in the fridge.
   b These events are dependent. If chocolate is on special, then it is more likely that John has enough money to buy chocolate.

6 {TMI, TIM, MTI, MIT, ITM, IMT}

7 a i Proportion of spinner that is red
   $\approx \frac{86}{300} \approx 0.287$ or 28.7%
   ii Proportion of spinner that is green
   $\approx \frac{300-132-86}{300} \approx 0.273$ or 27.3%
   b P(non-red) $= \frac{300-86}{300} \approx 0.713$

8 $n(U) = 7 + 3 + 12 + 2 = 24$
   a P(B) $= \frac{7+3}{24}$
   $= \frac{10}{24}$ or $\frac{5}{12}$
   b P(T but not B) $= \frac{12}{24}$ or $\frac{1}{2}$
   c P(T or B) $= \frac{7+3+12}{24}$
   $= \frac{22}{24}$ or $\frac{11}{12}$

   d P(T given B) $= \frac{3}{7+3}$
   $= \frac{3}{10}$

9 $p = 0.23$, $n = 76$
   ∴ we expect Katie to score $76 \times 0.23 \approx 17$ points next game.

10   Token 1   Token 2

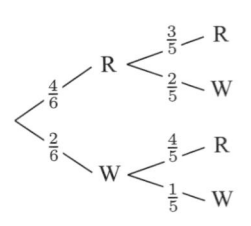

Tree diagram: R (4/6) → R (3/5), W (2/5); W (2/6) → R (4/5), W (1/5)

  a i P(two reds) $=$ P(RR)
   $= \frac{4}{6} \times \frac{3}{5}$
   $= \frac{2}{5}$
   ii P(red then white) $=$ P(RW)
   $= \frac{4}{6} \times \frac{2}{5}$
   $= \frac{4}{15}$
   iii P(red and white) $=$ P(RW or WR)
   $= \frac{4}{6} \times \frac{2}{5} + \frac{2}{6} \times \frac{4}{5}$
   $= \frac{8}{15}$
   iv P(at least one white) $= 1 -$ P(RR)
   $= 1 - \frac{2}{5}$ {from **a i**}
   $= \frac{3}{5}$

  b i P(the first is red given the second is red)
   $= \dfrac{\text{P(RR)}}{\text{P(RR)} + \text{P(WR)}}$
   $= \dfrac{\frac{2}{5}}{\frac{4}{6} \times \frac{3}{5} + \frac{2}{6} \times \frac{4}{5}}$
   $= \dfrac{\frac{2}{5}}{\frac{2}{3}}$
   $= \frac{3}{5}$
   ii P(the first is white given the second is red)
   $= 1 - \frac{3}{5}$
   $= \frac{2}{5}$

11 a
Brand	Frequency	Relative frequency
Sparkle	114	0.456
Fresh	68	0.272
Shine	45	0.18
Megawhite	23	0.092
Total	250	1.000

  b Sparkle   **c i** ≈ 0.092 **ii** ≈ 0.272

12 a 2 of the 6 faces are marked with a 2.
   ∴ P(2) $= \frac{2}{6}$ or $\frac{1}{3}$
   b i P(3) $= \frac{1}{6}$
   ∴ in 120 die rolls we would expect a 3 to occur $120 \times \frac{1}{6} = 20$ times.

ii $P(1) = \frac{3}{6} = \frac{1}{2}$

∴ in 120 die rolls we would expect a 1 to occur $120 \times \frac{1}{2} = 60$ times.

iii There are no 4s on the die, so we would not expect any 4s to occur.

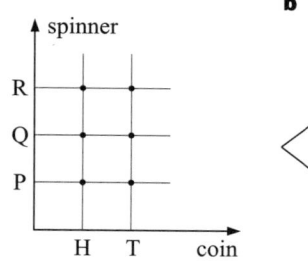

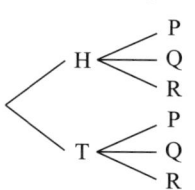

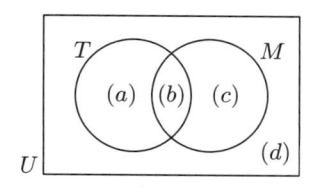

13 a spinner grid with points at (H,P), (H,Q), (H,R), (T,P), (T,Q), (T,R)

b coin–spinner tree: H→P,Q,R and T→P,Q,R

14 a Venn diagram with sets T and M, regions (a), (b), (c), and (d) outside.

$d = 4$ {4 horses are neither}
∴ $a + b + c = 32$
$a + b = 27$ {27 are thoroughbreds}
$b + c = 16$ {16 are mares}
So, $a = 16$, $b = 11$, $c = 5$.

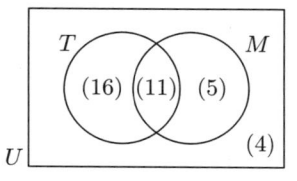

Venn diagram with values (16), (11), (5), (4).

b i $P(T \text{ and } M) = \frac{11}{36}$

ii $P(M \text{ but not } T) = \frac{5}{36}$

iii $P(T \text{ or } M) = \frac{16 + 11 + 5}{36} = \frac{32}{36} = \frac{8}{9}$

iv $P(T \text{ given } M) = \frac{11}{11 + 5} = \frac{11}{16}$

15 a Total number of students at the school
$= 3 + 1 + 312 + 247 + 24 + 15 + \ldots + 28 + 32$
$= 855$

b i $P(\text{girl}) = \frac{1 + 247 + 15 + 105 + 2 + 32}{855}$
$= \frac{402}{855}$
≈ 0.470

ii $P(\text{boy from Europe}) = \frac{86}{855}$
≈ 0.101

iii $P(\text{girl from the Americas}) = \frac{2 + 32}{855}$
≈ 0.0398

c $P(\text{the Asian student is a boy}) = \frac{312}{312 + 247}$
≈ 0.558

d $P(\text{the girl is from Australia or NZ}) = \frac{15}{402}$
≈ 0.0373

16 $n = 62$, $p = 0.045$
∴ Danny is expected to score $62 \times 0.045 = 2.79 \approx 3$ home runs in the season.

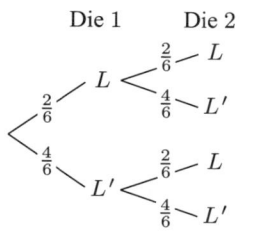

17 a Tree diagram: Die 1: L ($\frac{2}{6}$), L' ($\frac{4}{6}$); Die 2 from L: L ($\frac{2}{6}$), L' ($\frac{4}{6}$); Die 2 from L': L ($\frac{2}{6}$), L' ($\frac{4}{6}$).

b i $P(\text{both low}) = P(LL)$
$= \frac{2}{6} \times \frac{2}{6}$
$= \frac{1}{9}$

ii $P(\text{at least one not low}) = 1 - P(LL)$
$= \frac{8}{9}$

18 {HHHH, HHHT, HHTH, HTHH, THHH, HHTT, HTHT, HTTH, THHT, THTH, TTHH, HTTT, THTT, TTHT, TTTH, TTTT}

a $P(\text{at most 3 heads}) = 1 - P(4 \text{ heads})$
$= 1 - \frac{1}{16}$
$= \frac{15}{16}$

b $P(\text{two heads, given at least one head}) = \frac{6}{15}$
$= \frac{2}{5}$

c $P(\text{one tail, given at least two heads}) = \frac{4}{11}$

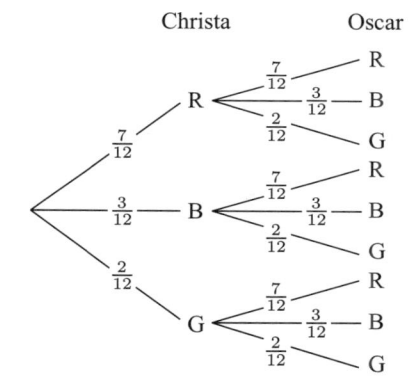

19 a Tree diagram Christa → Oscar:
Christa: R ($\frac{7}{12}$), B ($\frac{3}{12}$), G ($\frac{2}{12}$); Oscar from each: R ($\frac{7}{12}$), B ($\frac{3}{12}$), G ($\frac{2}{12}$).

b i $P(\text{red then green}) = P(RG)$
$= \frac{7}{12} \times \frac{2}{12}$
$= \frac{7}{72}$

ii $P(\text{two blue tiles}) = P(BB)$
$= \frac{3}{12} \times \frac{3}{12}$
$= \frac{1}{16}$

iii $P(\text{blue and green}) = P(BG \text{ or } GB)$
$= \frac{3}{12} \times \frac{2}{12} + \frac{2}{12} \times \frac{3}{12}$
$= \frac{1}{12}$

iv $P(\text{different colours})$
$= 1 - P(\text{same colour})$
$= 1 - P(RR \text{ or } BB \text{ or } GG)$
$= 1 - (\frac{7}{12} \times \frac{7}{12} + \frac{3}{12} \times \frac{3}{12} + \frac{2}{12} \times \frac{2}{12})$
$= \frac{41}{72}$

v P(no green tiles) = P(G'G')
$$= \tfrac{10}{12} \times \tfrac{10}{12}$$
$$= \tfrac{25}{36}$$

vi P(exactly one red) = P(RR' or R'R)
$$= \tfrac{7}{12} \times \tfrac{5}{12} + \tfrac{5}{12} \times \tfrac{7}{12}$$
$$= \tfrac{35}{72}$$

20 a These events are not dependent, since whether the home team wins will not affect whether Ben will forget his lunch.

b Let H be the event that the home team wins, and B be the event that Ben forgets his lunch.

  i P(H and B) = P(H) × P(B)
  $$= \tfrac{2}{3} \times \tfrac{1}{10}$$
  $$= \tfrac{1}{15}$$

  ii P(H or B') = P(H) + P(B') − P(H and B')
  $$= \tfrac{2}{3} + \tfrac{9}{10} - \tfrac{2}{3} \times \tfrac{9}{10}$$
  $$= \tfrac{29}{30}$$

  iii P(H' given B) = P(H')   {independent}
  $$= \tfrac{1}{3}$$

  iv P(B' given H') = P(B')   {independent}
  $$= \tfrac{9}{10}$$

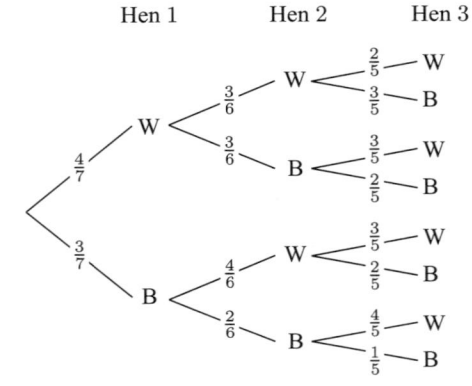

21 [Tree diagram: Hen 1, Hen 2, Hen 3]
Hen 1: W ($\tfrac{4}{7}$), B ($\tfrac{3}{7}$)
From W → Hen 2: W ($\tfrac{3}{6}$), B ($\tfrac{3}{6}$)
From B → Hen 2: W ($\tfrac{4}{6}$), B ($\tfrac{2}{6}$)
Hen 3 from WW: W ($\tfrac{2}{5}$), B ($\tfrac{3}{5}$)
Hen 3 from WB: W ($\tfrac{3}{5}$), B ($\tfrac{2}{5}$)
Hen 3 from BW: W ($\tfrac{3}{5}$), B ($\tfrac{2}{5}$)
Hen 3 from BB: W ($\tfrac{4}{5}$), B ($\tfrac{1}{5}$)

a i P(first 2 are white) = P(WW)
$$= \tfrac{4}{7} \times \tfrac{3}{6}$$
$$= \tfrac{2}{7}$$

  ii Each of the 7 hens are equally likely to be the last hen out.
  ∴ P(last hen out is brown) = $\tfrac{3}{7}$

b P(2nd hen white given 1st hen brown) = $\tfrac{4}{6}$ or $\tfrac{2}{3}$
  {from tree diagram}

c P(2nd and 3rd hens have same colour)
= P(WWW or WBB or BWW or BBB)
$$= \tfrac{4}{7} \times \tfrac{3}{6} \times \tfrac{2}{5} + \tfrac{4}{7} \times \tfrac{3}{6} \times \tfrac{2}{5} + \tfrac{3}{7} \times \tfrac{4}{6} \times \tfrac{3}{5} + \tfrac{3}{7} \times \tfrac{2}{6} \times \tfrac{1}{5}$$
$$= \tfrac{3}{7}$$

22 a A and B are not mutually exclusive, since A and B have a common outcome: the 2 of hearts.

b A and B are mutually exclusive, because a card cannot be both a picture card and a 7.

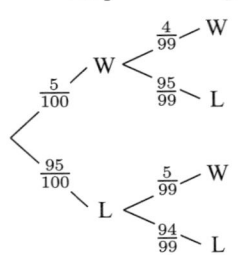

23 a [Tree diagram: 1st prize, 2nd prize]
1st prize: W ($\tfrac{5}{100}$), L ($\tfrac{95}{100}$)
From W → 2nd: W ($\tfrac{4}{99}$), L ($\tfrac{95}{99}$)
From L → 2nd: W ($\tfrac{5}{99}$), L ($\tfrac{94}{99}$)

b i P(both prizes) = P(WW)
$$= \tfrac{5}{100} \times \tfrac{4}{99}$$
$$= \tfrac{1}{495}$$

  ii P(second prize) = P(WW or LW)
  $$= \tfrac{5}{100} \times \tfrac{4}{99} + \tfrac{95}{100} \times \tfrac{5}{99}$$
  $$= \tfrac{1}{20}$$

  iii P(only second prize) = P(LW)
  $$= \tfrac{95}{100} \times \tfrac{5}{99}$$
  $$= \tfrac{19}{396}$$

c [Tree diagram: 1st prize, 2nd prize]
1st: K ($\tfrac{5}{100}$), P ($\tfrac{3}{100}$), N ($\tfrac{92}{100}$)
From K → 2nd: K ($\tfrac{4}{99}$), P ($\tfrac{3}{99}$), N ($\tfrac{92}{99}$)
From P → 2nd: K ($\tfrac{5}{99}$), P ($\tfrac{2}{99}$), N ($\tfrac{92}{99}$)
From N → 2nd: K ($\tfrac{5}{99}$), P ($\tfrac{3}{99}$), N ($\tfrac{91}{99}$)

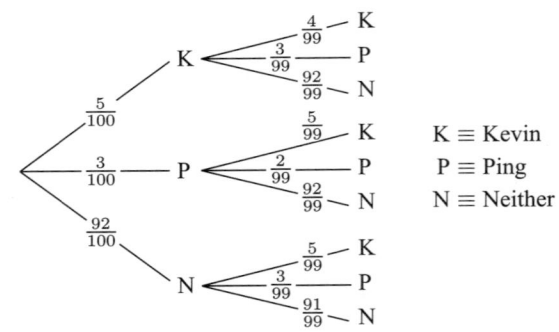

K ≡ Kevin
P ≡ Ping
N ≡ Neither

  i P(Kevin and Ping win prizes)
  = P(KP or PK)
  $$= \tfrac{5}{100} \times \tfrac{3}{99} + \tfrac{3}{100} \times \tfrac{5}{99}$$
  $$= \tfrac{1}{330}$$

  ii P(Ping but not Kevin wins a prize)
  = P(PP or PN or NP)
  $$= \tfrac{3}{100} \times \tfrac{2}{99} + \tfrac{3}{100} \times \tfrac{92}{99} + \tfrac{92}{100} \times \tfrac{3}{99}$$
  $$= \tfrac{31}{550}$$

24 a {ABCD, ABDC, ACBD, ACDB, ADBC, ADCB}

b i There are 2 ways A can sit opposite C {ABCD, ADCB}.
  ∴ probability = $\tfrac{2}{6}$ or $\tfrac{1}{3}$

  ii There are 2 ways D sits to the left of B {ABDC, ACBD, assuming clockwise labelling}.
  ∴ probability = $\tfrac{2}{6}$ or $\tfrac{1}{3}$

  iii There are 2 ways B sits between A and D {ABDC, ACDB}.
  ∴ probability = $\tfrac{2}{6}$ or $\tfrac{1}{3}$

c If A sits opposite D, it is impossible for C and B to sit together.
  ∴ probability = 0

d If B sits next to C, then A and D must sit together.
∴ probability = 1

e There are 2 ways B sits immediately to the right of A {ACDB, ADCB}.
A and C sit opposite each other in 1 of these {ADCB}.
∴ probability = $\frac{1}{2}$

25 a

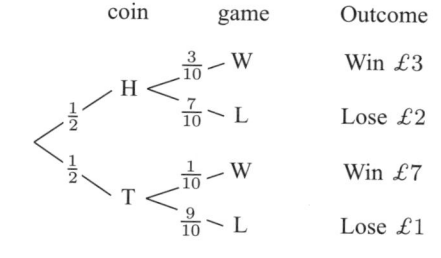

```
        coin      game    Outcome
               3/10 — W    Win £3
         H  
        /   7/10 — L       Lose £2
   1/2
        \   1/10 — W       Win £7
   1/2   T  
               9/10 — L    Lose £1
```

b **i** P(won £3) = P(HW)
$= \frac{1}{2} \times \frac{3}{10}$
$= \frac{3}{20}$

ii P(lost £1) = P(TL)
$= \frac{1}{2} \times \frac{9}{10}$
$= \frac{9}{20}$

iii P(won money) = P(HW or TW)
$= \frac{1}{2} \times \frac{3}{10} + \frac{1}{2} \times \frac{1}{10}$
$= \frac{1}{5}$

c Expected result
$= \frac{1}{2} \times \frac{3}{10} \times £3 + \frac{1}{2} \times \frac{7}{10} \times (-£2)$
$\quad + \frac{1}{2} \times \frac{1}{10} \times £7 + \frac{1}{2} \times \frac{9}{10} \times (-£1)$
$= -£0.35$

So, Ling is expected to lose £0.35 on average when she plays a game.

26 Probability that at least four of the boys scored
$= 0.25 \times 0.85 \times 0.80 \times 0.90 \times 0.85$ {A misses}
$+ 0.75 \times 0.15 \times 0.80 \times 0.90 \times 0.85$ {B misses}
$+ 0.75 \times 0.85 \times 0.20 \times 0.90 \times 0.85$ {C misses}
$+ 0.75 \times 0.85 \times 0.80 \times 0.10 \times 0.85$ {D misses}
$+ 0.75 \times 0.85 \times 0.80 \times 0.90 \times 0.15$ {E misses}
$+ 0.75 \times 0.85 \times 0.80 \times 0.90 \times 0.85$ {all score}
≈ 0.799

∴ P(Adrian scored, given at least 4 scored)
$\approx \frac{0.799 - (0.25 \times 0.85 \times 0.80 \times 0.90 \times 0.85)}{0.799}$
≈ 0.837

SOLUTIONS TO TOPIC 11: STATISTICS

1 Total sales $= 37 + 25 + 18 + 11 = 91$

Size	Sector angle
small	$\frac{37}{91} \times 360° \approx 146°$
medium	$\frac{25}{91} \times 360° \approx 99°$
large	$\frac{18}{91} \times 360° \approx 71°$
extra large	$\frac{11}{91} \times 360° \approx 44°$

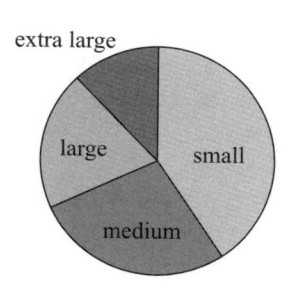

2 a discrete **b** continuous **c** categorical
d discrete

3 a Ages of club members

```
2 | 3 6
3 | 1 5 6
4 | 4 7 8 8
5 | 1 6 8 9
6 | 2 4 4       Key: 2 | 3 represents 23 years
```

b 7 of the 16 members are at least 50 years of age.
∴ percentage $= \frac{7}{16} \times 100\% = 43.75\%$

4 $\frac{54°}{360°} = 0.15$ or 15%
∴ 15% of 200 = 30 people gave the answer "conservative".

5 a census **b** sample

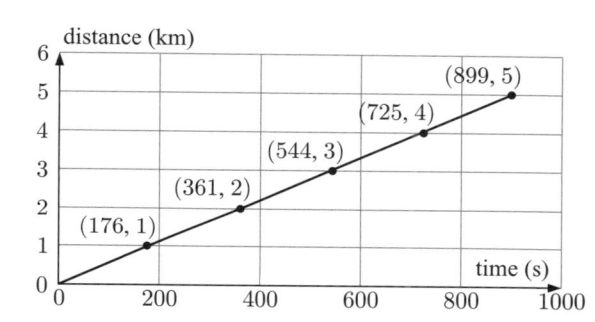

6 a
Graph showing distance (km) vs time (s) with points (176, 1), (361, 2), (544, 3), (725, 4), (899, 5).

b The data is continuous. It makes sense to join the points with straight line segments, as this shows Johann's progress throughout the race.

c Total time = 899 seconds
= 14 min 59 s

7 a Number of goals scored

b Discrete. You can only score a whole number of goals.

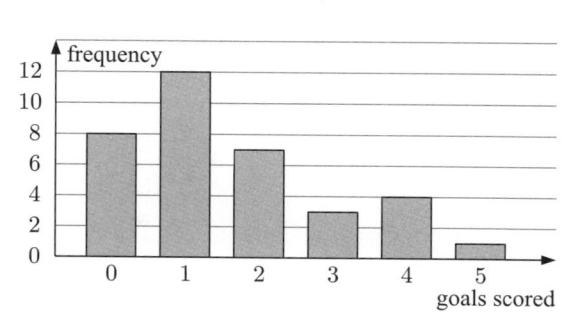

c Frequency bar chart of goals scored: 0→8, 1→12, 2→7, 3→3, 4→4, 5→1.

d **i** $\frac{8+12}{35} \times 100\% \approx 57.1\%$ of the footballers scored less than 2 goals.

ii $\frac{3}{35} \times 100\% \approx 8.57\%$ of the footballers scored 3 goals.

8 a If the police station was in a bad crime area, the policemen would be more likely to have a negative view about problems with crime.

b Dog and cat breeders will not be representative of the general population on the issue of pet registration laws.

c The survey can only be completed by people who have access to the internet.

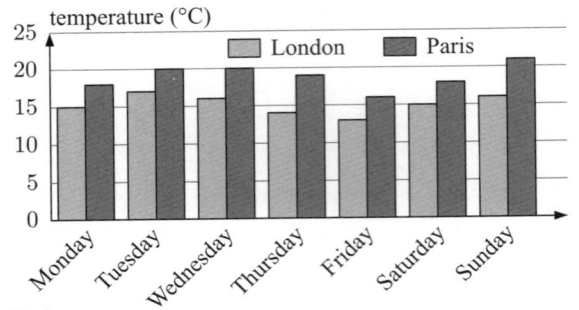

9 a

[Bar chart showing temperature (°C) for London and Paris from Monday to Sunday]

b Paris

c Both cities had a high on Tuesday, dropped in temperature until Friday and then warmed up on the weekend. Each day Paris was 3°C to 5°C warmer than London.

10 a i mean $= \dfrac{5+6+6+....+7+3}{10}$
$= \dfrac{56}{10}$
$= 5.6$

ii Since $n = 10$, $\dfrac{n+1}{2} = \dfrac{10+1}{2} = 5.5$

The ordered data set is
1 3 5 5 6 6 7 7 7 9
          ↑
       median

∴ median = average of 5th and 6th scores
$= \dfrac{6+6}{2}$
$= 6$

iii mode = 7 {most frequently occuring value}

b i mean $= \dfrac{0+0+0+1+....+6+7}{19}$
$= \dfrac{57}{19}$
$= 3$

ii Since $n = 19$, $\dfrac{n+1}{2} = 10$

∴ median = 10th score = 3

iii mode = 2 and 4

c i mean $= \dfrac{8.9+7.5+....+8.7+7.1}{10}$
$= \dfrac{82.8}{10}$
$= 8.28$

ii Since $n = 10$, $\dfrac{n+1}{2} = 5.5$

The ordered data set is
7.1 7.5 7.9 8.0 8.1 8.6 8.7 8.9 8.9 9.1
                     ↑
                  median

∴ median $= \dfrac{8.1+8.6}{2} = 8.35$

iii mode = 8.9

d i mean $= \dfrac{80+85+....+85+86}{11}$
$= \dfrac{907}{11}$
≈ 82.5

ii Since $n = 11$, $\dfrac{n+1}{2} = 6$

The ordered data set is
79 79 80 81 81 82 83 85 85 86 86
                         ↑
                      median

∴ median = 82

iii This data set has no mode.

11 a i negative   **ii** not linear   **iii** strong
b i positive   **ii** linear   **iii** moderate
c i no association   **ii** not linear   **iii** zero

12 $\dfrac{\text{sum of scores}}{18} = 27.5$

∴ sum of scores $= 27.5 \times 18$
$= 495$

13 Company Y, as on average it takes less time to deliver parcels than company X.

14 a mean $= \dfrac{7000+7000+....+29\,000+47\,000}{10}$
$= \dfrac{195\,000}{10}$
$= £19\,500$

Since $n = 10$, $\dfrac{n+1}{2} = 5.5$

The ordered data set is
£7000   £7000   £7000   £13\,000   £14\,000
£18\,000   £25\,000   £28\,000   £29\,000   £47\,000

∴ median $= \dfrac{14\,000+18\,000}{2}$
$= £16\,000$

The modal purchase price is £7000.

b No, the mode is the lowest value.

c The mean is affected by the extreme value of £47\,000.

15 $\dfrac{8+7+8+9+12+q+q+6}{8} = 7$

∴ $50 + 2q = 56$
∴ $2q = 6$
∴ $q = 3$

16 a

Without fertiliser		With fertiliser
9 9 7	1	
6 5 3 1	2	6 7 8
8 7 5 3 0 0	3	1 1 5 9
5 2 2 1 0	4	0 3 4 5 5 7
	5	1 4 8
	6	0 2

2 | 6 means 26 blueberries

b With fertiliser: 62   Without fertiliser: 45

c The plants with fertiliser have more blueberries on average.

∴ the fertiliser is effective.

17 a mode = 2 {highest frequency}

b

Number of sports	Frequency	Cumulative frequency
0	10	10
1	13	23
2	25	48
3	15	63
4	2	65
Total	65	

Since $n = 65$, $\frac{n+1}{2} = 33$

∴ median = 33rd score
= 2

c

Number of sports	Frequency	Product
0	10	0
1	13	13
2	25	50
3	15	45
4	2	8
Total	65	116

∴ mean = $\frac{116}{65}$
≈ 1.78

d range = maximum − minimum
= 4 − 0
= 4

18 a

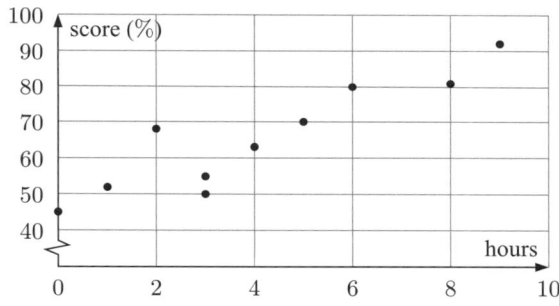

(scatter plot of score (%) vs hours, showing points increasing from about (0, 45) to (9, 90))

b There is a strong, positive, linear relationship between the hours of revision done and the exam score.

19 a *Height* can take any value on the number line between 2.0 and 8.0.

b The modal class is $6.0 \leqslant h < 7.0$ metres.

c

Height (h m)	Frequency	Midpoint	Product
$2.0 \leqslant h < 3.0$	2	2.5	5
$3.0 \leqslant h < 4.0$	8	3.5	28
$4.0 \leqslant h < 5.0$	11	4.5	49.5
$5.0 \leqslant h < 6.0$	10	5.5	55
$6.0 \leqslant h < 7.0$	12	6.5	78
$7.0 \leqslant h < 8.0$	7	7.5	52.5
		Total	268

mean ≈ $\frac{268}{50}$ ≈ 5.36

d i $\frac{10 + 12 + 7}{50} \times 100\% = 58\%$ of the trees sampled measured 5 metres or more.
∴ in the plantation, we would expect 58% of 5000 = 2900 trees to measure 5 metres or more.

ii $\frac{8 + 11}{50} \times 100\% = 38\%$ of the trees sampled measured between 3 and 5 metres.
∴ in the plantation, we would expect 38% of 5000 = 1900 trees to measure between 3 and 5 metres.

20 a Number of workers surveyed
= 3 + 5 + 9 + 11 + 10 + 12 + 9 + 8 + 5 + 2 + 1
= 75

b 9 + 11 + 10 + 12 = 42 workers attended between 2 and 5 meetings inclusive.

c $\frac{8 + 5 + 2 + 1}{75} \times 100\% \approx 21.3\%$ of workers attended more than 6 meetings.

21 Sum of measurements = $25 \times 27.4 + 15 \times 22.6$
= 1024

∴ mean of measurements = $\frac{1024}{40}$
= 25.6

22 a

No. of cyclists	Frequency	Product	Cumulative frequency
2	1	2	1
3	1	3	2
4	2	8	4
5	3	15	7
6	5	30	12
7	8	56	20
8	12	96	32
9	7	63	39
10	3	30	42
Total	42	303	

i mean = $\frac{303}{42} \approx 7.21$

ii Since $n = 42$, $\frac{n+1}{2} = 21.5$
∴ median = average of 21st and 22nd scores
= $\frac{8 + 8}{2}$
= 8

iii mode = 8 {highest frequency}

iv range = maximum − minimum
= 10 − 2
= 8

b

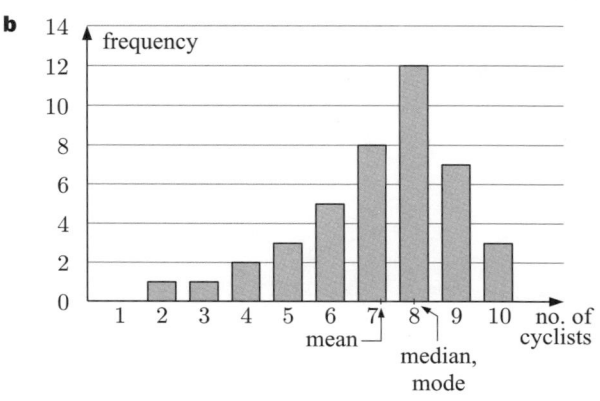

(bar chart of frequency vs no. of cyclists; bars at 2:1, 3:1, 4:2, 5:3, 6:5, 7:8, 8:12, 9:7, 10:3; mean marked between 7 and 8, median and mode marked at 8)

c The mean, as it takes the lower values into account.

23 a

Shoppers	Tally	Frequency
0 - 199	||||		6
200 - 399	|||| ||	7
400 - 599	|||| ||||		11
600 - 799	|||| ||	7
800 - 999	||	2
1000 - 1199	||||	4
1200 - 1399	||	2
1400 - 1599			1

b

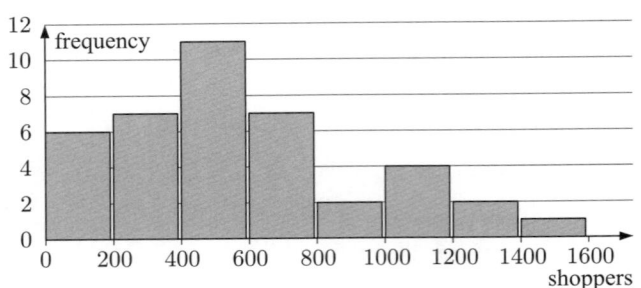

c The store received less than 400 visitors on
$6 + 7 = 13$ days.

d The modal class is 400 - 599 shoppers.

24 Since 9 is the mode, at least two of the numbers are 9.
∴ the six numbers are 9, 9, 7, 11, a, b.
Now, the mean $= 9$
∴ $\dfrac{9+9+7+11+a+b}{6} = 9$
∴ $36 + a + b = 54$
∴ $a + b = 18$
Also, the largest number is twice the smallest number.
∴ the only possibility is $a = 6$, $b = 12$.
∴ the smallest number is 6.

25

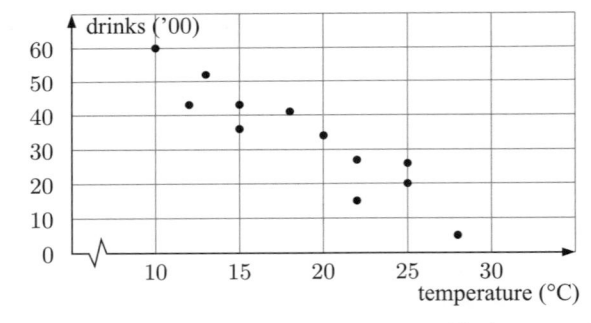

There is a strong, negative, linear relationship between the variables.

26 a i Since $n = 21$, $\dfrac{n+1}{2} = 11$
∴ median = 11th score = 10

ii We ignore the median and split the remaining data in two:
lower: 7 7 7 8 8 8 9 9 9 10
                    ↑
                    Q_1
∴ $Q_1 = \dfrac{8+8}{2} = 8$
upper: 10 10 11 11 11 11 11 12 13 13
                            ↑
                            Q_3
∴ $Q_3 = \dfrac{11+11}{2} = 11$

iii range = maximum − minimum
$= 13 - 7$
$= 6$

iv IQR $= Q_3 - Q_1$
$= 11 - 8$
$= 3$

b The ordered data set is
0 0 1 1 1 1 1 2 2 2 2 3 3 4 5 5 6 7 10
⎵_____lower_____⎵   ↑   ⎵_____upper_____⎵
                 median

i Since $n = 19$, $\dfrac{n+1}{2} = 10$
∴ median = 10th score = 2

ii Q_1 = median of lower half
$= 1$
Q_3 = median of upper half
$= 5$

iii range = maximum − minimum
$= 10 - 0$
$= 10$

iv IQR $= Q_3 - Q_1$
$= 5 - 1$
$= 4$

c The ordered data set is
87.9 88.1 88.6 89.5 89.7 89.9 89.9 90.9 92.7
⎵_____lower_____⎵   ↑   ⎵_____upper_____⎵
                    median

i Since $n = 9$, $\dfrac{n+1}{2} = 5$
∴ median = 5th score = 89.7

ii Q_1 = median of lower half
$= \dfrac{88.1 + 88.6}{2}$
$= 88.35$
Q_3 = median of upper half
$= \dfrac{89.9 + 90.9}{2}$
$= 90.4$

iii range = maximum − minimum
$= 92.7 - 87.9$
$= 4.8$

iv IQR $= Q_3 - Q_1$
$= 90.4 - 88.35$
$= 2.05$

27 a

Employees	Tally	Frequency
0 - 9			1
10 - 19	||	2
20 - 29	|||	3
30 - 39	||||	5
40 - 49	||||	5
50 - 59	||||		6
60 - 69	|||| |||	8
70 - 79	||||	5
	Total	35

b

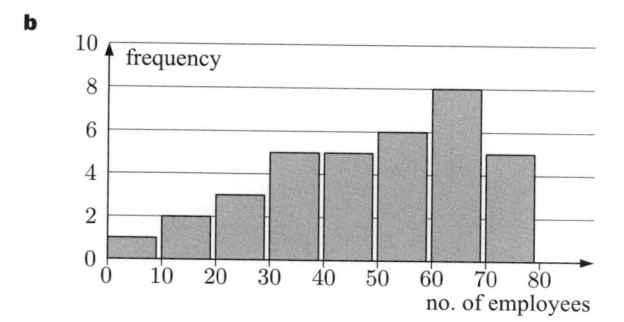

c $1 + 2 + 3 = 6$ businesses employed less than 30 people.

d $\dfrac{6 + 8 + 5}{35} \times 100\% \approx 54.3\%$ of businesses had at least 50 employees.

e The modal class is 60 - 69 employees.

28 a

Highest score	Frequency	Midpoint	Product
0 - 1999	25	999.5	24 987.5
2000 - 3999	32	2999.5	95 984
4000 - 5999	48	4999.5	239 976
6000 - 7999	15	6999.5	104 992.5
Total	120		465 940

mean $\approx \dfrac{465\,940}{120}$

≈ 3880 points

b The modal class is 4000 - 5999 points.

c Since $n = 120$, $\dfrac{n+1}{2} = 60.5$

There are $25 + 32 = 57$ scores below 4000 points, so the median is in the interval 4000 - 5999.

So, median $\approx L + \dfrac{N}{F} \times I$

where $L = 3999.5$, $N = 60.5 - 57 = 3.5$, $F = 48$, $I = 2000$

$\therefore$ median $\approx 3999.5 + \dfrac{3.5}{48} \times 2000$

≈ 4150 points

29 a The modal class is $750 \leqslant w < 800$.

b

Weight (w g)	Frequency	Midpoint	Product
$600 \leqslant w < 650$	2	625	1250
$650 \leqslant w < 700$	3	675	2025
$700 \leqslant w < 750$	7	725	5075
$750 \leqslant w < 800$	11	775	8525
$800 \leqslant w < 850$	2	825	1650
Total	25		18 525

mean $\approx \dfrac{18\,525}{25}$

≈ 741 g

c $\dfrac{2}{25} \times 100\% = 8\%$ of the zoo's meerkats weigh at least 800 g.

$\therefore$ in the colony, we would expect 8% of 50 = 4 meerkats to weigh at least 800 g.

d The estimate may not be reasonable if the diet and living conditions of the meerkats in Namibia are different from those of the meerkats in the zoo.

30 a minimum value = 100

b maximum value = 154

c $n = 25$ $\therefore$ $\dfrac{n+1}{2} = 13$

$\therefore$ median = 13th score = 126

d Q_1 = median of lower 12 scores
= average of 6th and 7th scores
= $\dfrac{113 + 115}{2}$
= 114

e Q_3 = median of upper 12 scores
= average of 19th and 20th scores
= $\dfrac{132 + 139}{2}$
= 135.5

f range = maximum − minimum
= 154 − 100
= 54

g IQR = $Q_3 - Q_1$
= 135.5 − 114
= 21.5

31 a 8 trains are represented by 4 squares.
$\therefore$ each square represents 2 trains.
There are 30 squares in total.
$\therefore$ sample size = $30 \times 2 = 60$.

b The modal class is $25 \leqslant t < 35$ minutes.
{highest frequency density}

c We use 1 square = 2 trains to find the frequency for each interval.

Time (t min)	Frequency	Midpoint	Product
$0 \leqslant t < 5$	4	2.5	10
$5 \leqslant t < 10$	8	7.5	60
$10 \leqslant t < 25$	18	17.5	315
$25 \leqslant t < 35$	20	30	600
$35 \leqslant t < 60$	10	47.5	475
Total	60		1460

mean $\approx \dfrac{1460}{60}$

≈ 24.3 minutes

32 a *Orchard A*:

Weight (w g)	Tally	Frequency
$200 \leqslant w < 210$	|||	3
$210 \leqslant w < 220$	||	2
$220 \leqslant w < 230$	|||| ||	7
$230 \leqslant w < 240$	|||| |||| |||	13
$240 \leqslant w < 250$	|||| |||| ||	12
$250 \leqslant w < 260$	|||	3
Total		40

Orchard B:

Weight (w g)	Tally	Frequency
$200 \leqslant w < 210$	|||	3
$210 \leqslant w < 220$	||||	4
$220 \leqslant w < 230$	|||| |||| ||||	15
$230 \leqslant w < 240$	|||| |||| ||	12
$240 \leqslant w < 250$	||||	5
$250 \leqslant w < 260$			1
Total		40

b

Orchard A

[histogram: weight (g) 200–260, frequency]

Orchard B

[histogram: weight (g) 200–260, frequency]

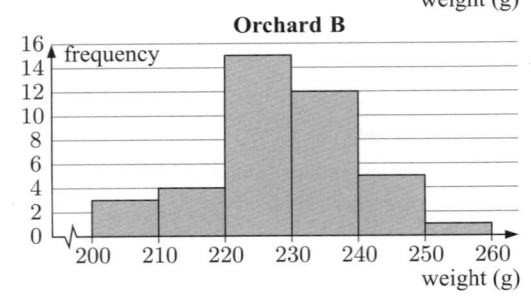

c The modal class for orchard A is $230 \leqslant w < 240$ g.
  The modal class for orchard B is $220 \leqslant w < 230$ g.

d The data for orchard A is skewed, while the data for orchard B is roughly symmetrical.

e Orchard A: range $= 258 - 201$
              $= 57$
  Orchard B: range $= 259 - 200$
              $= 59$

f Looking at the histograms, orchard A has more oranges at the higher end of the scale.
  So, orchard A produced heavier oranges.

33 a $\overline{x} = \dfrac{45 + 22 + 38 + \ldots + 28 + 55}{8}$
     $= \dfrac{287}{8}$
     ≈ 35.9 thousand pounds

  $\overline{y} = \dfrac{172 + 91 + 140 + \ldots + 162 + 271}{8}$
     $= \dfrac{1284}{8}$
     $= 160.5$ pounds

b, d, e

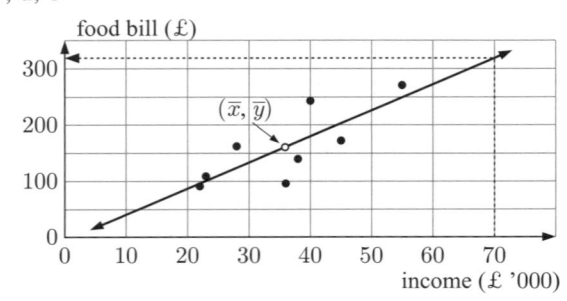

[scatter plot: food bill (£) vs income (£ '000), with $(\overline{x}, \overline{y})$ marked and line of best fit]

c There is a moderate, positive, linear association between *household income* and the *weekly food bill*.

f When $x = 70$, $y \approx 320$.
  So, the household's weekly food bill is approximately £320.

g This is outside the domain of the data, so the estimate may be unreliable, especially since the correlation is not very strong.

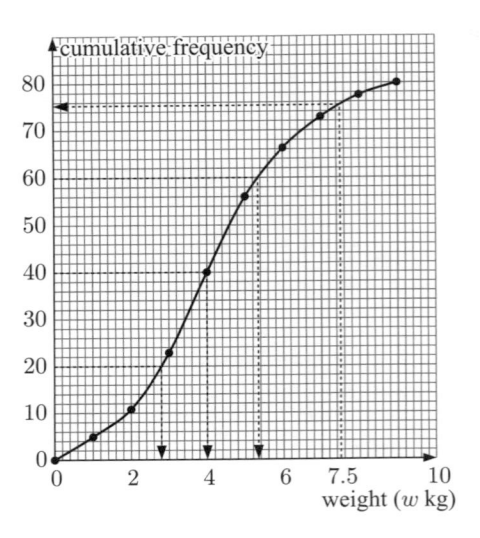

34

[cumulative frequency curve, weight (w kg) 0–10]

a 80 pumpkins

b 50% of 80 is 40.
  When CF $= 40$, $w = 4$.
  $\therefore$ the median is approximately 4 kg.

c When $w = 7.5$, CF ≈ 75
  $\therefore$ $80 - 75 = 5$ pumpkins are heavier than 7.5 kg, and will be entered in the local fair.

d IQR $= Q_3 - Q_1$
     $\approx (w$ when CF $= 60) - (w$ when CF $= 20)$
     $\approx 5.3 - 2.8$
     ≈ 2.5

35 a

Weight (w kg)	Frequency	CIW	Frequency density
$0 \leqslant w < 50$	2	50	0.04
$50 \leqslant w < 75$	13	25	0.52
$75 \leqslant w < 100$	19	25	0.76
$100 \leqslant w < 110$	36	10	3.6
$110 \leqslant w < 150$	10	40	0.25

[frequency density histogram, weight (kg)]

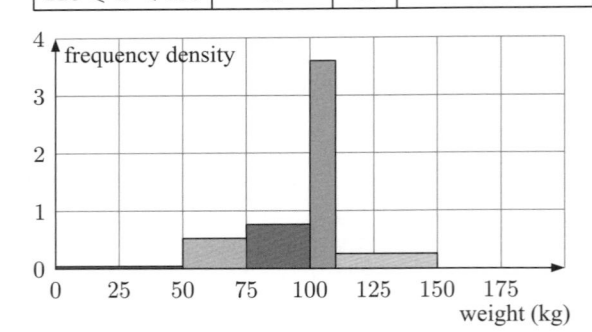

b The modal class is $100 \leqslant w < 110$ kg.
  {highest frequency density}

c

Weight (w kg)	Frequency	Midpoint
$0 \leqslant w < 50$	2	25
$50 \leqslant w < 75$	13	62.5
$75 \leqslant w < 100$	19	87.5
$100 \leqslant w < 110$	36	105
$110 \leqslant w < 150$	10	130

Using technology, mean ≈ 95.1 kg.

36 a Simon: mean $= \dfrac{50 + 15 + 1 + \ldots + 36 + 60}{12}$
            $= \dfrac{484}{12}$
            ≈ 40.3

The ordered data set is:
1 15 20 25 36 40 50 57 60 60 60 60
                        ↑
                     median

∴ median = $\dfrac{40+50}{2} = 45$

Barney: mean = $\dfrac{50 + 15 + 50 + ... + 16 + 57}{12}$
$= \dfrac{507}{12}$
$= 42.25$

The ordered data set is:
15 16 20 48 48 48 50 50 50 51 54 57
                        ↑
                     median

∴ median = $\dfrac{48+50}{2} = 49$

b Simon: range = $60 - 1$
$= 59$

Q_1 = median of lower 6 scores
$= \dfrac{20+25}{2}$
$= 22.5$

Q_3 = median of upper 6 scores
$= \dfrac{60+60}{2}$
$= 60$

∴ IQR $= 60 - 22.5$
$= 37.5$

Barney: range = $57 - 15$
$= 42$

Q_1 = median of lower 6 scores
$= \dfrac{20+48}{2}$
$= 34$

Q_3 = median of upper 6 scores
$= \dfrac{50+51}{2}$
$= 50.5$

∴ IQR $= 50.5 - 34$
$= 16.5$

c Barney is more consistent, he has a lower range and IQR.

37 a The independent variable is *cold days*, the dependent variable is *yield*.

b

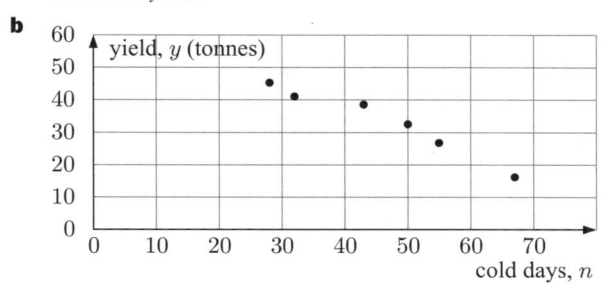

c $Y \approx -0.714n + 66.1$

d The gradient of -0.714 indicates that, for every additional cold day in the year, the yield decreases by 0.714 tonnes. The vertical intercept of 66.1 indicates that if there were no cold days, the yield would be 66.1 tonnes.

e i When $n = 40$, $Y \approx -0.714(40) + 66.14$
≈ 37.6

∴ if there are 40 cold days in a year, we would expect the yield to be 37.6 tonnes.

ii When $n = 100$, $Y \approx -0.714(100) + 66.14$
≈ -5.26

∴ if there are 100 cold days in a year, we would expect the yield to be -5.26 tonnes.

f The answer in **i** is within the domain, and appears to be reliable.
The answer in **ii** is outside the domain, and is clearly unreliable as the yield cannot be negative.

38 a

Time (t min)	Frequency	Cumulative frequency
$0 \leqslant t < 1$	3	3
$1 \leqslant t < 2$	11	14
$2 \leqslant t < 3$	19	33
$3 \leqslant t < 4$	15	48
$4 \leqslant t < 5$	7	55
$5 \leqslant t < 6$	5	60

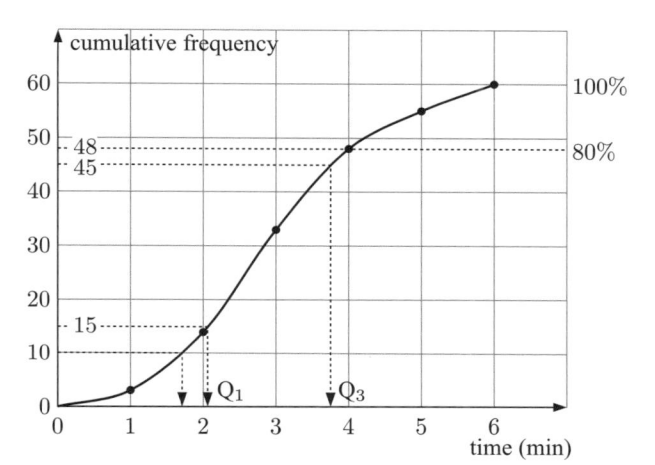

b When CF = 15, $t \approx 2.1$
∴ $Q_1 \approx 2.1$

When CF = 45, $t \approx 3.8$
∴ $Q_3 \approx 3.8$

∴ IQR $\approx 3.8 - 2.1$
≈ 1.7

This means that the middle half of the times were spread over 1.7 minutes.

c When CF = 80% of 60 = 48, $t = 4$.
∴ the 80th percentile is 4 minutes. This means that 80% of the group finished the puzzle in 4 minutes or less.

d When CF = 10, $t \approx 1.7$
∴ the cut-off time for selection is 1.7 minutes, or 1 minute 42 seconds.

39 a

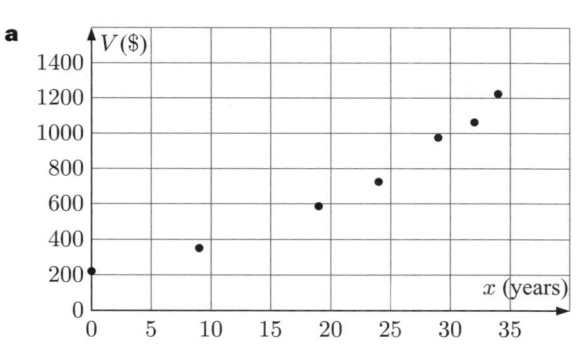

b $V \approx 28.9x + 127$

c There is a strong, positive association between the variables.
The linear model is not appropriate as the data does not lie in a straight line.

d We could draw a smooth curve through the points and extend it to $x = 49$.

SOLUTIONS TO PRACTICE EXAM 1

Paper 2

1 $\quad z = \dfrac{2}{3y-4}$

$\therefore \ z(3y-4) = \dfrac{2}{3y-4}(3y-4)$

$\therefore \ z(3y-4) = 2$

$\therefore \ \dfrac{z(3y-4)}{z} = \dfrac{2}{z}$

$\therefore \ 3y - 4 = \dfrac{2}{z}$

$\therefore \ 3y - 4 + 4 = \dfrac{2}{z} + 4$

$\therefore \ 3y = \dfrac{2}{z} + 4$

$\therefore \ \dfrac{3y}{3} = \dfrac{\frac{2}{z}+4}{3}$

$\therefore \ y = \dfrac{2}{3z} + \dfrac{4}{3}$

2 a $3^6 = 729 \Leftrightarrow \log_3 729 = 6$

b $5\sqrt{5} = 5^{1.5} \Leftrightarrow \log_5 5\sqrt{5} = 1.5$

3 a $\sin 60° = \dfrac{5}{x}$

$\therefore \ \dfrac{\sqrt{3}}{2} = \dfrac{5}{x}$

$\therefore \ \dfrac{\sqrt{3}}{2} \times 2x = \dfrac{5}{x} \times 2x$

$\therefore \ \sqrt{3}x = 10$

$\therefore \ \dfrac{\sqrt{3}x}{\sqrt{3}} = \dfrac{10}{\sqrt{3}}$

$\therefore \ x = \dfrac{10}{\sqrt{3}}$

b $\cos x° = \dfrac{3\sqrt{2}}{6}$

$= \dfrac{\sqrt{2}}{2} = \dfrac{1}{\sqrt{2}}$

$\therefore \ x° = \cos^{-1}\left(\dfrac{1}{\sqrt{2}}\right)$

$\therefore \ x = 45$

4 a 81, 80, 78, 75, 71, 66, 60 with differences $-1, -2, -3, -4, -5, -6$

b The difference table is:

n	1	2	3	4	5	6	7
u_n	81	80	78	75	71	66	60
$\Delta 1$		-1	-2	-3	-4	-5	-6
$\Delta 2$			-1	-1	-1	-1	-1

$\Delta 2$ values are constant, so the sequence is quadratic with general term $u_n = an^2 + bn + c$.

$2a = -1$, so $a = -\frac{1}{2}$

$3a + b = -1$, so $-\frac{3}{2} + b = -1 \quad \therefore \ b = \frac{1}{2}$

$a + b + c = 81$, so $-\frac{1}{2} + \frac{1}{2} + c = 81 \quad \therefore \ c = 81$

$\therefore$ the general term is $u_n = -\frac{1}{2}n^2 + \frac{1}{2}n + 81$

c ..., 60, 53, 45, 36, 26, 15, 3, -10 with differences $-7, -8, -9, -10, -11, -12, -13$

The first negative term is -10.
It is the 14th term.

5 Interest payable for 1 year $= 6\%$ of £1800
$= 0.06 \times £1800$

$\therefore$ interest payable for 40 months which is $\frac{40}{12}$ years
$= 0.06 \times £1800 \times \frac{40}{12}$
$= £360$

6 $g(x) = \dfrac{2-x}{x^3}$

a $g(-1) = \dfrac{2-(-1)}{(-1)^3} = \dfrac{2+1}{-1} = -3$

b $g(3) = \dfrac{2-3}{3^3} = -\dfrac{1}{27}$

c $g(-5) = \dfrac{2-(-5)}{(-5)^3} = \dfrac{2+5}{-125} = -\dfrac{7}{125}$

7 Let $x = 0.\overline{12}$
$= 0.121212....$

$\therefore \ 100x = 12.1212....$
$= 12 + x$

$\therefore \ 99x = 12$

$\therefore \ x = \dfrac{12}{99}$

$\therefore \ x = \dfrac{4}{33}$, so x is rational.

8 a $\{x \mid 0 \leqslant x \leqslant 8, \ x \in \mathbb{Z}\}$

b $\{x \mid x < -1 \text{ or } x \geqslant 2, \ x \in \mathbb{R}\}$

9 a Area $= \frac{1}{2} \times 12 \times 18 \times \sin 30°$
$= \frac{1}{2} \times 12 \times 18 \times \frac{1}{2}$
$= 54 \text{ cm}^2$

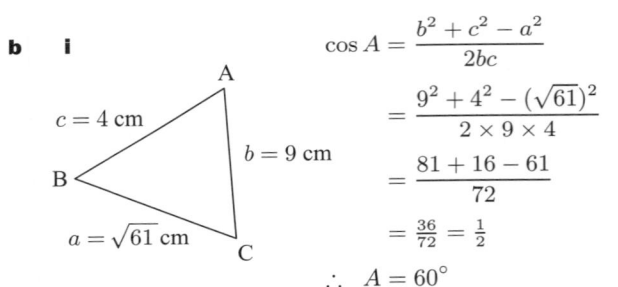

b i $\cos A = \dfrac{b^2 + c^2 - a^2}{2bc}$

$= \dfrac{9^2 + 4^2 - (\sqrt{61})^2}{2 \times 9 \times 4}$

$= \dfrac{81 + 16 - 61}{72}$

$= \dfrac{36}{72} = \dfrac{1}{2}$

$\therefore \ A = 60°$

Triangle with $c = 4$ cm, $b = 9$ cm, $a = \sqrt{61}$ cm, vertices A, B, C.

ii Area $= \frac{1}{2} \times 9 \times 4 \times \sin 60°$
$= 18 \times \dfrac{\sqrt{3}}{2}$
$= 9\sqrt{3} \text{ cm}^2$

10 a B **b** C **c** A

11 a $\overrightarrow{XZ} = \begin{pmatrix} -6-4 \\ -2-3 \end{pmatrix} = \begin{pmatrix} -10 \\ -5 \end{pmatrix}$

b $\overrightarrow{WY} = \begin{pmatrix} 2--2 \\ -1-2 \end{pmatrix} = \begin{pmatrix} 4 \\ -3 \end{pmatrix}$

c $\overrightarrow{VY} = \begin{pmatrix} 2-0 \\ -1--4 \end{pmatrix} = \begin{pmatrix} 2 \\ 3 \end{pmatrix}$

d $\overrightarrow{WZ} = \begin{pmatrix} -6--2 \\ -2-2 \end{pmatrix} = \begin{pmatrix} -4 \\ -4 \end{pmatrix}$

Paper 4

1

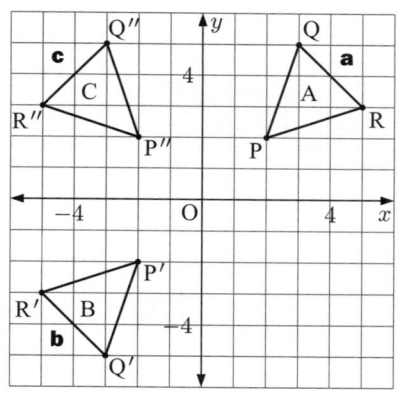

d a reflection in the y-axis

2 $f(x) = 3x^2 - 6x + 1$

a $f(3) = 3(3)^2 - 6(3) + 1$
$= 27 - 18 + 1$
$= 10$

b Cuts y-axis when $x = 0$
$f(0) = 1$, so y-intercept is 1.
Cuts x-axis when $y = 0$
$\therefore\ 0 = 3x^2 - 6x + 1$
$\therefore\ x = \dfrac{6 \pm \sqrt{36 - 4 \times 3}}{2 \times 3}$
$= \dfrac{6 \pm \sqrt{24}}{6}$
$= \dfrac{6 \pm 2\sqrt{6}}{6}$
$= 1 \pm \dfrac{\sqrt{6}}{3}$
$\therefore$ the x-intercepts are $1 - \dfrac{\sqrt{6}}{3}$ and $1 + \dfrac{\sqrt{6}}{3}$.

c The quadratic has a line of symmetry halfway between the x-intercepts.
The average of $1 - \dfrac{\sqrt{6}}{3}$ and $1 + \dfrac{\sqrt{6}}{3}$ is 1, so the line of symmetry is $x = 1$.
The line of symmetry passes through the vertex of the quadratic.
$\therefore$ the x-coordinate of the vertex is 1.
Now $f(1) = 3(1)^2 - 6(1) + 1$
$= -2$
$\therefore$ the vertex is $(1, -2)$.

d $f(x) = 3(x - 1)^2 - 2$

e $f(x)$ has vertex $(1, -2)$
Under $T\begin{pmatrix}1\\2\end{pmatrix}$, $g(x)$ has vertex $(2, 0)$.
$\therefore\ g(x) = 3(x - 2)^2$
$= 3(x^2 - 4x + 4)$
$= 3x^2 - 12x + 12$

3 $h = 3r$

a $V = \pi r^2 h$
$= \pi r^2 (3r)$
$= 3\pi r^3$

b $3\pi r^3 = 192\pi$
$\therefore\ 3r^3 = 192$
$\therefore\ r^3 = 64$
$\therefore\ r = 4$

c Surface area
$=$ area of curved surface $+$ area of 2 ends
$= 2\pi r h + 2\pi r^2$
$= 2\pi r(3r) + 2\pi r^2$
$= 6\pi r^2 + 2\pi r^2$
$= 8\pi r^2$
$= 8\pi \times 4^2$
≈ 402 cm^2

d Each cm^3 of lead weighs 11.37 g.
$\therefore$ using **b**, mass $= 11.37 \times 192\pi$
≈ 6860 g

e $V = \tfrac{4}{3}\pi r^3$
$\therefore\ 192\pi = \tfrac{4}{3}\pi r^3$
$\therefore\ r^3 = 192 \times \tfrac{3}{4}$
$\therefore\ r = \sqrt[3]{144}$
≈ 5.24
$\therefore$ the radius is approximately 5.24 cm.

4 a $2x + y = 3$ $2x + 3y = 15$

x	0	1.5
y	3	0

x	0	7.5
y	5	0

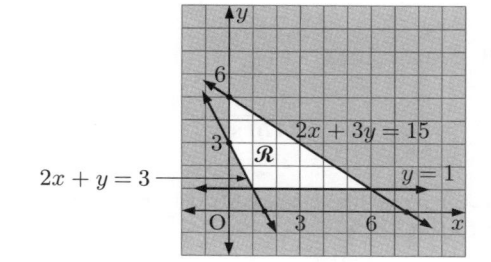

b

x	1	2	3	1	2	3	0
y	1	1	1	2	2	2	3
$x+2y$	3	4	5	5	6	7	6

$\rightarrow$ all other y values are too large

$(2, 1)$ is the only point with integer values such that $x + 2y = 4$.

5 a, d

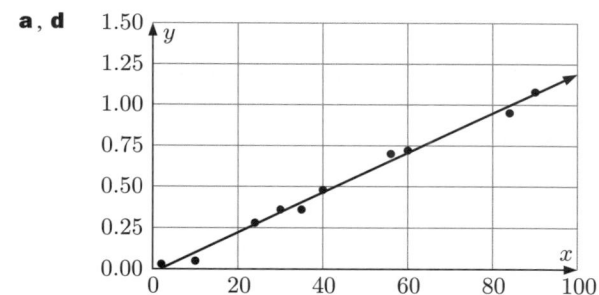

b very strong, positive correlation

c There is very strong, positive, linear correlation between the two variables, so as x increases, so does y. This indicates that the two variables are directly proportional.

d $y \approx 0.0121x - 0.0201$ (see above for graph)

e gradient ≈ 0.0121
For every extra female teacher on staff, a student will get an extra 0.0121 detentions a year.

f i 0.440 detentions   **ii** 1.37 detentions

The prediction for 38 teachers is reasonable, since the prediction is within the domain of the data and the variables are very strongly correlated.

The prediction for 115 teachers is not as reliable, as it is outside the domain of the data.

6 a 0.45 km = 450 m

   b i between 20 s and 50 s after starting

      ii average speed $= \dfrac{\text{distance}}{\text{time}}$

$= \dfrac{0.15 \text{ km}}{30 \text{ s}}$

$= \dfrac{150 \text{ m}}{30 \text{ s}}$

$= 5$ m/s

   c It happens 80 seconds after starting. He is stuck for 10 seconds.

   d i average speed for race $= \dfrac{0.45 \text{ km}}{170 \text{ s}}$

$= \dfrac{0.45 \text{ km}}{\frac{170}{60 \times 60} \text{ h}}$

$= \dfrac{0.45 \times 60 \times 60}{170}$ km/h

≈ 9.53 km/h

      ii average speed $= \dfrac{0.45 \text{ km}}{170 \text{ s}}$

$= \dfrac{450 \text{ m}}{170 \text{ s}}$

≈ 2.65 m/s

   e To complete the course Michel travels 0.45 km at 10.1 km/h,

so time taken $= \dfrac{0.45}{10.1} \times 3600 + 10$ s

$\approx 160.4 + 10$ s

≈ 170.4 s

So, Clinton finished in the quicker time.

7 a $\triangle$s ACE and BCD are congruent (SAS) because

      AC = BC   {given}

      CE = CD   {given}

      $A\widehat{C}E = B\widehat{C}D$ {vertically opposite}

      $\therefore$ AE = BD   {corresponding sides}

   b In $\triangle$s ABE, ADB,

      AB is common

      AE = BD   {part **a**}

      AC + CD = BC + CE   {AC = BC, CD = CE}

      $\therefore$ AD = BE

      $\therefore$ $\triangle$s ABE, ADB are congruent (SSS)

   c From **b**, $A\widehat{E}B = A\widehat{D}B$

But these are angles subtended by AB.

$\therefore$ ABDE is a cyclic quadrilateral.

8 a Data was recorded on $3 + 6 + 8 + 5 + 4 + 2$

$= 28$ days

b   Histogram of Surfing times

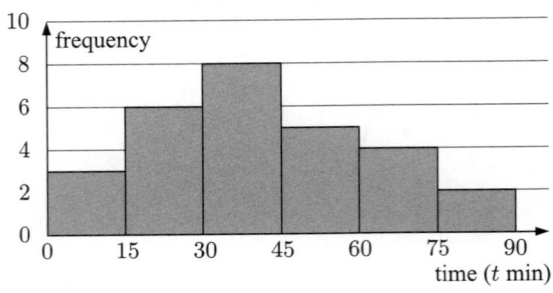

c The modal class is $30 \leqslant t < 45$ min.

d

Middle value (x)	Frequency (f)
7.5	3
22.5	6
37.5	8
52.5	5
67.5	4
82.5	2

The mean length of time ≈ 41.3 min.

e

Surfing time (mins)	Freq.	Cumul. Freq.
$0 \leqslant t < 15$	3	3
$15 \leqslant t < 30$	6	9
$30 \leqslant t < 45$	8	17
$45 \leqslant t < 60$	5	22
$60 \leqslant t < 75$	4	26
$75 \leqslant t < 90$	2	28

Cumulative frequency graph

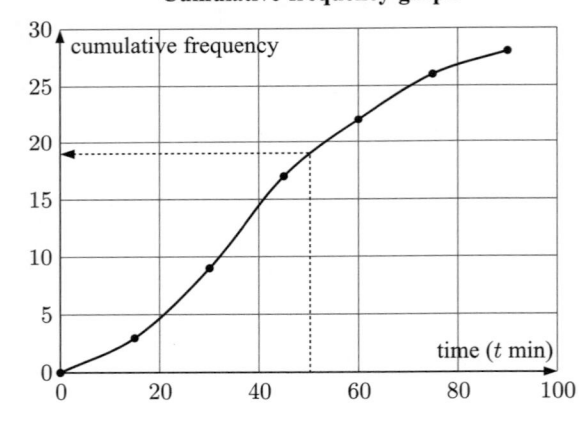

f Toby spends up to 50 minutes on approximately 19 days, so he spends more than 50 minutes on $28 - 19 = 9$ days.

$\therefore$ the probability that Toby 'fines' himself is $\frac{9}{28}$.

In a 30 day month he fines himself

$\frac{9}{28} \times 30 \approx 10$ days (nearest day).

$\therefore$ Toby will pay approximately

£5 $\times$ 10 = £50 in 'fines'.

9 $f(x) = 5 \times 4^{-x}$

   a i $f(-\tfrac{3}{2}) = 5 \times 4^{-(-\frac{3}{2})}$

$= 5 \times \left(2^2\right)^{\frac{3}{2}}$

$= 5 \times 2^3$

$= 40$

ii $f(x-1) = 5 \times 4^{-(x-1)}$
$= 5 \times 4^{-x} \times 4$
$= 20 \times 4^{-x}$
$4f(x) = 4 \times 5 \times 4^{-x}$
$= 20 \times 4^{-x}$
$\therefore\ f(x-1) = 4f(x)$

iii Domain $= \{x \mid x \in \mathbb{R}\}$
Range $= \{y \mid y > 0,\ y \in \mathbb{R}\}$

iv Asymptote is $y = 0$.

b By interchanging x and y, the inverse of $y = \dfrac{2x-1}{x+1}$
is $x = \dfrac{2y-1}{y+1}$
$\therefore\ x(y+1) = 2y - 1$
$\therefore\ xy + x = 2y - 1$
$\therefore\ xy - 2y = -x - 1$
$\therefore\ y(x-2) = -x - 1$
$\therefore\ y = \dfrac{-x-1}{x-2} = f^{-1}(x)$ or $f^{-1}(x) = \dfrac{x+1}{2-x}$

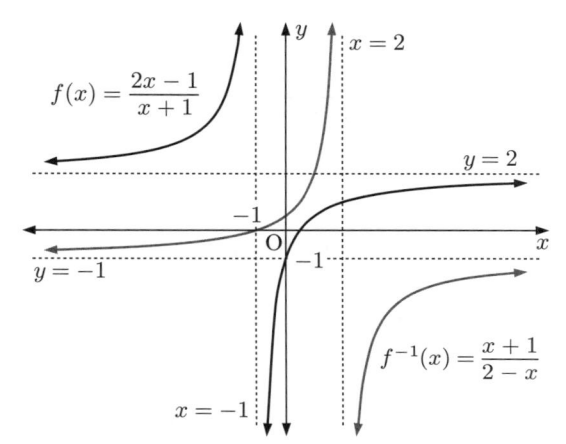

10 a $m_{PQ} = \dfrac{-1-(-5)}{6-(-1)} = \dfrac{-1+5}{6+1} = \dfrac{4}{7}$

$m_{PR} = \dfrac{1-(-5)}{2-(-1)} = \dfrac{6}{3} = 2$

$m_{QR} = \dfrac{1-(-1)}{2-6} = \dfrac{2}{-4} = -\dfrac{1}{2}$

$m_{QR} \times m_{PR} = -\dfrac{1}{2} \times 2 = -1$

$\therefore$ QR is perpendicular to PR.
$\therefore\ \triangle$PQR is right angled at R.

b midpoint X of QR is $\left(\dfrac{6+2}{2}, \dfrac{-1+1}{2}\right)$
$\therefore$ X is $(4, 0)$.
distance PX $= \sqrt{(4-(-1))^2 + (0-(-5))^2}$
$= \sqrt{5^2 + 5^2}$
$= \sqrt{50}$
$= 5\sqrt{2}$ units

c Let Y have coordinates (a, b).

For a parallelogram, the midpoints of the diagonals are the same point.

Midpoint of PQ is $\left(\dfrac{-1+6}{2}, \dfrac{-5+-1}{2}\right) = \left(\dfrac{5}{2}, -3\right)$

Midpoint of XY is $\left(\dfrac{a+4}{2}, \dfrac{b+0}{2}\right) = \left(\dfrac{a+4}{2}, \dfrac{b}{2}\right)$

$\therefore\ \dfrac{a+4}{2} = \dfrac{5}{2}$ and $\dfrac{b}{2} = -3$
$\therefore\ a + 4 = 5$ and $b = -6$
$\therefore\ a = 1$
So, Y is $(1, -6)$.

11 a We are given: $x + 2y = 12$ .... (1)
$x - 4y = -12$ .... (2)
$\therefore\ 2x + 4y = 24$ $\{(1) \times 2\}$
$\underline{x - 4y = -12}$
$3x \quad\quad = 12$ $\{$adding$\}$
$\therefore\ x = 4$
Substituting in (1) gives
$4 + 2y = 12$
$\therefore\ 2y = 8$
$\therefore\ y = 4$
So, $x = 4,\ y = 4$.

b Using technology,
$x = 4$ or $x = -12$
$\therefore\ x = 4,\ y = 4$ or $x = -12,\ y = 0$

12 Let Jo have €x.
$\therefore$ Gemma has €$3x$.
Together they have €x + €$3x$ = €$4x$
and $4x = 28.60$
$\therefore\ x = 7.15$
and $3x = 21.45$
So, Jo has €7.15 and Gemma has €21.45.

Paper 6

A. Investigation: The Hilbert curve

Part I

1 Using the base of the first pattern, the square is 1 unit wide. The second pattern has the same width. So 3 equal lengths make up 1 unit, and the length of each line segment is $\tfrac{1}{3}$ unit.

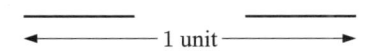

In the 3rd pattern, 7 lengths make up 1 unit, and so on.

Hilbert curve order	Length of line segments
1	1
2	$\tfrac{1}{3}$
3	$\tfrac{1}{7}$
4	$\tfrac{1}{15}$
5	$\tfrac{1}{31}$

2 The denominator is 2 raised to the power of the order number, then 1 is subtracted from it.

3 $u_n = \dfrac{1}{2^n - 1}$

For $n = 6$,
$u_6 = \dfrac{1}{2^6 - 1}$
$= \dfrac{1}{64 - 1}$
$= \dfrac{1}{63}$

For $n = 7$,
$u_7 = \dfrac{1}{2^7 - 1}$
$= \dfrac{1}{128 - 1}$
$= \dfrac{1}{127}$

4 As n becomes very large, the length becomes very small.

Part II

1 3   **2** 15   **3** $u_{n+1} = 4u_n + 3$

4

Hilbert curve order	Number of line segments
1	3 $(= 2^2 - 1)$
2	15 $(= 2^4 - 1)$
3	63 $(= 2^6 - 1)$
4	255 $(= 2^8 - 1)$
5	1023 $(= 2^{10} - 1)$

5 a $u_n = 2^{2n} - 1$ or $4^n - 1$
  b i $u_6 = 4^6 - 1 = 4095$
    ii $u_{10} = 4^{10} - 1 = 1\,048\,575$
    iii $u_{15} = 4^{15} - 1 = 1\,073\,741\,823$

6 $u_{n+1} = 4^{n+1} - 1$
$= 4 \times 4^n - 1$
$= 4 \times (4^n - 1) + 3$
$= 4u_n + 3$

7 As n becomes very large, the number of line segments becomes very large.

Part III

1

Hilbert curve order	Length of each line segment	Number of line segments	Total length
1	1	3	3
2	$\frac{1}{3}$	15	5
3	$\frac{1}{7}$	63	9
4	$\frac{1}{15}$	255	17
5	$\frac{1}{31}$	1023	33

2 $L_n = \dfrac{1}{2^n - 1} \times (2^{2n} - 1)$

from **Part I, 3**   from **Part II, 5**

$\therefore\ L_n = \dfrac{1}{2^n - 1} \times (2^n + 1)(2^n - 1)$
$= 2^n + 1$

a $L_6 = 2^6 + 1 = 65$ **b** $L_7 = 2^7 + 1 = 129$
c $L_{20} = 2^{20} + 1 = 1\,048\,577$

---

3 a As n gets larger, L_n becomes very large.
  b As n gets very large, the number of line segments gets extremely large, but the length of each line segment gets very small.
  However, in $L_n = \dfrac{1}{2^n - 1} \times (2^{2n} - 1)$,
  2^{2n} becomes larger much faster than 2^n.
  So, L_n becomes larger as n becomes larger.

B. Modelling: Hard Disk Storage

1

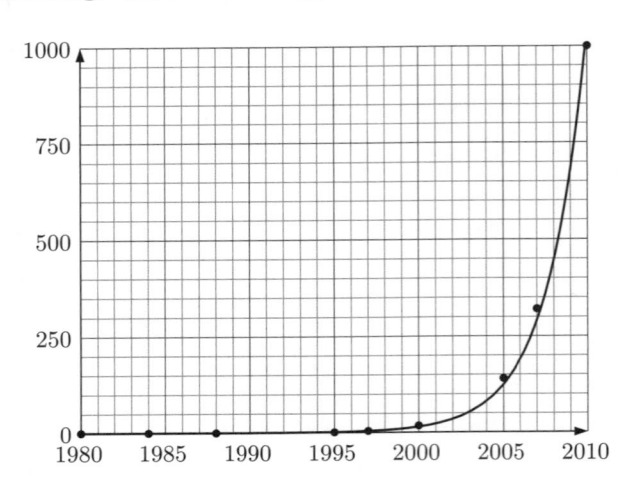

2 a i $S \approx 0.002\,76 \times 1.534^7$
    ≈ 0.0552 GB
  ii $S \approx 0.002\,76 \times 1.534^{12}$
    ≈ 0.469 GB

  b Using answers to **2 a i, ii**:
  $\left(\dfrac{0.469 - 0.0552}{0.0552}\right) \times 100\% \approx 750\%$ increase

  c If $S_t = 0.002\,76 \times 1.534^t$, then in 5 years,
  $S_{t+5} = 0.002\,76 \times 1.534^{t+5}$
  $= \underbrace{0.002\,76 \times 1.534^t}_{\text{this is } S_t} \times 1.534^5$
  $\approx 8.49 \times S_t$

  This corresponds to a 749% increase, so it will increase by approximately the same percentage every 5 years.

3 a $S = 6$ when $t \approx 17.96$ {using solver}
  The hard drive was made at the end of 1997 or start of 1998.
  b Our prediction should be fairly reliable, since it is within the domain of the data, and the variables have strong correlation.

4 a When $t = 40$, $S \approx 74\,804$ GB
  b Our prediction is outside the domain of the data, so it is not really reliable.

SOLUTIONS TO PRACTICE EXAM 2

Paper 2

1 a $-(-1)^3$
    $= -(-1)$
    $= 1$

b $\sqrt{28} - \sqrt{7}$
    $= \sqrt{4 \times 7} - \sqrt{7}$
    $= \sqrt{4} \times \sqrt{7} - \sqrt{7}$
    $= 2\sqrt{7} - \sqrt{7}$
    $= \sqrt{7}$

c $\log 9 + \log 15 - \log 5$
$= \log\left(\frac{9 \times 15}{5}\right)$
$= \log 27$
$= \log 3^3$
$= 3\log 3$

2 $-1, 2, -4, 8, -16$

a $-2^0, 2^1, -2^2, 2^3, -2^4, 2^5, -2^6$
The next two terms are $32, -64$.

b $u_n = -(-2)^{n-1}$

3 a C **b** A **c** B

4 $m = \frac{2-j}{3s}$

$\therefore 3s \times m = \frac{2-j}{3s} \times 3s$

$\therefore 3ms = 2 - j$

$\therefore 3ms + j = 2 - j + j$

$\therefore 3ms + j = 2$

$\therefore 3ms + j - 3ms = 2 - 3ms$

$\therefore j = 2 - 3ms$

5 a 111 minutes

b There are 20 scores, so the median is the average of the 10th and 11th scores.

$\therefore$ median $= \frac{93+95}{2} = 94$ minutes.

c The mode is the score that occurs most frequently.
$\therefore$ mode $= 93$ minutes.

d 5 scores are less than 90 minutes.
$\therefore$ percentage is $\frac{5}{20} \times 100\% = 25\%$

6 a $36^{\frac{3}{2}} = \left(6^2\right)^{\frac{3}{2}}$
$= 6^3$
$= 216$

b $5^{-3} = \frac{1}{5^3}$
$= \frac{1}{125}$

7 $2x + 5y = 11$ .... (1)
$3x - 2y = -12$ .... (2)
$4x + 10y = 22$ {multiply (1) by 2}
$15x - 10y = -60$ {multiply (2) by 5}
$19x = -38$ {adding}
$\therefore x = -2$ {$\div 19$}

Substituting this value into (1) gives $2(-2) + 5y = 11$
$\therefore -4 + 5y = 11$
$\therefore 5y = 15$
$\therefore y = 3$

Solution is $x = -2, y = 3$.

8 a $\overrightarrow{AB} = \begin{pmatrix} 5-1 \\ -2-4 \end{pmatrix} = \begin{pmatrix} 4 \\ -6 \end{pmatrix}$

b $C' = (-1+4, 2+-6) = (3, -4)$

c midpoint $= \left(\frac{1+5}{2}, \frac{4+-2}{2}\right) = (3, 1)$

9 $\frac{3}{x(x-1)} + \frac{x-4}{x-1} = \frac{3}{x(x-1)} + \frac{x(x-4)}{x(x-1)}$
$= \frac{3 + x(x-4)}{x(x-1)}$
$= \frac{3 + x^2 - 4x}{x(x-1)}$
$= \frac{x^2 - 4x + 3}{x(x-1)}$
$= \frac{(x-1)(x-3)}{x(x-1)}$
$= \frac{x-3}{x}$

10 Using the cosine rule,
$a^2 = 4^2 + 5^2 - 2 \times 4 \times 5 \times \cos 60°$
$= 16 + 25 - 40 \times \frac{1}{2}$
$= 41 - 20$
$= 21$
$\therefore a = \sqrt{21}$, as $a > 0$

11 Let $D =$ catch domestic flight
and $C =$ pay by credit card

a $P(C') = 1 - P(C)$
$= 1 - \frac{3}{4}$
$= \frac{1}{4}$

b $P(D$ and $C) = P(D) \times P(C)$
$= \frac{2}{5} \times \frac{3}{4}$
$= \frac{3}{10}$

Paper 4

1 $l = (r+2)$ cm

a Area of surface $=$ area of curved surface $+$ area of base
$= \pi r l + \pi r^2$
$= \pi r(r+2) + \pi r^2$
$= 2\pi r^2 + 2\pi r$ cm^2

b $\pi(2r^2 + 2r) = 220\pi$
$\therefore 2r^2 + 2r - 220 = 0$
$\therefore r^2 + r - 110 = 0$
$\therefore (r+11)(r-10) = 0$
$\therefore r = -11$ or 10
But $r > 0$, $\therefore r = 10$

c $l = r + 2 = 12$
$\therefore h = \sqrt{12^2 - 10^2}$
$= \sqrt{44} \approx 6.63$
Volume of cone
$= \frac{1}{3}\pi r^2 h$
$= \frac{1}{3}\pi \times 10^2 \times \sqrt{44}$
≈ 695 cm^3 (3 s.f.)

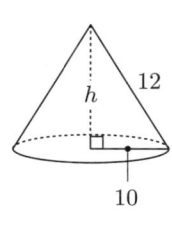

2 a frequency density $= \frac{\text{frequency}}{\text{class interval width}}$

Time (d min)	Freq.	CIW	Freq. density
$20 \leqslant d < 25$	6	5	1.2
$25 \leqslant d < 28$	10	3	$3.\overline{3}$
$28 \leqslant d < 30$	12	2	6
$30 \leqslant d < 32$	7	2	3.5
$32 \leqslant d < 36$	5	4	1.25

Histogram of drive times

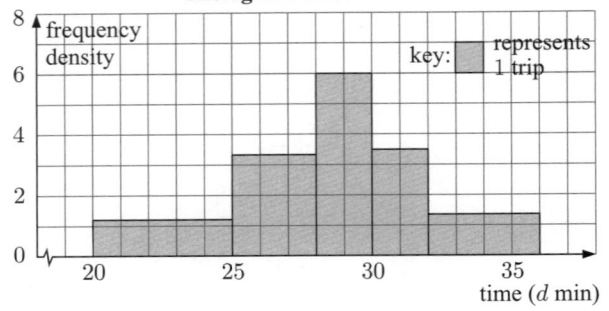

(graph: frequency density vs time (d min), key: represents 1 trip)

b Modal class is $28 \leqslant d < 30$. {highest frequency density}

c

Middle value (x)	Freq. (f)
22.5	6
26.5	10
29	12
31	7
34	5

Mean time ≈ 28.4 mins.

d

Time (d min)	Freq.	Cumul. freq.
$20 \leqslant d < 25$	6	6
$25 \leqslant d < 28$	10	16
$28 \leqslant d < 30$	12	28
$30 \leqslant d < 32$	7	35
$32 \leqslant d < 36$	5	40

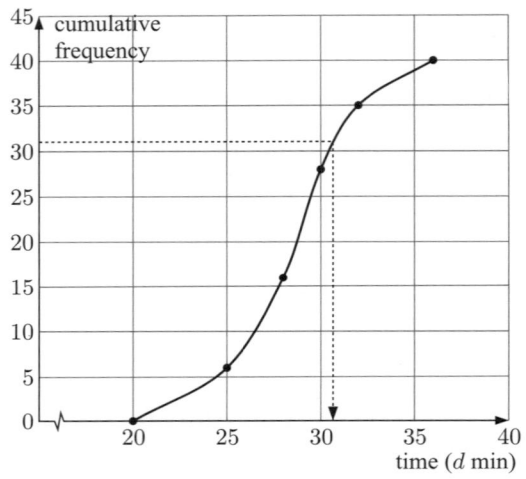

(cumulative frequency graph)

e 80% of 40 gives CF $= 32$, so from the graph, Peter should allow ≈ 31 minutes.

3 a The multiplier is $107.3\% = 1.073$

   i Value after 2 years   **ii** Value after 7 years
      $= £2500 \times (1.073)^2$ $= £2500 \times (1.073)^7$
      $\approx £2878.32$ $\approx £4093.91$

b $V = 2500 \times (1.073)^n$

c (graph of $V = 2500 \times (1.073)^n$, V(£) vs n (years))

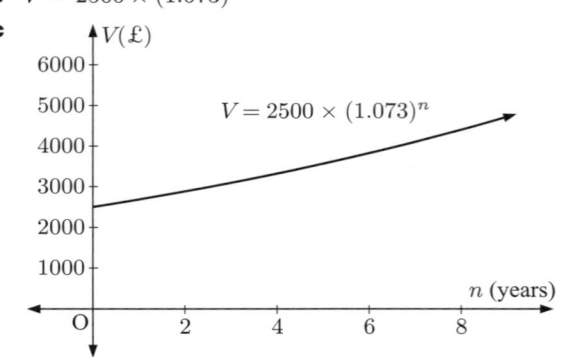

d Suppose the money is invested for t years.
Then $\quad 2500 \times (1.073)^t = 8900$
$$\therefore \ (1.073)^t = \tfrac{8900}{2500}$$
$$\therefore \ \log(1.073)^t = \log(\tfrac{8900}{2500})$$
$$\therefore \ t\log(1.073) = \log(\tfrac{8900}{2500})$$
$$\therefore \ t = \frac{\log(\tfrac{8900}{2500})}{\log 1.073}$$
$$\therefore \ t \approx 18$$
So, Dale's account is worth £8900 after 18 years.

4 a length OA $= \sqrt{2^2 + 4^2}$
$\qquad\qquad\quad = \sqrt{20}$
$\qquad\qquad\quad = 2\sqrt{5}$ units

b $m_{OA} = \dfrac{4-0}{2-0} = 2$ and OA passes through the origin.
$\therefore$ equation is $y = 2x$.

c A' is $(-2, 4)$

d $m_{OA'} = \dfrac{4-0}{-2-0} = -2$ and OA' passes through the origin.
$\therefore$ equation is $y = -2x$.

e Triangle OAA' has base 4 and height 4.
$\therefore$ area $\triangle OAA' = \tfrac{1}{2} \times 4 \times 4$
$\qquad\qquad\qquad = 8$ units^2

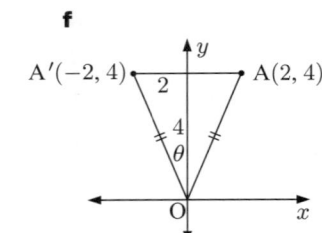

f (diagram showing $A'(-2, 4)$ and $A(2, 4)$ with angle θ at origin)

$\tan \theta = \tfrac{2}{4} = \tfrac{1}{2}$
$\therefore \theta = \tan^{-1}(\tfrac{1}{2})$
A' has to be rotated through 2θ to get to A, that is, $2\tan^{-1}(\tfrac{1}{2})$ or $\approx 53.1°$.

5 $f(x) = \dfrac{3}{x-2}$

a $f(4) = \dfrac{3}{4-2} = \tfrac{3}{2}$ or $1\tfrac{1}{2}$

b If $f(x) = 6$ then $\dfrac{3}{x-2} = 6$
$$\therefore \ \dfrac{3}{x-2} = \dfrac{6(x-2)}{(x-2)}$$
$$\therefore \ 3 = 6(x-2)$$
$$\therefore \ x - 2 = \tfrac{1}{2}$$
$$\therefore \ x = 2\tfrac{1}{2}$$

c Vertical asymptote is $x = 2$.
Horizontal asymptote is $y = 0$.

d The inverse of $y = \dfrac{3}{x-2}$ is $x = \dfrac{3}{y-2}$.
$$\therefore \ x(y-2) = \dfrac{3(y-2)}{y-2}$$
$$\therefore \ x(y-2) = 3$$
$$\therefore \ xy - 2x = 3$$
$$\therefore \ xy - 2x + 2x = 3 + 2x$$
$$\therefore \ xy = 3 + 2x$$
$$\therefore \ y = \dfrac{3 + 2x}{x}$$
$$\therefore \ f^{-1}(x) = \dfrac{3}{x} + 2$$

e

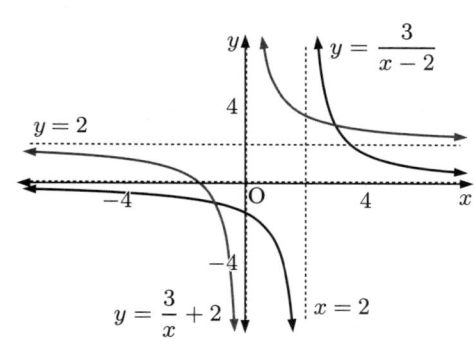

f Under the translation $\begin{pmatrix} 4 \\ -1 \end{pmatrix}$,

$x \to x - 4$ and $y \to y + 1$

$\therefore\ y = \dfrac{3}{x-2}$ becomes $y + 1 = \dfrac{3}{(x-4)-2}$

$\therefore\ y + 1 = \dfrac{3}{x-6}$

$\therefore\ y = \dfrac{3}{x-6} - 1$

6 a $\quad 4x - 5 \leqslant 3(x+2)$
$\therefore\ 4x - 5 \leqslant 3x + 6$
$\therefore\ 4x - 3x \leqslant 6 + 5$
$\therefore\ x \leqslant 11$

b $|x - 4| < -2x^2 + 5x + 3$

Using a graphics calculator:

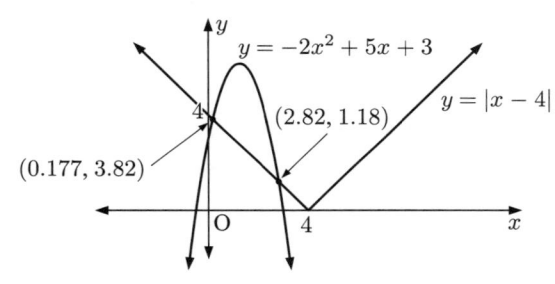

$\therefore\ 0.177 < x < 2.82$

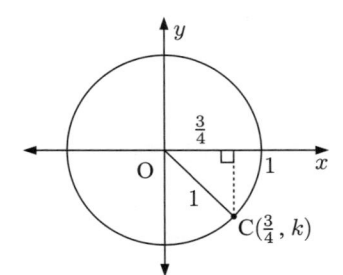

7 a

Using Pythagoras' theorem
$k^2 + \left(\tfrac{3}{4}\right)^2 = 1^2$
$\therefore\ k^2 + \tfrac{9}{16} = 1$
$\therefore\ k^2 = \tfrac{7}{16}$
$\therefore\ k = \pm\tfrac{\sqrt{7}}{4}$

But $k < 0$, so $k = -\tfrac{\sqrt{7}}{4}$

b i $m_{AC} = \dfrac{-\frac{\sqrt{7}}{4} - 1}{\frac{3}{4} - 0}$

$= \dfrac{-\sqrt{7}-4}{4} \times \dfrac{4}{3}$

$= \dfrac{-\sqrt{7}-4}{3}$

ii $m_{CB} = \dfrac{-1 - -\frac{\sqrt{7}}{4}}{0 - \frac{3}{4}}$

$= \dfrac{-4+\sqrt{7}}{4} \times \dfrac{-4}{3}$

$= \dfrac{4-\sqrt{7}}{3}$

c $m_{AC} \times m_{CB} = \dfrac{-\sqrt{7}-4}{3} \times \dfrac{4-\sqrt{7}}{3}$

$= \dfrac{-4\sqrt{7}+7-16+4\sqrt{7}}{9}$

$= -\tfrac{9}{9}$

$= -1$

$\therefore\ $ AC and CB are perpendicular.
$\therefore\ $ angle ACB is a right angle.

8 a Loss of $5200.

b

n	1	2	3	4	5
u_n	-5	-2.8	-0.2	2.8	6.2
$\Delta 1$	2.2	2.6	3.0	3.4	
$\Delta 2$		0.4	0.4	0.4	

The second differences are constant, so the sequence is quadratic with general term $u_n = an^2 + bn + c$.

$2a = 0.4$, so $a = 0.2$
$3a + b = 2.2$, so $0.6 + b = 2.2$
$\therefore\ b = 1.6$
$a + b + c = -5$, so $0.2 + 1.6 + c = -5$
$\therefore\ c = -6.8$

$\therefore\ $ the general term is $u_n = 0.2n^2 + 1.6n - 6.8$.

c For 6th year: $u_6 = 0.2 \times 6^2 + 1.6 \times 6 - 6.8 = 10$
Predicted profit is $10\,000$.

For 7th year: $u_7 = 0.2 \times 7^2 + 1.6 \times 7 - 6.8 = 14.2$
Predicted profit is $14\,200$.

d The predictions will only be reliable if the business trends continue in the same way. A drought, for example, could affect this business.

9 a i $\overrightarrow{AM} = \tfrac{1}{3}\mathbf{b}$ $\quad$ **ii** $\overrightarrow{BN} = \tfrac{1}{3}\mathbf{b} - \tfrac{1}{3}\mathbf{a}$

iii $\overrightarrow{MN} = -\tfrac{1}{3}\mathbf{b} + \mathbf{a} + \tfrac{1}{3}\mathbf{b} - \tfrac{1}{3}\mathbf{a} = \tfrac{2}{3}\mathbf{a}$

b ABNM is a trapezium $\{\overrightarrow{MN} \parallel \overrightarrow{AB}\}$

c $\triangle$s ABC and MNC are similar
$\{A\widehat{C}B$ is common, $C\widehat{N}M$ and $C\widehat{B}A$ corresponding$\}$

$\dfrac{AB}{MN} = \dfrac{3}{2}$

$\therefore\ $ area $\triangle ABC = \left(\tfrac{3}{2}\right)^2 \times 40$ cm^2

$\therefore\ $ area of ABNM $= \left(\tfrac{3}{2}\right)^2 \times 40 - 40$

$= 50$ cm^2

10 a Midyear $\quad$ mean ≈ 71.3 marks

median $= \dfrac{73 + 74}{2} = 73.5$ marks

IQR $= Q_3 - Q_1$

$= \dfrac{79 + 81}{2} - \dfrac{61 + 66}{2}$

$= 80 - 63.5$

$= 16.5$ marks

Final mean ≈ 73.3 marks
median = 77 marks
IQR = $Q_3 - Q_1$
= 85 − 61
= 24 marks

b The mean and median for the final results were higher than for the midyear, but the final results had greater spread.

c [scatter plot of final score vs mid-year score]

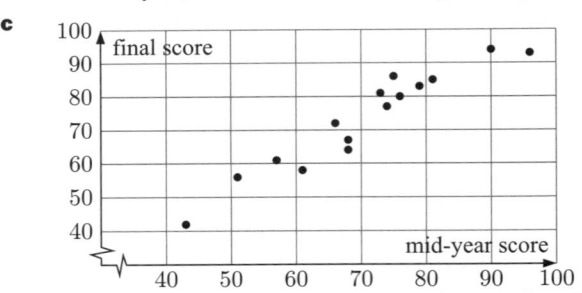

d There is a strong positive linear correlation between the two sets of exam scores.

e $y \approx 1.03x + 0.969$

f 86. This is the value of y (rounded to the nearest integer) that we obtain when we substitute $x = 83$ into the equation in **e**.
This is a sensible prediction as it is within the domain of the data and the variables are strongly correlated.

11

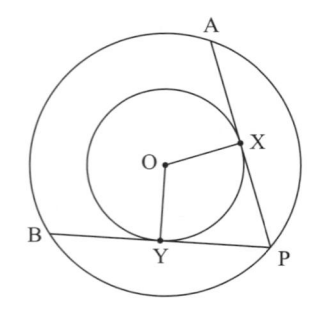

a PX and PY are both tangents to the small circle.
∴ PX = PY
{tangents from an external point}

b In the large circle, X is the midpoint of AP {radius of a circle perpendicular to a chord, bisects the chord}
∴ PX = XA

c Using similar reasoning to **b**,
PY = YB and PX = PY from **a**
∴ PA = PB
∴ △PAB is isosceles and $\hat{BAX} = \hat{ABY}$.

d △OXY is isosceles {OX = OY, equal radii}
∴ $\hat{OXY} = \hat{OYX}$

e $\hat{OXA} = \hat{OYB} = 90°$ {radius-tangent}
∴ $\hat{AXY} = 90° + \hat{OXY}$
= $90° + \hat{OYX} = \hat{XYB}$ {using **d**}
Now, $\hat{BAX} + \hat{AXY} + \hat{XYB} + \hat{YBA} = 360°$ {∠s in quad.}
∴ $2\hat{BAX} + 2\hat{XYB} = 360°$ {using **c**}
∴ $\hat{BAX} + \hat{XYB} = 180°$
∴ AXYB is a cyclic quadrilateral.
{opposite angles supplementary}

12 a $P = \{2, 3, 5, 7, 11, 13, 17, 19\}$
$C = \{4, 6, 8, 9, 10, 12, 14, 15, 16, 18, 20\}$
$M = \{5, 10, 15, 20\}$

b P and C. A number can not be both prime and composite.

c [Venn diagram with sets P and C within universe U, showing M overlapping; P contains 2, 3, 7, 11, 13, 17, 19; intersection P∩C contains 5 (in M); C contains 4, 6, 8, 9, 12, 14, 18 and 10, 15, 20 (in M); 16 in C∩M; 1 outside both]

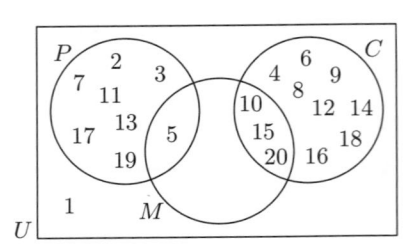

d i P(C then P) = P(C) × P(2nd is P | 1st is C)
= $\frac{11}{20} \times \frac{8}{19}$
= $\frac{22}{95}$

ii P(C and M then C and M) = $\frac{3}{20} \times \frac{2}{19}$
= $\frac{3}{190}$

iii P(2nd is prime given that it is a multiple of 5) = $\frac{1}{4}$

Paper 6

A. Investigation: Freya Sums

Part I

1 $4 = 4^1$

2 a [grid: 4 4 / 4 4]   **b** $4 \times 4 = 16 = 4^2$

3 a 3×3 **b** [grid: 4 4 4 / 4 4 4 / 4 4 4]   **c** [grid: 16 16 / 16 16]
$4 \times 16 = 64 = 4^3$

4 $4 \times 64 = 256 = 4^4$

5 Freya sum for an $n \times n$ grid is 4^{n-1}.

6 Freya sum is $4^{17-1} = 4^{16}$
= 4 294 967 296

Part II

1 a [grid: 1 0 / 0 0] sum = 1
b [grid: 1 1 / 0 0] sum = 2
c [grid: 0 1 / 0 0] sum = 1
d [grid: 1 0 / 1 0] sum = 2
e [grid: 1 1 / 1 1] sum = 4

2 A 3×3 grid with 1 at a corner and all other entries 0 will have Freya sum 1, because the 1 is in only one box.
A 3×3 grid with 1 at the beginning or end of a 2nd row or column and all other entries 0 will have Freya sum 2, because the 1 is in 2 boxes.
A 3×3 grid with 1 at the centre and all other entries 0 will have Freya sum 4, because that 1 is in all 4 boxes.

3 [grid: 1 2 1 / 2 4 2 / 1 2 1]

4 a [grid: 1 1 / 0 1] sum = 3
b [grid: 1 1 / 0 1] sum = 3
c [grid: 2 2 / 2 1] sum = 7

5 a 1 (corner) + 2 (end of 2nd row) = 3
   b 2 (start of 2nd column) + 1 (corner) = 3
   c 2 (start of 2nd column) + 4 (middle) + 1 (corner) = 7

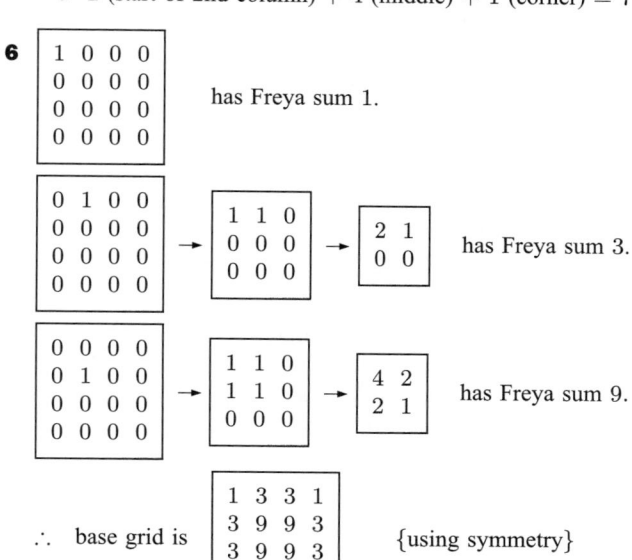

6
$\begin{array}{|cccc|}\hline 1 & 0 & 0 & 0 \\ 0 & 0 & 0 & 0 \\ 0 & 0 & 0 & 0 \\ 0 & 0 & 0 & 0 \\ \hline\end{array}$ has Freya sum 1.

$\begin{array}{|cccc|}\hline 0 & 1 & 0 & 0 \\ 0 & 0 & 0 & 0 \\ 0 & 0 & 0 & 0 \\ 0 & 0 & 0 & 0 \\ \hline\end{array}$ → $\begin{array}{|ccc|}\hline 1 & 1 & 0 \\ 0 & 0 & 0 \\ 0 & 0 & 0 \\ \hline\end{array}$ → $\begin{array}{|cc|}\hline 2 & 1 \\ 0 & 0 \\ \hline\end{array}$ has Freya sum 3.

$\begin{array}{|cccc|}\hline 0 & 0 & 0 & 0 \\ 0 & 1 & 0 & 0 \\ 0 & 0 & 0 & 0 \\ 0 & 0 & 0 & 0 \\ \hline\end{array}$ → $\begin{array}{|ccc|}\hline 1 & 1 & 0 \\ 1 & 1 & 0 \\ 0 & 0 & 0 \\ \hline\end{array}$ → $\begin{array}{|cc|}\hline 4 & 2 \\ 2 & 1 \\ \hline\end{array}$ has Freya sum 9.

∴ base grid is $\begin{array}{|cccc|}\hline 1 & 3 & 3 & 1 \\ 3 & 9 & 9 & 3 \\ 3 & 9 & 9 & 3 \\ 1 & 3 & 3 & 1 \\ \hline\end{array}$ {using symmetry}

7 Freya sum is $1 + 9 + 9 + 3 + 9 + 3 = 34$.

Part III

1 $\begin{array}{|ccc|}\hline 1 & 2 & 3 \\ 4 & 5 & 6 \\ 7 & 8 & 9 \\ \hline\end{array}$ → $\begin{array}{|cc|}\hline 12 & 16 \\ 24 & 28 \\ \hline\end{array}$ → 80

2 $\begin{array}{|ccc|}\hline 1 & 2 & 1 \\ 2 & 4 & 2 \\ 1 & 2 & 1 \\ \hline\end{array}$ $\begin{array}{|ccc|}\hline 1 & 2 & 3 \\ 4 & 5 & 6 \\ 7 & 8 & 9 \\ \hline\end{array}$
base grid

Multiplying corresponding elements and adding gives
$1 + 4 + 3 + 8 + 20 + 12 + 7 + 16 + 9 = 80$ which checks.

3 $\begin{array}{|ccc|}\hline 1 & 5 & 2 \\ 6 & 9 & 7 \\ 3 & 8 & 4 \\ \hline\end{array}$ Sum is
$1 + 10 + 2 + 12 + 36 + 14 + 3 + 16 + 4 = 98$

4 $\begin{array}{|cccc|}\hline 1 & 3 & 3 & 1 \\ 3 & 9 & 9 & 3 \\ 3 & 9 & 9 & 3 \\ 1 & 3 & 3 & 1 \\ \hline\end{array}$ $\begin{array}{|cccc|}\hline 1 & 2 & 3 & 4 \\ 5 & 6 & 7 & 8 \\ 9 & 10 & 11 & 12 \\ 13 & 14 & 15 & 16 \\ \hline\end{array}$
base grid

Multiplying corresponding elements and adding gives
$1 \times 1 + 3 \times 2 + 3 \times 3 + \ldots + 3 \times 14 + 3 \times 15 + 16 = 544$

5 a When $n = 1$, we have $2^{2-3}(1 + 1^2)$
$= \frac{1}{2} \times 2 = 1$ ✓
When $n = 2$, we have $2^{4-3}(1 + 2^2)$
$= 2 \times 5 = 10$
$= 1 + 2 + 3 + 4$ ✓
When $n = 3$, we have $2^{6-3}(1 + 3^2)$
$= 8 \times 10 = 80$ ✓
When $n = 4$, we have $2^{8-3}(1 + 4^2)$
$= 2^5 \times 17$
$= 32 \times 17 = 544$ ✓

b For a 5×5 grid, $n = 5$.
So, the Freya sum will be $2^{10-3}(1 + 25)$
$= 2^7 \times 26$
$= 128 \times 26$
$= 3328$

B. Modelling: Wind Turbines

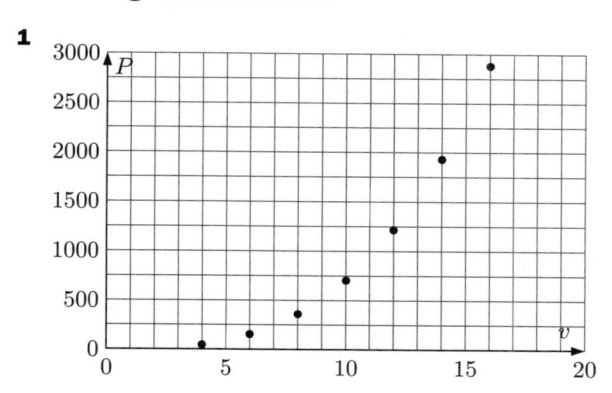

1

2 When $v = 4$, $P = 45$
so $45 = k \times 4^n$ .... (1)
When $v = 16$, $P = 2880$
so $2880 = k \times 16^n$ .... (2)
∴ $\frac{2880}{45} = \frac{k \times 16^n}{k \times 4^n}$ {(2) ÷ (1)}
∴ $\left(\frac{16}{4}\right)^n = 64$
∴ $4^n = 64$
∴ $n = 3$
From (1), $45 = k \times 4^3$
∴ $k = \frac{45}{64} \approx 0.703$

3 Using $P = 0.703 \times v^3$:
a i $P \approx 297$ **ii** $P \approx 2373$
b $\left(\frac{2373 - 297}{297}\right) \times 100\% \approx 699\%$ increase
c Yes. Doubling the wind speed increases power output by a factor of $8 \,(= 2^3)$, which is an increase of 700%.

4 $k \approx 0.703$ (from **2**)
So, $0.703 = c \times 31^2$
∴ $c = \frac{0.703}{31^2} \approx 7.32 \times 10^{-4}$

5 a $c \approx 7.32 \times 10^{-4}$, $r = 40$
∴ $k \approx 7.32 \times 10^{-4} \times 40^2 \approx 1.17$
So, $P = 1.17 \times v^3$
∴ at 10 m/s, $P = 1.17 \times 10^3$
$= 1170$ kW

b Find v such that $0.703v^3 = 1170$
Using the model, and technology, $v \approx 11.85$ m/s.

6 Power and wind speed are linked by a cubic. So, a small increase in speed leads to a larger increase in power generated. Conversely, this means that half the power output over a year will come from much less than half the operating time.

SOLUTIONS TO PRACTICE EXAM 3

Paper 2

1 a
$$\frac{2x+3}{x^2-3x-4} - \frac{2}{x+1}$$
$$= \frac{2x+3}{(x+1)(x-4)} - \left(\frac{2}{x+1}\right)\left(\frac{x-4}{x-4}\right)$$
$$= \frac{2x+3-2(x-4)}{(x+1)(x-4)}$$
$$= \frac{2x+3-2x+8}{(x+1)(x-4)}$$
$$= \frac{11}{(x+1)(x-4)}$$

b
$$\frac{1+\sqrt{2}}{3-\sqrt{2}}$$
$$= \left(\frac{1+\sqrt{2}}{3-\sqrt{2}}\right)\left(\frac{3+\sqrt{2}}{3+\sqrt{2}}\right)$$
$$= \frac{3+4\sqrt{2}+2}{9-2}$$
$$= \frac{5+4\sqrt{2}}{7}$$

2 a $m_{AB} = \frac{3-(-2)}{7-(-3)} = \frac{5}{10} = \frac{1}{2}$

b gradient of perpendicular to AB is -2
midpoint of AB is $\left(\frac{-3+7}{2}, \frac{-2+3}{2}\right) = (2, \frac{1}{2})$
∴ equation of perpendicular bisector is
$2x + y = 2(2) + 1(\frac{1}{2})$
∴ $2x + y = \frac{9}{2}$

c Substituting $(4, -3)$ into $2x + y$ gives $2(4) + -3 = 5$ which does not equal $\frac{9}{2}$.
∴ C does not lie on the perpendicular bisector.

d $\overrightarrow{AC} = \begin{pmatrix} 4-(-3) \\ -3-(-2) \end{pmatrix} = \begin{pmatrix} 7 \\ -1 \end{pmatrix}$

e $|\overrightarrow{AC}| = \sqrt{7^2 + (-1)^2}$
$= \sqrt{49+1}$
$= \sqrt{50}$
$= 5\sqrt{2}$ units

3 a $-8, -3, 2, 7, 12, 17$ (+5 each)
The next two terms are 12 and 17.

b $u_n = 5n - 13$

4 a $A = \{9, 10, 11, 12, 13, 14, 15, 16, 17, 18, 19, 20\}$

b $B = \{9, 10, 12, 14, 15, 16, 18\}$

c $A \cap B = \{9, 10, 12, 14, 15, 16, 18\}$

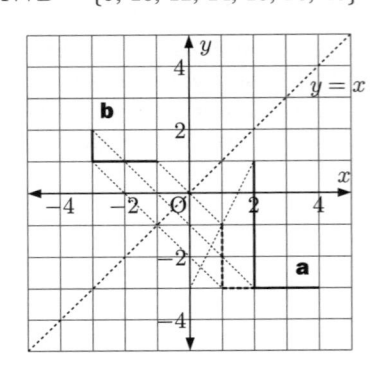

5 [graph with lines $y = x$ and labels a, b]

6 Let a small bucket have capacity s litres and a large bucket have capacity l litres.
Then $2s + 3l = 18$ .... (1)
and $7s + 2l = 29$ .... (2)
∴ $4s + 6l = 36$ $\{(1) \times 2\}$
and $-21s - 6l = -87$ $\{(2) \times -3\}$
∴ $-17s = -51$ {adding}
∴ $s = 3$
Substituting in (1) gives
$2(3) + 3l = 18$
∴ $3l = 12$
∴ $l = 4$
∴ a large bucket contains 4 litres.

7 $(x-4)^2 = 5(14-x)$
∴ $x^2 - 8x + 16 = 70 - 5x$
∴ $x^2 - 3x - 54 = 0$
∴ $(x-9)(x+6) = 0$
∴ $x = 9$ or -6

8 The amplitude is 3. ∴ $a = 3$
The period is 180°. ∴ $\frac{360°}{b} = 180°$
∴ $b = 2$

9 The known angles in the quadrilateral are 123°, 104° {vertically opposite} and 63° {angles on a line}.
Sum of angles of a quadrilateral is 360°.
∴ $x + 123 + 104 + 63 = 360$
∴ $x + 290 = 360$
∴ $x = 70$

10 a 4 dolphins {most frequent}

b skewed

c total sightings are whales (81) and dolphins (125)
∴ more likely to see a dolphin.

Paper 4

1 a $\triangle OAB$ is isosceles {equal radii}
∴ $O\hat{A}B = O\hat{B}A = \alpha$
∴ $2\alpha + 110° = 180°$ {angle sum of $\triangle$}
∴ $2\alpha = 70°$
∴ $\alpha = 35°$

b $A\hat{D}B = \frac{1}{2}A\hat{O}B$
{angle at centre $= 2 \times$ angle at circle subtended by same arc}
∴ $A\hat{D}B = \frac{1}{2} \times 110° = 55°$

c $O\hat{B}D = B\hat{D}C = \alpha$ {alt. angles, parallel lines}
∴ $B\hat{D}C = 35°$
∴ $A\hat{D}C = 55° + 35° = 90°$
∴ $A\hat{D}C$ is a right angle.

d Since $A\hat{D}C$ is a right angle {from **c**}, AC must be a diameter of the circle.
∴ if OB, which is a radius, is 5 cm, then AC = 10 cm.

e △OBD is isosceles {equal radii}
∴ O$\widehat{B}$D = O$\widehat{D}$B = α = 35°
∴ B$\widehat{O}$D = 110° {angle sum of triangle}
∴ D$\widehat{O}$A = 360° − 110° − 110°
= 140° {sum of angles about a point}
△OAD is isosceles {equal radii}
∴ D$\widehat{A}$O = 20° {equal base angles}

f In △ACD, $\sin D\widehat{A}O = \dfrac{DC}{10}$
∴ DC = 10 sin 20°
≈ 3.42 cm

2 a Area of Birdsville
= area of square + area of trapezium
$= 1^2 + \left(\dfrac{2.5 + 4}{2}\right) \times 2$
= 1 + 6.5
= 7.5 km²

b i P(a questionnaire goes to Birdsville)
$\approx \dfrac{\text{area of Birdsville}}{\text{area of northeast suburbs}}$
$\approx \dfrac{7.5}{\frac{1}{4} \times \pi \times 10^2}$
≈ 0.0955
∴ P(BB′ or B′B)
= P(B) × P(B′) + P(B′) × P(B)
≈ 2 × 0.0959 × 0.9045
≈ 0.173

ii We are assuming that the area of the suburb is proportional to the number of households it contains.

3 a i In the first 4 minutes. {graph is steepest}

ii $\text{speed} = \dfrac{\text{distance}}{\text{time taken}}$
$= \dfrac{2.5 \text{ km}}{4 \text{ min}}$
$= \dfrac{2.5 \text{ km}}{\frac{4}{60} \text{ h}}$
$= 2.5 \times \frac{60}{4}$ km/h
= 37.5 km/h

b Wei stops after 9 minutes. It takes 3 minutes to get directions.

c i $\text{average speed} = \dfrac{\text{distance}}{\text{time taken}}$
$= \dfrac{5 \text{ km}}{14 \text{ min}}$
≈ 0.357 km/min

ii $\text{average speed} = \dfrac{5000 \text{ m}}{14 \times 60 \text{ s}}$
≈ 5.95 m/s

d $\text{speed} = \dfrac{\text{distance}}{\text{time}}$
∴ $\text{time} = \dfrac{\text{distance}}{\text{speed}}$
Time taken by Hao $= \dfrac{5.6 \text{ km}}{42 \text{ km/h}}$
$= \dfrac{5.6 \times 60}{42}$ min
= 8 min
So, Hao arrives 5 + 8 = 13 minutes after Wei starts.
Wei takes 14 minutes.
So, Hao arrives first.

4 a Volume of cylinder $= \pi r^2 l$
$= \pi r^2 \times 2r$
$= 2\pi r^3$ cm³

b mass of wood + mass of copper
$m = (\frac{4}{3}\pi R^3 - 2\pi r^3) \times 1.2 + 2\pi r^3 \times 9$
$= \dfrac{4 \times 1.2}{3}\pi R^3 - 2.4\pi r^3 + 18\pi r^3$
$= 1.6\pi R^3 + 15.6\pi r^3$

c $15.6\pi r^3 = m - 1.6\pi R^3$
∴ $r^3 = \dfrac{m - 1.6\pi R^3}{15.6\pi}$
∴ $r = \sqrt[3]{\dfrac{m - 1.6\pi R^3}{15.6\pi}}$

d Substituting R = 6 and m = 1120,
$r = \sqrt[3]{\dfrac{1120 - 1.6 \times \pi \times 6^3}{15.6\pi}}$
≈ 0.888
The cylinder has radius ≈ 0.888 cm and height ≈ 1.775 cm.

5 a Value after 1 year = \$15 000 × 1.08
Value after 2 years = \$15 000 × 1.08 × 0.98
Value after 3 years = \$15 000 × 1.08 × 0.98 × 1.13
= \$17 939.88

b Suppose the rate of compound interest is i%.
After 3 years compound interest
$15\,000(1 + i)^3 = 17\,939.88$
∴ $(1 + i)^3 = \dfrac{17\,939.88}{15\,000}$
∴ $1 + i = \sqrt[3]{\dfrac{17\,939.88}{15\,000}}$
≈ 1.0615
∴ $i \approx 0.0615$
∴ $i \approx 6.15\%$
∴ average compound change in value ≈ 6.15%.

c $I = Prn$
∴ $I = \$17\,939.88 \times 0.062 \times 3$
≈ \$3336.82
So \$17 939.88 generates \$3336.82 interest in 3 years.
Using the average compound rate,
value after 3 years $= \$17\,939.88 \times (1.0615)^3$
≈ \$21 457.52
∴ interest earned ≈ \$3517.64
Melissa should not take the offer.

6 a

[Graph showing lines $2y - x = -2$, $x + 3y = 12$, $y + 2x = 9$ with region $\mathcal{R}$ having vertices $(3, 3)$, $(6, 2)$, $(4, 1)$]

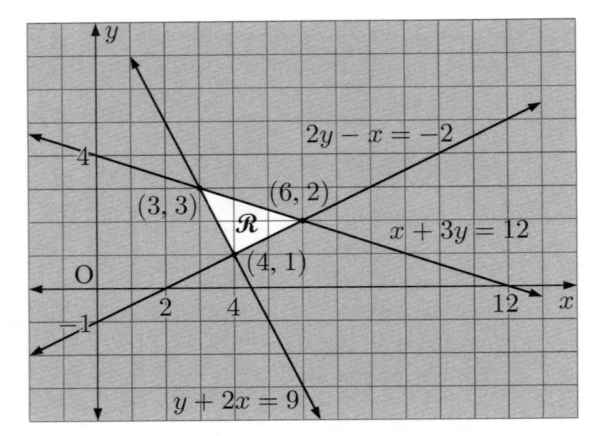

b

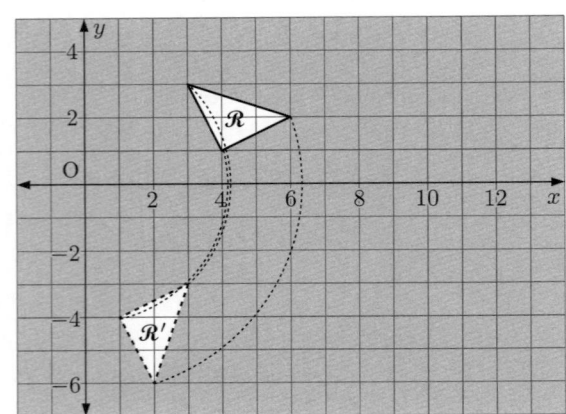

[Graph showing region $\mathcal{R}$ and its image $\mathcal{R}'$ after transformation]

7 Let T_e = students who earned over €10 per hour and T_w = students who earned less than €12 per hour.

a $n(T_e \cup T_w) = n(T_e) + n(T_w) - n(T_e \cap T_w)$
$\therefore \quad 30 = 15 + 23 - n(T_e \cap T_w)$
$\therefore \quad n(T_e \cap T_w) = 38 - 30$
$= 8$

[Venn diagram with T_e containing (7), intersection (8), T_w containing (15), universe U]

b 8 students

c Mean wage $\approx \dfrac{7 \times €13 + 8 \times €11 + 15 \times €9}{30}$
$\approx €10.47$

8 $f(x) = -x^2 - 10x - 21$

a i When $y = 0$, $-x^2 - 10x - 21 = 0$
$\therefore \quad x^2 + 10x + 21 = 0$
$\therefore \quad (x + 3)(x + 7) = 0$
$\therefore \quad x = -3$ or -7
$\therefore$ x-intercepts are -3 and -7.

ii When $x = 0$, $y = -21$
$\therefore$ y-intercept is -21.

iii The axis of symmetry lies midway between the x-intercepts.
Since the average of -3 and -7 is -5, the axis of symmetry is $x = -5$.
$\therefore$ the x-coordinate of vertex is -5
Now $f(-5) = -25 + 50 - 21 = 4$
$\therefore$ the vertex is $(-5, 4)$.

b, c

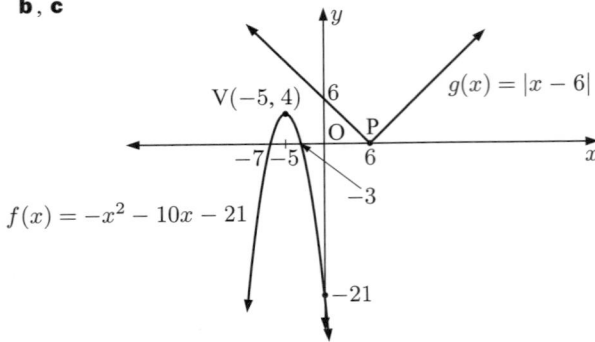

[Graph showing $f(x) = -x^2 - 10x - 21$ with vertex V(-5, 4), and $g(x) = |x - 6|$ with point P at 6]

d P is 6 units from the origin.
Distance OV $= \sqrt{(-5 - 0)^2 + (4 - 0)^2}$
$= \sqrt{25 + 16}$
$= \sqrt{41}$
> 6
$\therefore$ P is closer to the origin.

e i $m_{\text{VP}} = \dfrac{0 - 4}{6 - -5} = -\dfrac{4}{11}$
$\therefore$ equation of VP is $4x + 11y = 4 \times 6 + 11 \times 0$
$\therefore \quad 4x + 11y = 24$

ii midpoint of VP is $\left(\dfrac{-5 + 6}{2}, \dfrac{4 + 0}{2}\right) = \left(\dfrac{1}{2}, 2\right)$

9 a Using technology:
i median $= 38$ **ii** range $= 52 - 26 = 26$
iii IQR $= 42 - 35 = 7$

b, d Using technology, $(\overline{x}, \overline{y}) \approx (38.4, 2858)$

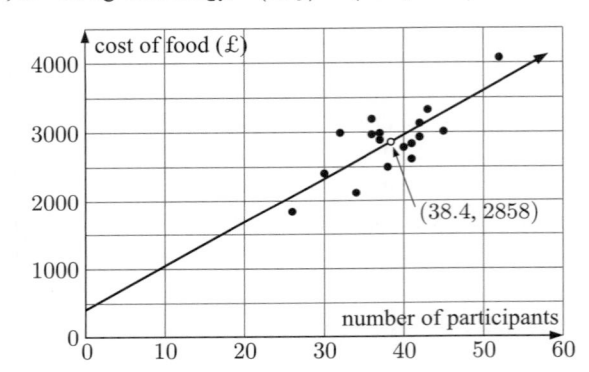

[Scatter plot of cost of food (£) vs number of participants showing point (38.4, 2858) and regression line]

c moderate, positive linear correlation

d (see graph above)
Let participants $= x$
food costs $= y$
Line of regression is $y \approx 63.8x + 410$

e gradient ≈ 63.8
For each extra participant in the camp, food costs will increase by £63.80.

f i £2643 **ii** £5514
The prediction in part **i** would be more reliable as it is within the domain of the data.
The prediction in part **ii** would not be as reliable as it is outside the domain of the data.
Also, the variables are only moderately correlated.

10 a $AC^2 = (\sqrt{3})^2 + 1^2 = 4$
$\therefore \quad AC = 2$ cm as $AC > 0$
Now $\dfrac{AC}{AB} = \dfrac{2}{\sqrt{3}}$
$\therefore \quad AC = \dfrac{2}{\sqrt{3}} AB$

But △s ABC and ACD are similar {given}

$\therefore$ AD $= \frac{2}{\sqrt{3}}$AC

$= \frac{2}{\sqrt{3}} \times \frac{2}{\sqrt{3}}$AB

$= \frac{4}{3} \times \sqrt{3}$

$= \frac{4}{\sqrt{3}}$ cm

Similarly, $\frac{AE}{AD} = \frac{2}{\sqrt{3}}$

$\therefore$ AE $= \frac{2}{\sqrt{3}}$AD

$= \frac{2}{\sqrt{3}} \times \frac{4}{\sqrt{3}}$

$= \frac{8}{3}$ cm

b **i** Sequence is $\sqrt{3}, 2, \frac{4}{\sqrt{3}}, \frac{8}{3}, \ldots$

which is geometric, with $u_{n+1} = \frac{2}{\sqrt{3}} \times u_n$.

ii $u_n = \sqrt{3} \times \left(\frac{2}{\sqrt{3}}\right)^{n-1}$

c $\tan\theta = \frac{1}{\sqrt{3}}$ $\therefore$ $\theta = 30°$

d There are $\frac{360°}{30°} = 12$ triangles

$u_{13} = \sqrt{3} \times \left(\frac{2}{\sqrt{3}}\right)^{12}$

$= \frac{2^{12}}{3^{\frac{11}{2}}} \approx 9.732$

Now AB ≈ 1.732 cm, so this hypotenuse is approximately 8.00 cm longer than AB.

11 a

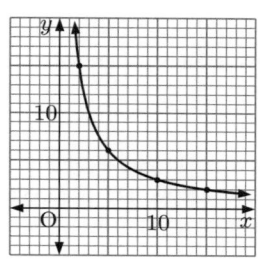

b For $x > 0$, $y = \frac{30}{x}$ $\therefore$ $f(x) = \frac{30}{x}$

y is inversely proportional to x.

c $g(x) = 5 \times 3^x$

i $g(4) = 5 \times 3^4$
$= 405$

ii $g(x+2) = 5 \times 3^{x+2}$
$= 5 \times 3^x \times 3^2$
$= 9 \times (5 \times 3^x)$
$= 9\,g(x)$

iii Domain $= \{x \mid x \in \mathbb{R}\}$
Range $= \{y \mid y > 0,\ y \in \mathbb{R}\}$

iv Horizontal asymptote is $y = 0$.

d **i** $f(g(x)) = \frac{30}{5 \times 3^x} = \frac{6}{3^x}$

$= 2 \times 3^1 \times 3^{-x}$

$= 2 \times 3^{1-x}$

ii $y = 5 \times 3^x$ has inverse $x = 5 \times 3^y$

$\therefore 3^y = \frac{x}{5}$

$\therefore \log(3^y) = \log\left(\frac{x}{5}\right)$

$\therefore y\log 3 = \log x - \log 5$

$\therefore 0.4771y \approx \log x - 0.6990$

$\therefore g^{-1}(x) \approx \frac{\log x}{0.477} - 1.46$

12 a A stretch with invariant line $x = 1$ and scale factor 3.

b $\overrightarrow{AB} = \begin{pmatrix} 3-1 \\ 1-3 \end{pmatrix} = \begin{pmatrix} 2 \\ -2 \end{pmatrix}$

$\overrightarrow{AC} = \begin{pmatrix} 2-1 \\ 6-3 \end{pmatrix} = \begin{pmatrix} 1 \\ 3 \end{pmatrix}$

c Let D have coordinates (a, b).

$\overrightarrow{AB} \parallel \overrightarrow{CD}$ and $|AB| = |CD|$

$\therefore \begin{pmatrix} 2 \\ -2 \end{pmatrix} = \begin{pmatrix} a-2 \\ b-6 \end{pmatrix}$

$\therefore a - 2 = 2$ and $b - 6 = -2$

$\therefore a = 4$ and $b = 4$

So, D is $(4, 4)$.

d D is 3 units from $x = 1$.

$\therefore$ D′ is $3 \times 3 = 9$ units from $x = 1$

So, D′ is $(10, 4)$.

e midpoint of AD′ is $\left(\frac{1+10}{2}, \frac{3+4}{2}\right) = \left(\frac{11}{2}, \frac{7}{2}\right)$

midpoint of B′C′ is $\left(\frac{7+4}{2}, \frac{1+6}{2}\right) = \left(\frac{11}{2}, \frac{7}{2}\right)$

$\therefore$ AB′D′C′ is a parallelogram
{diagonals bisect each other}

Paper 6

A. Investigation: Rolling Dice

1

Alec

		1	2	3	4	5	6
Bethany	1	D	A	A	A	A	A
	2	B	D	A	A	A	A
	3	B	B	D	A	A	A
	4	B	B	B	D	A	A

2 a P(Alec winning) $= \frac{14}{24} = \frac{7}{12}$

b P(Bethany winning) $= \frac{6}{24} = \frac{1}{4}$

c P(draw) $= \frac{4}{24} = \frac{1}{6}$

3 a

Alec

		1	2	3	4	5	6	7	8
Bethany	1	D	A	A	A	A	A	A	A
	2	B	D	A	A	A	A	A	A
	3	B	B	D	A	A	A	A	A
	4	B	B	B	D	A	A	A	A

b **i** P(Alec winning) $= \frac{22}{32} = \frac{11}{16}$

ii P(Bethany winning) $= \frac{6}{32} = \frac{3}{16}$

iii P(Draw) $= \frac{4}{32} = \frac{1}{8}$

4

Alec

		1	2	3	4	5	6	7	8
Bethany	1	D	A	A	A	A	A	A	A
	2	B	D	A	A	A	A	A	A
	3	B	B	D	A	A	A	A	A
	4	B	B	B	D	A	A	A	A
	5	B	B	B	B	D	A	A	A

P(Alec winning) $= \frac{25}{40} = \frac{5}{8}$

P(Bethany winning) $= \frac{10}{40} = \frac{1}{4}$

P(Draw) $= \frac{5}{40} = \frac{1}{8}$

5 Alec has an x-sided die and Bethany has a y-sided die, $x > y$. From **1**, **3 a**, and **4**:

- the number of Draws $= y$ .... (1)
- the number of Bs is the sum of the first $(y-1)$ integers
$$= \frac{y(y-1)}{2} \quad \text{.... (2)}$$
- the number of As is xy minus (1) and (2)

So the number of As is $\quad xy - y - \dfrac{y(y-1)}{2}$

$$= xy - y - \frac{y^2}{2} + \frac{y}{2}$$
$$= xy - \frac{y}{2} - \frac{y^2}{2}$$

$\therefore \ \text{P(Draw)} = \dfrac{y}{xy} = \dfrac{1}{x}$

$\text{P(Bethany winning)} = \dfrac{y(y-1)}{2xy} = \dfrac{y-1}{2x}$

$\text{P(Alec winning)} = \dfrac{xy - \frac{y}{2} - \frac{y^2}{2}}{xy}$

$\qquad = 1 - \dfrac{1}{2x} - \dfrac{y}{2x}$

6 It is fair if Alec's die has 1 more side than Bethany's die.

7 Bethany rolls one 12-sided die, so she could obtain each of the results 1 to 12 with probability $\frac{1}{12}$.

Alec rolls two 6-sided dice, so he could obtain each of the results 2 to 12, with different probabilities.
We summarise in the following table:

k	P(Alec gets k)	P(Bethany loses)	P(draw)	P(Bethany wins)
2	$\frac{1}{36}$	$\frac{1}{12}$	$\frac{1}{12}$	$\frac{10}{12}$
3	$\frac{2}{36}$	$\frac{2}{12}$	$\frac{1}{12}$	$\frac{9}{12}$
4	$\frac{3}{36}$	$\frac{3}{12}$	$\frac{1}{12}$	$\frac{8}{12}$
5	$\frac{4}{36}$	$\frac{4}{12}$	$\frac{1}{12}$	$\frac{7}{12}$
6	$\frac{5}{36}$	$\frac{5}{12}$	$\frac{1}{12}$	$\frac{6}{12}$
7	$\frac{6}{36}$	$\frac{6}{12}$	$\frac{1}{12}$	$\frac{5}{12}$
8	$\frac{5}{36}$	$\frac{7}{12}$	$\frac{1}{12}$	$\frac{4}{12}$
9	$\frac{4}{36}$	$\frac{8}{12}$	$\frac{1}{12}$	$\frac{3}{12}$
10	$\frac{3}{36}$	$\frac{9}{12}$	$\frac{1}{12}$	$\frac{2}{12}$
11	$\frac{2}{36}$	$\frac{10}{12}$	$\frac{1}{12}$	$\frac{1}{12}$
12	$\frac{1}{36}$	$\frac{11}{12}$	$\frac{1}{12}$	$\frac{0}{12}$

$\text{P(Alec wins)} = \frac{1}{36} \times \frac{1}{12} + \frac{2}{36} \times \frac{2}{12} + \frac{3}{36} \times \frac{3}{12} + \frac{4}{36} \times \frac{4}{12}$
$\qquad + \frac{5}{36} \times \frac{5}{12} + \frac{6}{36} \times \frac{6}{12} + \frac{5}{36} \times \frac{7}{12} + \frac{4}{36} \times \frac{8}{12}$
$\qquad + \frac{3}{36} \times \frac{9}{12} + \frac{2}{36} \times \frac{10}{12} + \frac{1}{36} \times \frac{11}{12}$
$\qquad = \frac{216}{432} = \frac{1}{2}$

$\text{P(draw)} = \frac{1}{12}$

$\text{P(Bethany wins)} = \frac{1}{36} \times \frac{10}{12} + \frac{2}{36} \times \frac{9}{12} + \frac{3}{36} \times \frac{8}{12} + \frac{4}{36} \times \frac{7}{12}$
$\qquad + \frac{5}{36} \times \frac{6}{12} + \frac{6}{36} \times \frac{5}{12} + \frac{5}{36} \times \frac{4}{12}$
$\qquad + \frac{4}{36} \times \frac{3}{12} + \frac{3}{36} \times \frac{2}{12} + \frac{2}{36} \times \frac{1}{12}$
$\qquad = \frac{180}{432} = \frac{5}{12}$

B. Modelling: Weight by Age

1 By inspection, we see that the quadratic model is the best fit.

2 $0.2054 \times (7\frac{1}{4})^2 - 0.3346 \times (7\frac{1}{4}) + 14.643 \approx 23.0$ kg

3 Power: 0
Quadratic: 14.6 kg
Exponential: 10.2 kg

4 **a** The power model was closest, but a birth weight of 0 kg is not a sensible prediction.

b $\overline{x} \approx 6.43$, $\overline{y} \approx 21.3$

c $\text{slope} = \dfrac{21.3 - 3.2}{6.43 - 0}$
$\qquad \approx 2.815$
$y\text{-intercept} = 3.2$
$\therefore$ equation is $y \approx 2.815x + 3.2$

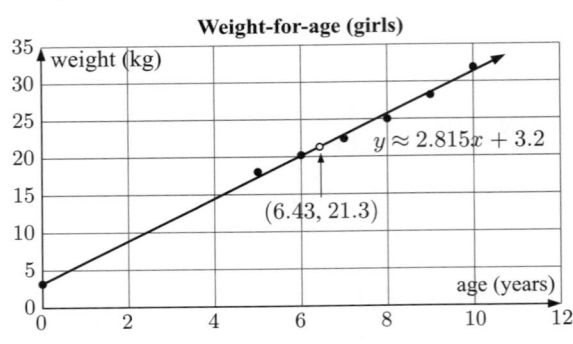

Weight-for-age (girls)

(graph showing weight (kg) vs age (years) with line $y \approx 2.815x + 3.2$ through point $(6.43, 21.3)$)

d It implies that a girl will continue to gain weight, at a constant rate from birth, indefinitely. This is not very appropriate given that a human's growth rate is different at different life stages.

5 Weight ≈ 79 kg
In 'real-life' terms a 16 year old girl weighing 79 kg seems plausible. However, given that we are extrapolating beyond the upper pole it is not a reliable estimate.

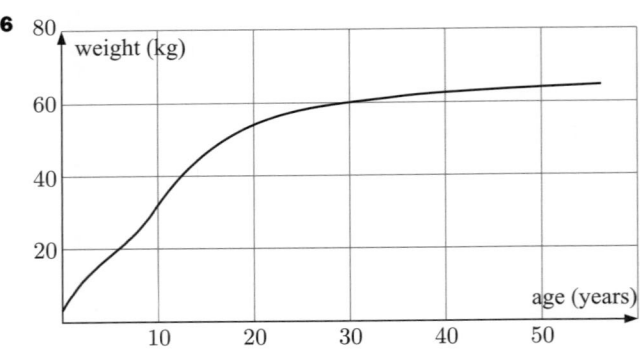

6 (graph showing weight (kg) vs age (years) levelling off near 60-65 kg)

Given that most people will attain a particular weight and then maintain it (with minor fluctuations), none of the models are appropriate because they all have weight continuing to increase indefinitely. A different model that takes this into account is needed.

NOTES

NOTES